Maize Genetics and Breeding in the 20th Century

Maize Genetics and Breeding in the 20th Century

Editors

Peter A. Peterson
Iowa State University
Ames, Iowa

Angelo Bianchi
Istituto Sperimentale Cerealicoltura
Rome

World Scientific
Singapore • New Jersey • London • Hong Kong

Published by

World Scientific Publishing Co. Pte. Ltd.

P O Box 128, Farrer Road, Singapore 912805

USA office: Suite 1B, 1060 Main Street, River Edge, NJ 07661

UK office: 57 Shelton Street, Covent Garden, London WC2H 9HE

Library of Congress Cataloging-in-Publication Data
Peterson, Peter A.
    Maize genetics and breeding in the 20th century / Peter
A. Peterson, Angelo Bianchi.
      p. cm.
    Includes bibliographical references and index.
    ISBN 981022866X (alk. paper)
    1. Corn -- Breeding.   2. Corn -- Genetics.   3. Corn -- Breeding -
- Research -- United States.   4. Corn -- Genetics -- Research -- United
States.   5. Plant breeders -- United States -- Biography.   6. Plant
geneticists -- United States -- Biography.   I. Bianchi, Angelo.
II. Title.
SB191.M2P4175   1999
633.1'52--dc21   98-38059
                                                    CIP

British Library Cataloguing-in-Publication Data
A catalogue record for this book is available from the British Library.

Printed in Singapore by Uto-Print

Preface

Upon my arrival in Cold Spring Harbor in the fall of 1947 to join the Demerec carcinogen project, I (*PAP*) was introduced to a *Drosophila* project for which I had near zero acquaintance. But with the patient tutelage of Dr. M. Demerec and Dr. Bruce Wallace, I soon became familiar with the materials and quickly became adept at handling the *Drosophila* cultures and finding mutants. In fact it became quite routine, but with my lab colleagues, the interest was enlivened by discussions—especially so with the continual flow of national and international visitors to the Cold Spring Harbor laboratories which generated more interest. This was 1947, and the effort to understand the gene was quite simplistic in the light of the revelations that would come in six years.

But the study of genetics had many aspects in the several laboratories at Cold Spring Harbor. There were the MacDowell mouse studies along with the B. Kaufmann and the Demerec *Drosophila* laboratories. In addition, both Demerec and E. Witkin had bacterial laboratories on locus dissection and repair enzymes.

But the plant genetic studies carried on in the small nursery next to the Carnegie library and the small watery inlet of the Harbor appeared to be more appealing. Of course, one could not separate the methodology of the maize nursery from the enthusiasm displayed by Barbara McClintock.

In any case, when I decided to embark on formal genetic training, it was not difficult to choose the organism for genetic study. That it became maize genetics and remained so for one-half a century is a testament to a rightful decision in 1948 as Sally and I headed for Illinois to join M. M. Rhoades. Yet, during the following three decades, I did have an opportunity to try other areas (a bacterial laboratory with P2 phage in Sweden in 1968, followed by mitochondrial laboratories in Vienna and in Cambridge, England in 1972–1973, and then again in a bacterial laboratory with the IS2 phage in Freiburg, Germany in 1977). Yet, maize was always a magnet beckoning me back to the field nursery.

v

When I began the editorship of *MAYDICA* for the Americas, I became acquainted with the maize community as I handled submitted papers and sought diverse reviewers for the various maize related papers. Then the editorial board, and especially Francesco Salamini, suggested we begin commemorating noted maize breeders and geneticists who have contributed to maize as a major breeding accomplishment and genetic tool. This has led to a collection of papers especially generated by the enthusiasm and interest of the students and colleagues of those commemorated. Without this very active interest, the collection would not have been possible.

Through all this editing and activity, I have maintained my teaching and research activity along with generating desired mutants with the collaboration with the Max Planck Institute in Cologne for nearly two decades.

But I find it important to note that this pursuit in maize genetics and related activities would never have been possible without the supporting understanding of my wife, Sally. Only someone in research or especially in a maize program with the seven-day-week summers could appreciate the support that wives provide in their spouse's academic programs. For over half a century, Sally has not only supported but has provided an understanding ear to the ever-evolving developments that accompany an academic life. All the wives that have patiently sustained such a program deserve a sincere vote of thanks.

Peter A. Peterson
Ames, Iowa

Acknowledgment

Peter A. Peterson would like to acknowledge the dedicated effort by Ms. Wendy Henerlau in the attention and care that she has applied to the development of this book.

Contents

Maize Genetics and Breeding in the 20th Century

PART A

DEVELOPMENT OF MAIZE GENETICS AND BREEDING

by Peter A. Peterson*

*Department of Agronomy, Iowa State University, Ames, IA 50011.

Introduction

Today's prolific maize production is the culmination of at least 10,000 years of prodigious selection by America's first breeders. As the maize plant developed from its *teosinte* origin, many major and minor genomic events took place (Doebley *et al.*, 1990; 1995). Abetted by an alert, subsistence-oriented observer, the Indian of the Americas—during the Neolithic period—captured heritable changes that led to the maize plant we know today. The changes were rapid from the very small ear of progenitor maize, the standard ear seen today emerged. Such rapidity in change would be expected if present day rapid changes isolated by today's current plant breeder are an indicator. Thus, this volume is a tribute to these anonymous breeders, tillers of the soil, who might have looked like William Henry Jackson's (1902) thoughtful caretaker of the maize genome (Fig. 1).

Fig. 1. Hopi Indian with his hoe in a cornfield. Photograph by William Henry Jackson, copyright 1902. (Detroit Photographic Company)

The rich heritage of the maize genome is a bequest from the past, handed down to the early 20th century breeder and geneticist. As in a track relay race, the geneticist, breeder, and pathologist have carried the baton through the 20th century into the 1990s and have passed it to the molecular biologists (see Freeling and Walbot, 1994. for an extensive listing) who are embarking on the molecular dissection of the maize genome.

Scientific advance is essentially a beneficiary of past endeavors rarely acknowledged. The scientific historian need only look at the record, in this case of maize genetics, to give documentation and structure to the early discoveries and then to rectify history (Dunn, 1965; Rhoades, 1984). Who will follow the

3

molecular biologists? Most likely, the proteins involved in gene expression will be analyzed by employing new technologies. X-ray crystallographs will provide an insight into how small changes affect gene expression (Nielsen *et al.,* 1991; Wickelgren, 1995).

Over the past 100 years, maize has attracted an outstanding group of scientists who have applied themselves to the genetics and breeding of this very important crop plant (Crabb, 1947; 1992; Wallace and Brown 1988; Rhoades, 1984; Lee, 1984). Maize is one of the most important feed-grain crops in the world. In the United States, it represents a massive export commodity (approximately 2.4 billion bushels annually) and is a basis for many other products. In 1977, in an effort to recognize outstanding members of the maize community, the international journal *MAYDICA*, devoted to maize and its relatives, initiated a series of commemorative volumes to laud some of the various achievers in this field. Their biographies and their relations to others are summarized in the following pages.

The scientists commemorated in this volume are the scientific progeny of these forerunners of the emerging field of genetics, and have been some of the field's important contributors. The "scholarly" progeny of these early researchers have been involved for the past 100 years in research and subsequent maize improvement and will conduct the maize research program of the future (Fig. 24).

1 Flowering of Maize Genetics

The early decades of the 20th century saw a surge of activity in the study of the inheritance of newly discovered traits. This was before the era of hybrid corn, and as a consequence, researchers attention focused on the progeny of crosses of newly uncovered traits. How exciting to see a changed or altered form that could be proven to be a heritable mutant in successive crosses. One of the many prolific investigators in the emerging study of genetics in the early decades of the 20th centur was G. N. Collins (1909; 1917) and his co-worker, J. H. Kempton. (Collins and Kempton, 1920; 1927). During the first two decades following the uncovering of Mendel's laws (1900), they published on the inheritance of readily observed traits such as flowering, tillering, grain texture, morphological changes, and several others. These studies and many others (cited in Emerson *et al.,* 1935) established the solid foundation of maize genetics by the mid-1930s. Linkages among the traits soon were established and different linkage groups were identified. The chromosome relationships of these genetic linkages were to be identified as soon as proper chromosome handling techniques (staining, etc.) were developed. In other laboratories, *"the chromosome theory of heredity"* was clarified and finally proven (Bridges, 1916) and raised the question of how the genetic linkage groups could be related to the maize chromosomes.

These maize chromosomes were elusive. The Merle T. Jenkins promotional activity while heading the corn program in the United States Department of Agriculture (USDA) of funding basic studies to augment corn breeding programs included establishment of a new initiative—a USDA satellite laboratory at Cornell University under L. E. Randolph. But Randolph was trained in the "root-tip school" of chromosome studies. This technique used paraffin-embedded root-tips and microtome cutting followed by slide-making and analysis of serial sections. He needed help, and an assistant was provided. This was the young graduate student, Barbara McClintock from Brooklyn, New York, who came to Cornell to study botany. She asked, wasn't there a better way to look at maize chromosomes?

How she came across Belling's efficient acetocarmine method is not clear (though this might be expected since Cornell University was one of the most advanced centers for genetic and cytological studies of maize at that time), but it soon became apparent that these maize chromosomes could be defined better if Belling's (1927) "squash" acetocarmine technique was applied. The plant researchers' cytological techniques are difficult to assess, because the chromosome smear method was used as early as 1900 by Sutton (1900, 1902) in the study of the grasshopper, *Brachystola/magnor*. Soon, with the study of triploids and the derivative trisomics with the chromosome smear method, McClintock and her colleague, Henry E. Hill (McClintock and Hill, 1931) were able to relate genetic linkage groups to individual chromosomes. Thus, in the few short years from 1926 to 1933, the cytogenetics of maize was firmly established as a potent genetic tool, with chromosomes identified and coupled with their genetic organization from the longest (chromosome 1) to the shortest (chromosome 10). Maize genetics was soon to explode.

The genes then described through three decades of inheritance studies could be assigned to an established linkage group, and these could be identified with the cytologically described individual chromosomes. This was synergism at its best. It was not long before the Emerson group correlated the extensive body of accumulated genetic knowledge with linkage relationships of all the chromosomes by using both two-point and three-point linkage tests; they related this to maize cytogenetics in their Cornell memoirs (Emerson *et al.*, 1935). These memoirs became a focus for many working with maize genetics and were a significant document in the progress of this field.

What this clearly illustrates is the pyramid developing, where successive investigators build on the findings of their predecessor (Table 2). Similarly, PCR and RAPD technologies were rapidly developing and incorporated in the further analysis of genetic studies (Table 1).

Table 1. Definitions and glossary of terms.

PCR	Polymerase Chain Reaction
RFLP	Restriction-Fragment-Length Polymorphism
Xenia	Influence of pollen on the embryo and endosperm
Double fertilization	The two sperms in the pollen tube – one fusing with the egg ➔ embryo; the other with the two polar nuclei ➔ endosperm.
P-allele	Controls the pericarp phenotype which is a product of the female genotype.

Table 2. Some of the described genes in maize;
their initial discovery and the span of years before final molecular study.

Gene	Original Genetic Description	Molecular Development
$a1^*$	Emerson, 1918	O'Reilly et al., 1985
$a2^*$	Jenkins, 1932	Menssen et al., 1990
$a3^*$	Lindstrom, 1935	Robinett et al., 1995
$ae1$	Vineyard and Bear, 1952	Stinard et al., 1993
$brn1$	Robertson, 1984	Stinard, 1994
$bt1^*$	Mangelsdorf, 1926; Wentz, 1926	Sullivan et al., 1991
B-Peru	Emerson, 1921	Chandler et al., 1989
$bz1^*$	Rhoades, 1952	Fedoroff et al., 1984 Ralston et al., 1988
$bz2^*$	Neuffer, 1954	
$c1^*$	East and Hayes, 1911	Paz-Ares et al., 1986 Cone et al., 1986
$c2^*$	Brink and Greenblatt, 1954	Wienand et al., 1986

Table 2. (Continued)

Gene	Original Genetic Description	Molecular Development
Dt1	Rhoades, 1935, 1936	a^{dt} -Brown *et al.*, 1989
P[*]	Emerson, 1914, 1929 Lock, 1906	Lechelt *et al.*, 1989
Kn[*]	Bryan and Sass, 1941	Veitt *et al.*, 1990
vp1[*]	Eyster, 1924, 1931	McCarty *et al.*, 1991 Hattari *et al.*, 1992 McCarty, 1992
w3	Lindstrom *et al.*, 1923	
gl5[*], *gl6*[*], *gl7*[*], *gl8*[*], *gl9*[*]	Emerson *et al.*, 1935 Sprague (Unpublished)	Schnable *et al.*, 1995 Moose and Sisco, 1994
in1[*]	Fraser, 1924	Burr *et al.*, 1996
o2[*]	Singleton and Jones, 1935 (see Emerson *et al.*, 1935) Salamini, 1980	Schmidt *et al.*, 1990
r1[*]	East and Hayes, 1911	Dellaporta *et al.*, 1988
rf2[*]	Duvick, 1956	Schnable and Wise, 1994
Self incom- patibility	East and Mangelsdorf, 1925	Haring *et al.*, 1990 Franklin-Tong, 1993
sod1	Baum and Scandalios, 1982	
Transposons		
En/Spm[*]	Peterson, 1953	Pereira *et al.*, 1985
Pvv Mp (Ac)[*]	Emerson, 1914, 1917	Fedoroff *et al.*, 1983
Mu[*]	Robertson, 1977	Chomet *et al.*, 1991 and several labs
Uq	Friedmann and Peterson, 1982	Pisabarro *et al.*, 1991

2 Relevant Genetic Milestones in the Late 19th and Early 20th Centuries

Mendel's discovery and clear elucidation of hereditary units in 1866 at the monastery in Brünn, Austria, began a propitious and exciting era. Little did Mendel know that at about that same time some 1,000 kilometers to the north in Sweden, Alfred Noble (Fant, 1993) was perfecting the utilization of dynamite, and only two years later, in 1868, the United States Patent Office granted him a patent. Given the level of communication in the mid-19th century, the world did not universally know of these two events: dynamite utilization and of course, eventually smokeless gun powder; and, the birth of genetics. Both were to become significant and revolutionary events—the impact of world wars and the changes leading to the genetic revelations of the later decades of 1900s.

Certainly, the rediscovery of Mendelian principles was a most significant event (Correns, 1902; de Vries, 1901). Mendel, though a practical breeder, was probably the first to utilize maize genetics (Rhoades, 1984). Though a number of noted botanists (Carl von Naegeli, Weismann, and deVries) in the late 19th century did study hereditary units, they did not test their ideas with thoughtful strategies. Mendel, however, broke with the past. He practiced precise and deliberate experimental strategies by noting clear and distinct "phenotypes" and more importantly, *recording* these results. This feature is a hallmark of Mendelian analysis, a legacy that is practiced today by every experimental geneticist.

Also in the 19th century, a significant political event stimulated the growth of plant science. This was the development of the land-grant colleges, created under the Morrill Act, signed by President Lincoln following the American Civil War. The Act allowed states to use income from grants of federal land to fund public institutions. This Congressional Act emphasized practical matters such as agriculture and "mechanic arts," and this initiative opened new avenues of research into important crop plants.

At the beginning of the 20th century, the relationship of chromosomes to heredity made genetics an objective scientific endeavor (reviewed by Wilson, 1925). Were the chromosomes the carriers of the hereditary material? Very early, genes were found to be linked on chromosomes with a genetic linkage that could be broken and with a measurable distance between the genes (Sturtevant, 1913). The extensive late 19th century revelations in plant science, such as sperm development, heterofertilization, double fertilization, xenia, (Nawaschin, 1899; Guignard, 1899) and other significant botanical features, also were important to maize research.

What insoluble problems would have confronted a maize geneticist if double fertilization was not understood? In pursuing transmission genetics and following hereditary units, it was important to know whether gametes were coupled (as Bateson professed) or whether genetic elements were coupled, or whether it was a case of replication preference (Bateson, 1916; 1930). [Note that Correns (1899; 1901) unknowingly observed linkage early in the 20th century in segregating progenies of sugary (*su*) outcrosses. It was not until two decades later that this linkage, expressed as "non-Mendelian aborted ratios," caused by a *su-ga* (gamete factor) linkage was reported by Emerson (1934); the original observation was made during a class exercise at the University of Nebraska in which students were attempting to verify Mendelian experiments. How distressing to get an aborted ratio!].

Of course, credit is due to those early observers, who noticed that a parent's traits were coupled in the progeny (Bateson and Punnett, 1911). But without the chromosome relation, how could this be envisaged? What elements in a maize cross were the product of the cross? Could one see the genotype of the male parent expressed among the phenotypes of the progeny of a cross (xenia)? If one was keenly observant, wouldn't one be surprised that the phenotype of the male appeared on the female ear after a cross? Certainly, one would need to know of the workings of double fertilization. Today, xenia is being exploited to add qualitative features such as oil and carbohydrates in hybrid progeny (Dyer, 1994). Further, the relevant features of gamete coupling are being explored (Faure *et al.*, 1994) by scanning electron microscopy.

The chromosome theory of inheritance was developed in the early part of the 20th century (McClung, 1902; Morgan, 1910), followed quickly by the confirmation of chromosome linkage and the knowledge that genes from a parent need not be coupled but could be independent (Bateson and Punnett, 1911;

Bateson, 1930; Wilson, 1925). Principles of heredity were falling into place. Possibly the "blood theory of inheritance" was laid to rest during this period, though it is still prevalent in "present day common parlance."

During this early period of the 20th century, researchers studied quantitative inheritance. G. H. Shull (1909) looked at heterotic phenomena, and his early experiments were a significant beginning for future studies in corn breeding. At approximately the same time, East (1910) and Emerson (Emerson and East, 1913) were experimenting with heredity that seemed to be "continuous." Now, in the 1990s, we are convinced that genes of quantitative traits are apparent, but we are less sure of the exact nature of this control (Peterson 1992; 1993a, b). Researchers currently working to identify quantitative trait loci (QTL) still seek the same answers. Many models are offered to bring molecular findings into plant breeding programs (Jansen, 1993; Jansen and Stam, 1994; van Ooijen and Jansen, 1994).

3   The Maize Plant

The achievements in maize research are possible because of the qualities of the maize plant and maize ear (Figs. 2–5), which can easily be stored, preserved, and scored for genetic analysis. The maize plant's separate male and female organ activities are easily controlled and readily accessible in a crossing program. Further, maize has an unparalleled number of interesting mutants (Neuffer *et al.,* 1968; 1996).

The maize plant is "user-friendly." The individual maize plant (Fig. 2) offers experimenting field researchers in a field-crossing program a "window" of approximately five days of available pollen. This is not to say that there are five days' leeway in attending to a plant, because "nicking" problems often present a window of crossing opportunity of only one day depending on the available receptive silks to be used in the cross (silk receptivity is not everlasting). Thus, the field researcher expects to attend to the crossing nursery continuously through the pollinating period (in the Midwest, approximately July 10–August 15). This could last four to six weeks, with an early low-activity period building to a plateau of heavy activity, followed by another period of low activity in August. The seven-day week is a shocking surprise to those not having "interned" in such a program during their graduate days.

To the corn breeder the corn plant offers a bountiful opportunity for manipulating the genome to a "customized need." The Salazar and Hallauer (1986) experiment to obtain long and short ears in a mass-selection program illustrates this very well (Figs. 4 and 5).

Fig. 2. The corn plant.

13

Fig. 3. Maize ears illustrating the arrangement of rows that have made this plant a unique genetic tool. The kernel mutants as color, or morphological as sugary (*su*) or floury (*fl*), are easily scored to validate Mendelian ratios.

Fig. 4. Hallauer's selection of long and short cobs by mass selection. From this original ear, selection yielded the two extreme types (Salazar and Hallauer, 1986).

Fig. 5. Distribution of the extreme types.

4   Emergence of Corn

From Sustainable Agriculture to Hybrid Corn

Fig. 6. G.H. Shull (Photo courtesy of *Genetics Biographics.*)

Maize genetics began as a food utility. The significance of maize to the early natives of North and South America is emphasized in their art and religion (Townsend, 1992). The early manipulation of maize produced germplasm that was successfully used in the improvement of maize for utilitarian purposes. From its beginning at least 10,000 years ago (Wallace and Brown, 1988), maize has become one of the most productive food plants known. With the introduction of hybrid maize in the early-to-middle part of the 20th century, farmers throughout the world adopted this crop.

Following the initial groundwork on hybrid maize by George Harrison Shull (Figs. 6 and 12) in 1909 (Mangelsdorf, 1955), stimulus for the commercial propagation of maize in the 1920s came from a number of leaders, notably Henry A. Wallace (Figs. 7 and 11) and his allies (Crabb, 1992; Lee, 1984), including Wallace's colleague and plant breeder Raymond F. Baker (Fig. 12). Wallace's faith in the potential of hybrid maize is well-documented (Wallace and Brown 1988). From this beginning, maize research institutes originated throughout the world, including Zemun, Yugoslavia, under

Fig. 7. Henry Wallace of Pioneer Hi-Bred International visiting with Professor Fenaroli at the Institute Cerealcultura in Bergamo, Italy, 1958.

Dr. Milorad Piper in 1945 and Bergamo, Italy, under Professor Luigi Fenaroli (Fig. 7) (which H. Wallace visited in 1958).

The interested reader will want to seek out Crabb's anecdotal history of the "early corn makers" for a description of corn-breeding activity in the early 20th century. As Crabb (1992) relates, corn breeding enthusiasts were on the trail of Shull's pure-line breeding suggestion by 1913. Numerous corn breeders began selling hybrid seed, and many companies evolved that remain well-known today, though a number have been absorbed by international combines led by pharmaceutical and chemical companies.

5 The Bussey Institute and Its Role in the Development of Plant Science

What was the plant science scene in the United States in the early part of the 1900s? Especially prominent in the Midwest was the Biblical version of creation. Therefore Midwestern researchers were slow to bring genetics into mainstream thinking and application although R. A. Emerson (Figs. 8 and 14) (1911, 1914) in the first decade of the 20th century was actively pursuing genetic studies at the University of Nebraska at Lincoln. [Though Emerson's publication on the variegation in mutable pericarp appeared in 1914, this writer estimates, based on the crossing protocol in that paper, that Emerson began his experimentation in the early part of the

Fig. 8. R.A. Emerson (Photo courtesy of *Genetics Biographics*.)

first decade of the 20th century (Peterson, 1995)]. Before the Land Grant Act (1862), universities rarely gave the study of plants a place in their curricula. Perhaps the elitist schools considered botany and other plant sciences too vocational or hobbyist. The overseers of Harvard, however, took the initiative, though somewhat conservatively, and only after a bequest from a prominent donor—namely, Benjamin Bussey, who gave an appropriate endowment of land and other resources. The center for plant science investigation in the United States in the first two decades of this century was the Bussey Institution under Professor Edward Murray East (Fig. 9).

An Illinois native, East worked as an agricultural chemist at the University of Illinois, where he received his advanced degrees. As the Land Grant Act gained momentum and additional resources became available to Agricultural Experiment Stations, East learned of an opportunity for a position in Connecticut.

He moved from the University of Illinois to the Connecticut Agricultural Experiment Station near New Haven. In 1909, E. M. East was invited to take a position at the Bussey Institution (on the recommendation of Professor W. Bateson, then the Silliman lecturer at Yale University). East continued at the Bussey Institution until his death in 1938. In 1909, Shull demonstrated the importance of inbreeding in isolating desirable germplasm, followed by crossing to maximize vigor and productiveness. East (1910) continued along these lines to develop systems of maize seed production.

Fig. 9. E. M. East (Photo courtesy of *Genetics Biographics.*)

The East group at "the Bussey" became the focal point for plant sciences and genetics on the east coast and, essentially, for the United States. Many of the geneticists who will be honored and discussed on the following pages are direct or indirect "progeny" of the Bussey group under Professor East (Fig. 24).

A note on the development of the Bussey Institution is appropriate here, because succeeding developments in maize as well as other areas of genetic research have their scientific roots there (Fig. 24). The Bussey Institution in Jamaica Plains, Massachusetts, about 10 miles from Cambridge, began as a School of Agriculture and Horticulture at Harvard University. The gray stone building, originally built as a private mansion, served as an undergraduate school of husbandry and gardening from 1871-1908 (Anonymous, 1873; Weir, 1994). This Institute was established in 1835 by the will of Benjamin Bussey of Roxbury, Massachusetts. It was Mr. Bussey's desire in bequeathing *"Woodland Hill, consisting of over two acres of land...to establish there a course of instruction in practical agriculture..."* (Anonymous, 1873). The original mandate of the bequest was to give systematic instruction in agriculture and in useful and ornamental horticulture. Obviously, the manicured grounds around Woodland Hill influenced this gift. A sufficient endowment was established, and several professional appointments were made by Harvard University in the early 1870s, including agricultural chemistry, horticulture, entomology, and other disciplines. [An interesting sidelight is the story of Jim Dole. Because Harvard did not offer hands-on practical botany, Dole, while an undergraduate student at Harvard, received his first exposure to agriculture at the Bussey in the late 1890s. This exposure eventually led to the development of the Dole Pineapple empire in Hawaii where massive acreages of the upland region of Oahu show endless row on row of pineapples (Dole and Porteous, 1990).]

Professor East was made Professor of Experimental Morphology in 1914. At that time, it was still too early for the Harvard faculty to acknowledge genetics as a scientific discipline. (Note that it was only a few years after Morgan revealed the chromosome theory of inheritance.) It was not until 1926 that East was given the title of Professor of Genetics. Obviously, the Harvard group that became a Biology Department in 1926 was quite conservative. If only they could have envisaged the latter part of the 20th century, with the emergence of molecular genetics and Harvard's eminent role in advanced genetic findings (Science, 1995).

The Bussey group in the Boston area under the leadership of E. M. East had a far-reaching influence. East attracted many students who aspired to learn more about this emerging science of genetics. He trained a large number of early genetics researchers (Fig. 24), including D. F. Jones (Figs. 10 and 11), who went on to make significant contributions with his practical applications in maize breeding (Nelson, 1993; Mangelsdorf, 1970), and R. A. Brink, with his extensive group of students investigating such questions as paramutation and imprinting, prominent subjects at the end of the 20th century.

Fig. 10. D. F. Jones at the Connecticut Experiment Station near New Haven in the 1940s.

Note that these East-trained students, although educated in the Harvard curriculum, went on to pursue practical considerations in plant breeding. Certainly, D. F. Jones' emphasis on breeding problems (Jones, 1918) had a significant impact on commercial maize development (Nelson, 1993). R. A. Brink also embarked on the improvement of alfalfa in his early period at Wisconsin. Paul Mangelsdorf and Edgar Anderson, also of the early Bussey group, went on to study the evolution of maize. Another of East's students, E. Sears, led research into wheat cytogenetics (McFadden and Sears, 1947).

Though these pages focus on the development of maize genetics, the Bussey group influenced other areas of plant genetics that generated more

Fig. 11. D.F. Jones, Henry Wallace, and Paul Mangelsdorf discussing their favorite subject in a cornfield in the 1940s. (Nelson, 1993; Mangelsdorf, 1975)

genetic discoveries. This impact came largely from one of East's first students, Karl Sax. Sax himself uncovered much on the origin of wheat (Sax and Sax, 1924). One of Sax's students, C. "Charlie" Rick, went to the University of California at Davis in the 1940s and made important discoveries relative to the tomato and his students brought tomato genetics into molecular investigation. The current effort by Steve Tanksley (a Rick student) at Cornell University in relating the various members of the *Solanaceae* is part of the thread (Sax followed by Rick) that began in the East era at the Bussey Institution.

Another of Sax's students was Norman Giles. From his early studies with *Tradescantia* in the 1940s, he quickly adopted the newly developed molecular tools, embarking on gene organization and expression studies in *Neurospora*. Giles attracted a large number of students to Yale. One was G. Fink, now of the Whitehead Institute in Cambridge, Mass, who undertook the study of yeast genetics and uncovered a number of interesting aspects, including the *Ty* transposon. Of course, in the late 20th century, Fink and many others assimilated *Arabidopsis* into their genetic studies.

The reader may want to examine Weir's (1994) descriptions of the growth and development of the Bussey. Yet, the major trunk of the East pedigree tree (Fig. 24) leads to R. A. Emerson.

6   The Legendary School of
Maize Genetics Under Emerson

Among the group from the Bussey Institution that had the greatest impact on the growth of maize genetics is Professor R. A. Emerson (Figs. 8 and 24), who left the University of Nebraska for Cornell University in 1915 to head the Department of Plant Breeding. There, he initiated the "legendary school of maize genetics," (Beadle, 1950). Emerson brought with him to Cornell two young graduate students in maize research, E. G, Anderson and E. W. Lindstrom—the beginning of the Emerson group. In his latter years, E.G. Anderson went to the California Institute of Technology to lead an extensive analysis of the maize material exposed to the Bikini Islands atomic bomb test, along with A. E. Longley and numerous visitors. [Ironically, this A-bomb material was pivotal in advancing maize genetics, as it provided the maize community with numerous chromosome translocations, inversions and an unusual and bountiful assortment of mutants (Anderson *et al.*, 1949)].

Meanwhile, following the completion of his training in 1922, E. W. Lindstrom (Fig. 12), went to Iowa State College at Ames (later, to become Iowa State University) to establish a Department of Genetics—a unit which lasted 70 years before being dismantled in the early 1990s. At ISU, Lindstrom

Fig. 12. At the Heterosis Conference at Iowa State College (Ames, 1952) from left to right: John Gowen, E.W. Lindstrom, G.H. Shull, I. Johnson, and Raymond F. Baker.

21

was pivotal in bringing genetics to maize breeding programs and generating interest among Iowa's commercial corn breeders.

In the second and third decades of this century, the Emerson group was most prominent in laying the broad foundation of maize genetics by collating the accumulated maize mutants, assembling that information, and bringing usable genetic information to the scientific audience (Emerson *et al.*, 1935). Emerson was a very effective leader, bringing together maize geneticists scattered across the country both at annual meetings and through correspondence. Most importantly, however, he was a significant beacon, attracting a coterie of outstanding students anxious to engage in plant genetics.

Among other milestones, the development of the maize chromosome idiotype early in the 1930s by Barbara McClintock quickly established the cytogenetics of maize (McClintock, 1929), following earlier studies by Kuwada (1911; 1915). Emerson's (1921) own studies on plant color genes greatly influenced maize researchers. The Emerson *et al.*, 1935 summary of linkage studies in maize formed the basis for the significant collection for the study of the genetics of mutants (Emerson *et al.*, 1935) and laid the foundation for much of what eventually evolved into the computerized versions of mutants and Zea notes of the 1990's maize newsletters. The reader will find the 1935 publication, with its detailed descriptions and two- and three-point linkage tests (and their often widely divergent values) interesting reading, showing how the great number of early geneticists struggled with and analyzed what are now considered simple concepts. The zeal of George Beadle for this project early in the growth of maize as a genetics tool raised the stature of maize genetics through this publication. Not only was Professor Emerson the leader of this effort, but he was significant in generating and training an enthusiastic number of students who became prominent in the field of maize genetics. His presence with his students in the maize summer nursery at the "Hole" at Cornell University must have encouraged them in their attachment to these maize projects (Beadle, 1950).

Though we are concentrating on the maize "progeny", one should not neglect Emerson's other students; for example, Milislav Demerec (Fig. 13), a young arrival from Croatia in the former Yugoslavia. After leaving Cornell University, he went to Cold Spring Harbor at Long Island, N.Y. and later became the Director of the Carnegie Institution laboratories, as well as the Long Island Biological Laboratory which later evolved into the Cold Spring

Harbor Laboratory (Watson, 1991). Demerec became prominent in the genetics of *Drosophila* and bacteria (Watson, 1991). He also fostered the growth of the world-renowned annual Cold Spring Harbor Symposia, which focused on significant advanced biological problems.

George Beadle, following his maize genetics studies under Emerson, looked to other organisms for further questions on heredity and with B. Ephrussi pursued thoughtful questions on gene interaction and development in *Drosophila*. Not satisfied with *Drosophila* to answer his probing questions on gene action, Beadle went on to other organisms; such as *Neurospora* with E. Tatum.

Fig. 13. M. Demerec at the Carnegie laboratory, Cold Spring Harbor, N.Y.

George Beadle was visionary; soon after hearing C. C. Lindegren outline the advantages of *Neurospora* in a lecture at Stanford in the late 1930s Beadle embarked on one-gene, one-enzyme studies (with E. Tatum), also attracting students and fostering their development. How many Professors other than Emerson have had two students (Beadle and McClintock) awarded Nobel Prizes in medicine?

The fifth generation of Emerson-influenced students (see Schwartz in Part B, Chapter 31 and Fig. 24) are represented in scientific publications of the 1990s and can be expected to continue research into the next century. A number of the geneticists and breeders honored here are students of Emerson or students of those who spent time in Emerson's laboratory.

The reader will find three lines of maize workers in these pages who are commemorated in *MAYDICA* volumes. These groups are not separate but overlapping, as illustrated by the "professional tree" (Fig. 24). The genetics and cytogenetics line began with Burnham, Rhoades, McClintock, Brink, and Stadler; they were followed by Schwartz, Peterson, Laughnan, Nelson, Patterson, Coe, Neuffer, and later by Walden and Smith. Those who concentrated on maize evolution studies, also originating with East, included Mangelsdorf and E. Anderson, who led to Iltis, then to B. Benz and J. Doebley. The plant breeding group includes those also originating with East, leading to R. A. Emerson, H. K. Hayes, and D. F. Jones, who then trained Russell, Brown, Sprague, Hallauer, Gardner, Zuber, and Hooker. There are a number of "crossovers;" both Russell (Burnham) and Sprague (Emerson) had training in genetic-cytogenetic laboratories, and Nelson and Walden studied at a plant-breeding laboratory at the Connecticut Experiment Station with D. F. Jones. These crossovers were significant in enriching researchers' outlooks and scientific strategies.

7   The Early Maize Cytogenetics Group

Burnham, Brink, Rhoades, McClintock, Laughnan

Cytogenetics is a correlative discipline in which genetic findings are confirmed with cytological methods. (It has been succeeded by today's molecular investigation of maize mutants.) The discipline was most significantly demonstrated by McClintock in her studies of cytological crossing-over associated with the exchange of genetic material (Creighton and McClintock,1931). Yet, cytogenetics in the United States had an earlier beginning in the Kansas wheat fields, when McClung (1902) confirmed that grasshopper males had a different chromosomal makeup from females. This and other similar studies were carried on very early in the 20th century (Wilson, 1925).

These observations of grasshopper chromosomes dealt with the idiotype, and should have given credence to the chromosome theory of heredity; however they lacked a gene-related phenotype such as the red-white eye phenotype that *Drosophila* provided (Morgan, 1910). When McClintock perfected the Belling-type staining methods in maize cytology, a new era of maize cytogenetics emerged. Signposts (knobs, chromomeres, centromeres, chromosome length, short-arm, and long-arm traits) on the chromosome could be used as markers, so that chromosome events such as crossover exchanges could be confirmed and related to genetic findings (Creighton and McClintock, 1931). Chromosome breakage seen genetically as marker loss or exchange could be confirmed cytologically. Even fractured telomeres could be related to a mutant phenotype (McClintock, 1944).

Pursuing the same cytogenetic tradition, M. M. Rhoades with E. Dempsey (1953) investigated an inversion in chromosome 3, a truly classic study in cytogenetic analysis. (As one student in a cytogenetics course remarked, "It isn't something you want to read with your feet up on the desk.") Similarly, the extraordinary genetic effects (preferential gene transmission) of abnormal 10 would have been hidden without the "window" provided by cytology. The cytogenetic legacy has been largely replaced in the 1990s by molecular technology with RFLP, PCR, and RAPD (Restriction Fragment Length Polymorphisms, Polymerase Chain Reaction, and Random Amplified Polymorphic DNA, respectively) tools. Cytogenetic activity has lessened for several reasons, such as the allure of molecular technology and a lack of training and familiarity with classical cytogenetic tools. A perusal of the posters and reports in the last decades of the 20th century of the Annual Maize Genetics Meetings provide an example of this trend, although the 1996 meeting witnessed a mild resurgence of interest in maize cytogenetics.

The reader will find an uneven treatment of each of the scientists. In some cases, there is a greater personal familiarity by the author. In other cases, there is an attempt to portray the "genetic atmosphere" in the period—a background that bears on the discovery element of the investigator.

Among the early maize cytogeneticists was Dr. "Charlie" Burnham. Though he was not the first honored in the *MAYDICA* series, he was one of the Emerson group of four who are so well-known in the famous photo of Emerson, his students, and the dog in the Cornell field (Fig. 14). We see in this photograph that George Beadle used a pollinating apron when operating in his maize nursery, whereas M. M. Rhoades, Charlie Burnham, and R. A. Emerson used the "belt technique" to hold their pollinating bags. (Today, most field geneticists have not altered this modest and inexpensive field apparatus.) Though these early workers

Fig. 14. The Emerson group in 1929 at the "Hole" at Cornell University, Ithaca, New York, left to right (standing) C. Burnham, M.M. Rhoades, R.A. Emerson and B. McClintock. George Beadle with dog. (from M.M. Rhoades early 1930s)

used such simple implements, their work and experiments were quite creative. The Emerson group of Burnham, Rhoades, McClintock, and Beadle was exceptional, and they found the genetic materials propitious for their use. They aggressively analyzed these genetic materials. Burnham had a long history in cytogenetics and spent a period in Emerson's field. He is scientifically and academically related to another member of the *MAYDICA* series—one of his students, W. A. Russell—and also to R. A. Brink, under whom he earned his doctoral degree. Burnham assembled a vast amount of information on cytogenetics in the volumes he published, and worked diligently to bring to the forefront much of what is now known about inversions, translocations, and other aberrations related to maize cytogenetics. Many of his students have gone on to carry his banner. Burnham's students and his line of scholarly descent represent one of the few remnants of true cytogenetic effort. Most prominent among the Burnham students is Ron Phillips (University of Minnesota), who has in turn trained a number of people in this area. Efforts in cytogenetics also are continued by Wayne Carlson (a Rhoades student) at the University of Iowa and by Marjorie McGuire (under Lowell Randolph at Cornell University) at the University of Texas in Austin, Texas.

After his "retirement" (1972-1992), Burnham devoted his energy to additional maize genetics experiments, including interdependent rings-of-four, use of interchange heterozygosity for microwaveable popcorn, and ornamental corn. His principal focus following his retirement, however, was on restoring the American chestnut tree (*Castanea dentata*). He started the non-profit American Chestnut Foundation, which now boasts almost 2,000 members, maintained a research farm in West Virginia, had several state chapters, and awarded grants for chestnut research. Burnham initiated a backcross breeding program in chestnuts; the first backcross trees are now in the nursery. He passed away in 1995.

R. Alexander Brink spanned an interesting era of genetics. He was very active in the early post-Mendelian period (the 1920s) and continued into the era of molecular genetics. Many of his early studies, such as on the *wx* locus, found their way into molecular investigations many decades later. In his Bussey days, Brink initiated studies on pollen which eventually led him to discriminate *Wx* from *wx* genotypes with iodine staining. His seminal studies with the *wx* locus became the focus of study of a number of laboratories, most prominently the Susan Wessler group at the University of Georgia (Athens, Georgia), utilizing the vast array of *wx* mutants with their numerous insertions to uncover the molecular polymorphism so rampant in the maize genome. Brink's studies on genetic crossovers related to cytologically observable cytological events were parallel to the Creighton and McClintock demonstration (Brink and Cooper, 1935). Brink did not originate in the Emerson school, but, as indicated earlier, he was trained under E. M. East at the Bussey Institution. After earning his doctoral degree at the Bussey Institution, Brink went to Wisconsin, where he initiated the development of a genetics department with an outstanding group of geneticists.

When R. A. Brink retired, a number of papers were assembled to honor him, and a large number of his former students traveled to Madison, Wisconsin, for the occasion. The papers represent the diversity of his students, who are now contributing to the fund of maize discoveries worldwide. Brink, with diligent effort, brought out a number of interesting findings with his studies of the *P-vv* allele. Though the allele "gestated" for four decades after its early description by Emerson (1914; 1917), it was finally cloned and given further study by a number of groups (Lechelt *et al.,* 1986; 1989; the S. Dellaporta group at Yale University; T. Peterson, now at Iowa State University; and Erich Grotewald at Cold Spring Harbor New York). (Eight decades after the first genetic studies, the *P* allele was brought into the mainstream of heterologous relationships where it was found to be a *myb* homologous protein.) The paramutation story and further effort with the *R* locus currently are being pursued most notably by Brink's student Jerry Kermicle and by the Chandler group at Oregon (Patterson and Chandler, 1995). Equally important are his earlier studies with differential pollen-tube growth, plant reproduction with alfalfa breeding, maize breeding, and disease resistance in alfalfa. Brink passed away in 1984.

Marcus M. Rhoades: The *MAYDICA* dedication for Marcus M. Rhoades came in a second dedication in his 80th year when a large number of his students spanning a 30-year period assembled at Bloomington, Indiana in 1983. (The first dedication, at his 70th year, was published in *Theoretical and Applied Genetics.*) He undertook his graduate studies in the Emerson school at Cornell University (Fig. 14). The E. Dempsey (1994) biographical sketch, an excellent source of M. M. Rhoades material, portrays the life of a devoted scientific collaborator. This early group from Cornell University, especially B. McClintock and M. M. Rhoades, continued to meet and discuss mutual problems and interests in maize genetics well after their Cornell days while both were in the New York area (Rhoades at Columbia University and McClintock 35 miles away at Cold Spring Harbor). With that beginning in Emerson's laboratory, Rhoades, his students, and his students' students, have contributed greatly to maize genetics.

Rhoades always was meticulous and vigorous in his analytical procedures, probing the analyses of his maize crosses and passing judgment, and imparted to his students a clear dedication to excellence though there was a "sink or swim" quality to his training. This is evident in the productivity of his "scholarly progeny" (Fig. 24). He never hesitated to analyze his own material in a most dedicated manner often with his colleague of more than four decades, Ellen Dempsey.

Rhoades' move to Illinois in 1948 from Columbia University contributed to the growth of maize cytogenetics. This was the re-entry of cytogenetics and especially maize genetics to the Midwest in the "corn belt." The Illinois site was unique because of its aggressive recruitment of top faculty. In the Department(s) of Chemistry and Biochemistry, Irwin C. "Gunny" Gunsalus was joined by S. Spiegelman and Salvador Luria. This group developed molecular biology in the late 1940s and early 1950s. An extensive array of speakers invited to the Illinois campus by Gunsalus, Luria, and Spiegelman generated excitement about genetics on the Illinois campus and among the array of diverse students gathered in joint seminars and by this, enriched each other's courses (for an example, Alan Campbell from Spiegelman's group in his genetics course in Botany). Thus, was the tempo of the Illinois campus in that period.

In the Botany Department at the University of Illinois, Rhoades taught his well-known cytogenetics course beginning in 1949. He patiently outlined, among other landmark papers, the basic and classical demonstration of four-strand crossing over and the chromosome theory of heredity (Bridges, 1916) in *Drosophila* to an eager group of students. This was prior to the discovery of the DNA double-helix (Watson and Crick, 1953) or later events such as the discovery of genes with introns, exons, promoters, and codigenic sequences (Singer and Berg, 1991). Imagine the excitement of concentrating on the classics in basic genetic concepts, a focus that is now absent among currently trained geneticists. Such creative efforts on gene dissection (the A^b complex by John Laughnan), Bithorax in *Drosophila* by Ed Lewis and the numerous lozenge alleles by Mel Green were some of the pre-DNA-era highlights of the 1940s. Such was the "genetic world" that Rhoades encompassed and unraveled to his students when he arrived in Illinois (1948). Intracistronic exchange and bacterial conjugation were yet to be uncovered (Benzer, 1955) (Lederberg *et al.*, 1951). Although the Avery, McCloud and McCarty (1944) experiment on pneumococcus transformation was discussed, it had yet to be fixed in genetic thinking. Surprisingly, the Avery, McCloud, and McCarty experiments on the transforming principal (*tp*) were not readily connected with the Hershey and Chase (1952) blender experiments (not cited by Hershey and Chase, 1952) that finally culminated in the description of the double helix. Only after the fact was *tp* connected to the thread leading to the DNA discovery. Rhoades passed away in December, 1991. As a fitting tribute from his colleagues, the 1992 Asilomar meeting was dedicated to his contributions to maize genetics.

Barbara McClintock's early successes (1931, 1933) were with the Emerson group. Her time in the Emerson laboratory overlapped with that of Marcus Rhoades and George Beadle. [Though formerly under the tutelage of the plant cytologist Professor L. W. Sharp (known for his universally used textbook *Introduction to Cytology*), in the Botany Department at Cornell University, her mentoring in Genetics was in the Plant Breeding Department under R. A. Emerson. With a minor in Genetics with A. C. Fraser, it is evident that McClintock had her experimental experience in Emerson's field nursery (Fig. 14).] One could say that her presence at that time in Cornell was the ultimate in synergism. Here was maize, a genetic tool with an abundance of mutants; eager colleagues; and an intellectually receptive environment. Everyone at Cornell recognized McClintock's talents even at the beginning of her career in the 1920s (Rhoades 1984). To the Emerson group's already-developed genetic basis, she brought her remarkable and disciplined observational skills which developed into an innovative cytological talent. Probably, one of her major assets was the ability to maintain a focus. In essence, her efforts marked the beginning of maize cytogenetics, in which the chromosome order (short-to-long-arm) and chromosome length (longest to shortest) would be established. When McClintock was born in 1902, two years had passed since the rediscovery of Mendel's laws and the relation of chromosomes to the sex phenotype in grasshoppers (McClung 1902). By the time she arrived at Cornell from Brooklyn, ten years had passed since Bridges

(1916) had demonstrated, and thus established, the chromosome theory of heredity with exacting cytogenetic thinking.

McClintock was a steadfast and prodigious worker during those early Cornell days, and this attitude continued throughout her career. Her early-morning forays in the maize nursery at Cold Spring Harbor in those humid summer days situated between the Carnegie Institute Library building and the watery inlet of Cold Spring Harbor on the northern shore of Long Island Sound, were a common sight during growing seasons from the 1940s to the 1970s. Her white shirt and field outfit were fresh and spotless, a remarkable contrast to many of us who spend days (and nights) in a genetics nursery. She worked late hours in the laboratory, analyzing her cytological specimens in order to confirm the genotypes prior to the next day's crossing program. This is the ultimate in cytogenetic efficiency—the nightly cytological results dictating the next day's genetic crossing strategy. But, McClintock knew her limits and how to relax during those busy summer days. Cold Spring Harbor is an idyllic setting (Watson 1991), and she often could be seen stretched out on the sun-drenched floating dock in the Harbor, to recover her strength before proceeding back into the laboratory. Such was the soothing effect of the Cold Spring Harbor waters in the mid-summer and especially on Indian-Summer fall days. (The author would like to note here that these personal notes of McClintock come from a one-year informal association while the author was an assistant with M. Demerec at Cold Spring Harbor and occupied lodgings on the lower floor of the same building as McClintock's laboratory).

McClintock knew how to relax in other ways. She found jazz music soothing. She would often listen to stations coming across Long Island Sound from stations on the southern Connecticut shoreline. She was also a fan of Marlene Dietrich's singing, probably as a result of McClintock's short stay in Germany. She did have a small record collection of Dietrich's songs. Barbara's mother (a painter, I believe water colors) and father, as well as her sister and nephew, would visit her at the Cold Spring Harbor Laboratory during the late 1940s. The author and his wife, Sally, often would converse with her on the portico of the Animal Building (now the McClintock Laboratory) about things in general and exciting goings-on in her maize field during that 1948 summer.

Cold Spring Harbor (The Carnegie Institution of Washington Laboratory (CIW) and the Long Island Biological Laboratory (LIBL)) during the late 1940s and 1950s was (and still is) a center of cutting-edge concepts in biology.

There were many visitors: Max Delbruck from the California Institute of Technology; Salvador Luria from Indiana University, coming for the summer of 1948 with his students Jim Watson and Richard Siegel (three years before Jim Watson would embark on his scientific pursuits in Copenhagen and later in Cambridge) (Watson 1980); and Mark Adams from New York University, with his eagerly attended phage course. McClintock enjoyed the discussions of advancing topics of genetics with eager young visiting researchers as well as those attending the significant Cold Spring Harbor Symposia, especially during lunches at the venerable Bancroft Hall. The Cold Spring Harbor visitors were an exciting group, and as Watson relates, it was a summer at Cold Spring Harbor that "turned him on." But the summer would end, and all the visitors would return to their campuses. The Carnegie laboratory would return to the winter season quiet-time with four to five principal researchers and their assistants. During those days in the 1940s, McClintock would continue with her analysis, while the symposia attendees and summer visitors returned to their laboratories throughout the United States and the world and pursued their area of research. There is little evidence that these numerous visitors discussed her current effort (transposons) in the late 1940s; if they did, they could hardly have put it into a proper context. The confirmation that transposons are universal, not necessarily related to a feature only in a maize geneticist's nursery, came with the discovery of the IS2s in *E. coli* in 1967 (Jordan *et al.*, 1968; Shapiro, 1969) still nearly two decades away. Only then did the universality of transposable elements resonate among the biological community, and McClintock's contribution was recognized (Peterson, 1995).

McClintock's first major contribution was the demonstration that chromosomes are individually recognizable by their relative lengths and arm ratios (Fig. 17)—and this is still basic today in maize genetics. This was preceded by her adaptation and meticulous development of Belling's acetocarmine staining techniques. McClintock experimented with and perfected laboratory routines for the sound development of maize genetics and cytogenetics so that genetic maps with proper short-arm centromere long-arm orientation could be developed (Fig. 18). (She was also a perfectionist in photographing her specimens.) Further, distinct patterns were associated with knobs and chromomeres at characteristic positions. She clearly developed the concept that gene linkage can be broken by exchange and can be proven by cytogenetic markers (Creighton and McClintock, 1931); it was an elegant example of chromosome manipulation in maize cytogenetics that was simultaneously demonstrated in *Drosophila* (Stern, 1931). This finding was associated with McClintock's attempt to improve the capacity to analyze chromosomes and her efforts to make maize a successful cytogenetic tool. Her attempts to make staining and chromosome identity techniques readily available to other researchers aided the development of maize cytogenetics. Even her photos were taken and developed by her because only she could capture what she saw under the microscope.

The Carnegie Institution of Washington in the first half of this century was an idyllic environment for genetics and for research in general. It was most suitable for Barbara McClintock—in fact, it was probably the only setting in which she could proceed unencumbered by the common distractions of most universities. There at Carnegie, largely answerable only to her own initiatives and direction, she could proceed with innovative ideas without answering to funding agencies. [In fact, one director at the National Science Foundation (NSF) has indicated that McClintock would have had difficulty with that agency's demands regarding proposal approval or peer review panels. It is difficult to imagine that a typical NSF panel would have found the early transposon investigation worth supporting. Outlining prospective detailed experiments and providing anticipated answers was not possible in the early transposon experiments. Most of the exciting results were serendipitous.] No. Dogma in science is difficult to overcome as is the case for many findings, reverse transcriptase, for example. Thus, this Long Island site was most appropriate for McClintock's unencumbered efforts in maize cytogenetics.

How many maize workers have observed the same material with variegation that McClintock saw when she first uncovered transposable elements? Her insight into the workings of maize quickly unraveled basic genetic concepts which have now been confirmed (Peterson, 1994). She had enough confidence in her observations to go against the prevailing tide, and she was not diverted by non-believers. The discovery of transposons in the 1940s opened up a vast field of study that will carry investigators into the 21st century. Much of her transposon research was conducted during the mid-1940s during the United States' participation in World War II. Despite the intensity with which she worked single-handedly in the laboratory and genetics nursery, she avidly read the *Christian Science Monitor* and the *New York*

Times for accounts of the activity and progress of World War II. She was very concerned about world affairs and the dangers of irradiation.

She was also an active correspondent. She typed responses to many who wrote to her to discuss concepts or request seed (Figs. 19–21). She often initiated correspondence to relate her views and expressed admiration to authors on interesting findings. Her notations in the margins of her copies of the *Maize Genetics Cooperation Newsletter* attest to her intense interest in numerous findings. McClintock could be called an "activist" in things genetic. Following her retirement from active research, she actively "recruited" visiting scientists to Cold Spring Harbor to discuss the latest findings in a wide assortment of subjects. This writer remembers the Berkeley geneticist Gerry Rubin visiting the Cold Spring Harbor Laboratory in the mid-1980s and discussing genetics with McClintock until the early morning hours. Few question that her attention to detail and to critical analysis catapulated maize into its present position as a genetic tool of great prominence. Rhoades writes of McClintock's many contributions in this volume (see Part B, Chapter 26). The reader may find Regina Kahmann's (1992) account of some of McClintock's daily life interesting. Evelyn Fox Keller (1983) captured highlights of her life and work in the McClintock biography, *A Feeling for the Organism*. McClintock passed away in September, 1992.

John Laughnan: During the 1940s, possibly because of World War II, there was a dearth of developing maize geneticists. But at the University of Missouri, under Louis Stadler's tutelage, H. Roman, S. Fogel, L. Smith, J. Laughnan and others were undertaking graduate studies in maize genetics. Laughnan did not come from the Emerson group, though his mentor, namely, Stadler (Fig. 15) spent a brief time at Cornell with Emerson [according to M. M. Rhoades (1984), Stadler wasn't welcomed, possibly due to his "laid-back" nature]. Stadler was trained elsewhere. Yet, the training under Stadler was rigorous. Two of Stadler's students,

Fig. 15. L. J. Stadler (Photo courtesy of *Genetics Biographics.*)

Hershel Roman and Seymour Fogel, with the background and rigorous training in maize genetics, went on to make impressive contributions in pioneering studies in yeast genetics. Could it be that a maize thinking-framework could be applied to yeast and develop yeast into a viable genetics? A look at current issues of *Genetics* clearly illustrates that yeast is a very actively researched genetic organism.

Laughnan, trained in the Missouri Agronomy Department, brought a plant-breeder's eye to a genetics nursery as he blocked out ranges (vs. the traditional serpentine nursery arrangement) for massive crosses to recover that rare crossover in $A1^b$. He vigorously pursued a successful strategy for attacking the dissection of complex loci. He "nursed" his one or two recently discovered *sh2* plants (Mains, 1949) for use in his $A1^b$ crossover investigations in 1948, even "milking" late-shedding pollen after sunset using his truck headlights to see his way in the nursery. His early work with the anthocyanin loci and the unraveling of the origin of derivatives arising from the complicated $A1$ locus, at a time when the gene concept was still abstract (late 1940s, early 1950s), was at the forefront of genetics, along with *Drosophila* workers (M. Green with *lz* and E. Lewis with *Bi*) in the pre-DNA period of the dissection of genes. The separation by crossing over of the $A1^b$ allele into α and ß components was a focal point for plant gene study. Could all genes be tandemly duplicate and be so separated? Genetics was the only "window" into the gene content of chromosomes. In the late 1940s, legitimate questions were: How were genes arranged—into separable components? Always? What was the space, if any, between genes? This was the time when the dark- and light-staining bands seen in salivary glands of *Drosophila* were the visual representation of gene concepts.

Puzzled by nonreciprocal exchanges in the origin of derivatives, Laughnan was not content with the limited opportunity presented by maize chromosomes for observing non-reciprocal events. Was there a universal mechanism to explain these events? To investigate this, Laughnan needed visible and recoverable cytological events; the salivary glands of *Drosophila* were the ideal tool. Because of this, Laughnan went on to test his concepts in *Drosophila* (with a sabbatical leave at the California Institute of Technology), quickly adapting *Drosophila* technology using salivary gland analysis at the Bar gene complex as a more useful tool, with which he attempted to test the basis of the origin of non-crossover derivatives. But maize genetic studies beckoned him back to the cornfield (Fig. 16). His research studies with the cytoplasmic male sterile in the *S* system are significant in showing that reversions were associated with changes in the nuclear genome. Obvious throughout Laughnan's work is his thoughtful analysis and his pursuit of every last detail in his quest to resolve a research problem. Above all he was a very sympathetic listener and willingly explained and described any concept that would clarify a point. Laughnan was in this way, much like Bruce Wallace (the author's supervisor during the Cold Spring Harbor days in the late 1940s) who derived great pleasure in describing genetic events to the novice. Laughnan's numerous students are a testimony to his influence as a teacher and his efforts to bring quality assessment to genetic analysis.

Fig. 16. The Illinois group in the 1970s at the University of Illinois experimental plots at Urbana, Illinois. Kneeling from left to right: Bob Lambert, Art Hooker, D.E. Alexander, and John Laughnan. Standing (left to right): Earl Patterson, G.F. Sprague, and John Dudley.

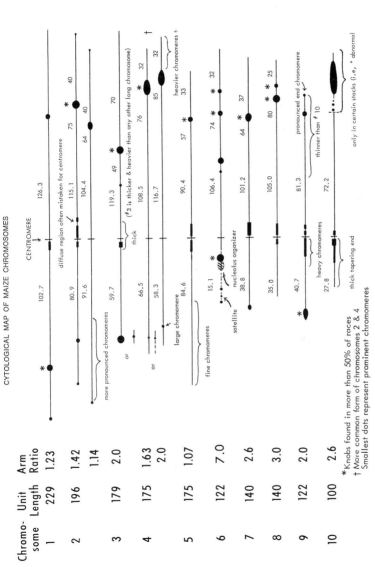

Fig. 17. Cytological Map of Maize Chromosomes. Drawn to scale with major distinguishable characteristics (based on data from McClintock and Longley, also Longley and Kato, 1965). Reproduced from *Mutants of Maize*, 1968, p. 4, by Neuffer, Jones and Zuber, by permission of the Crop Science Society of America.

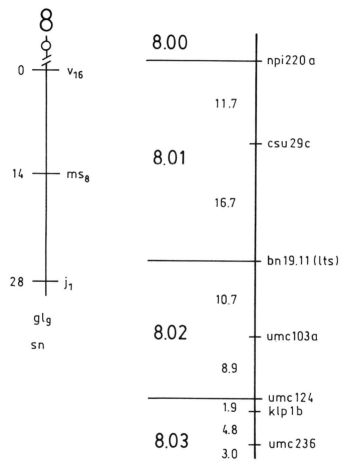

Fig. 18. Map of Chromosome 8. Left—genetic map; right—portion of map illustrating the bins with molecular markers.

CARNEGIE INSTITUTION OF WASHINGTON

DEPARTMENT OF GENETICS

COLD SPRING HARBOR, LONG ISLAND, N. Y.

March 22, 1951

Mr. Peter Peterson
308 Natural History Bldg.
University of Illinois
Urbana, Illinois

Dear Peter:

Your letter came this morning. I was much interested to learn of your problem and the manner in which it is progressing. It appears as if you had made considerable progress.

This past fall, much of the mutable chlorophyll material was thrown out as I realized it could not be followed if other studies were to continue. I believe some ears of the mutable luteus stock and possibly some of the mutable pale green were saved. The mutable luteus is autonomous but the pale green is controlled by another factor (not Ac). If you wish to test for Ac, tester stocks are available and I should be glad to send some to you. Will send seed to you shortly.

Please give my best to Sally. I hope all has been well with the family.

Sincerely,

Barbara

Barbara McClintock

Fig. 19. McClintock letters to Peterson illustrating her selfless effort to extend advice and help to young maize geneticists. Copy of a typical McClintock response to a question and request for seed or exchange of ideas.

CARNEGIE INSTITUTION OF WASHINGTON

DEPARTMENT OF GENETICS

COLD SPRING HARBOR, LONG ISLAND, N. Y.

April 23, 1951

Mr. Peter Peterson
Department of Botany
University of Illinois
Urbana, Illinois

Dear Peter:

     I am sending you some seed that I hope will have
what you need. They are not just what I would like to send;
but as I wrote before, most of the seed of var. lu and var.
light green has been discarded.

     Packages #1 and #2 should segregate lu^S and lu^{var}.
The variegation is intermediate. You may see little
variegation until the plants are quite well grown (Lu → lu
type). These plants should be used in crosses as they will
carry lu^S.

     Packages #3 and #4 have seed that may segregate light
green. I have not made tests. They come from the self-
pollination of F_1 plants derived from crosses of normal x
variegated light-green. I have not tested this seed for the
presence of light-green. May I suggest that you make a sample
test in the greenhouse before putting seed in the fields. If light-
green is present, grow plants first in greenhouse and transplant
to the field when they are fairly large.

     Packages #5 and #6 contain seed for Ac-testers. #5 is
homozygous for C Sh wx Ds and no Ac. It can be used against c.
#6 is homozygous I Sh Wx Ds, no Ac. It can be used against
colored kernels to give I-C variegated when Ac is present.

     All good luck on your problem. Please give my best
to Sally.

Sincerely yours,

Barbara McClintock

Fig. 20. Other exchanges.

CARNEGIE INSTITUTION OF WASHINGTON

GENETICS RESEARCH UNIT

BOX 200

COLD SPRING HARBOR, NEW YORK 11724

August 21, 1973

Dear Peter,

My apologies for this long delay in answering your several letters. The one written on May 29 requesting use of the photos that I had sent to you, arrived here much after your June 7 visit. Of course, you could have used these photos and I hope that you did. I assumed, however, that your paper had already been submitted by the time of your visit here.

I will gladly send you kernels with wx-m7 and also with wx-m9 for comparison. If you let me know what you have in mind for wx-m7 I can select the proper states for your purposes. wx-m7 is reported and illustrated in the Carnegie Institution Year Book number 64, issued December 1965. Photos are also shown in Figure 1 of my paper in the Brookhaven Symposium no. 18, (1965).

Thank you for inviting me to attend the ceremonies associated with presentation to Marcus Rhoades of the volume commemorating his 70th birthday. Because of unusually heavy pressures on me during this summer, I developed a disturbing degree of arrhythmia (irregular heart beat). Although in my case it is psychosomatically induced, it does require that I take care to avoid serious consequences. Thus, I am attempting to avoid all extra strains until the intense fatigue and restlessness this summer has imposed has subsided. Nevertheless, I do hope that the celebration is a happy one.

Best regards,

Sincerely,

Barbara

Fig. 21. In 1973, McClintock discloses her developing arrhythmia.

8 The Maize Breeding Group

Sprague, Russell, Hallauer, Brown, Gardner, Zuber, Stuber

Certainly one of the great achievements in plant breeding is the remarkable success of hybrid maize, begun with the experiments in the early part of this century by G. H. Shull (1909) at the Carnegie Institution of Washington at Cold Spring Harbor, and by many private plant breeders. (Many of the latter created seed corn companies, which were absorbed in the multinational acquisitions of the 1970s and 1980s.) The annual incremental rise in maize yields worldwide is a testimony to the efforts of a number of plant breeders.

W. A. Russell: "Russ," a plant breeder at Iowa State University, was most influential in capturing the inbred lines developed from the Iowa (B) Stiff Stalk Synthetic (BSSS) populations, initiated by G. Sprague in his "pooling method" (soon to be discussed). One of the most successful maize breeding lines that Russell uncovered—namely, the B73 line—was developed from BSSS and is used in many maize breeding programs throughout the world. Russell, a diligent worker, excelled at field activity (enduring hot, dry summers, and in the fall, wind, snow, and sleet until the last harvest was "in the bin") coupled with rigorous analysis of data to bring forth so many desirable inbred lines. Certainly, significant were his untiring efforts to select for greater resistance to the devastating effects of the European corn borer and for improved stock quality. He took part in the "team effort" of Iowa State University and the USDA throughout the best part of the 20th century. He has influenced many students, a number of whom are now successful maize breeders and leaders in the maize industry. An outpouring of friends, family, and appreciative industry colleagues came to his retirement party in 1988. Following his retirement, he still continues to apply himself daily to questions of maize breeding, writing reviews, and reviewing journal articles.

George Sprague (Figs. 16 and 24) came out of the genetics program at Cornell University and was exceptional at combining genetics and plant breeding in this century, in the tradition of E. M. East. He has diligently applied genetic techniques, technology, and thinking into his plant breeding efforts. In the 1930s, Sprague swam against the tide and decided to "pool" his genetic resources (his best lines that contained stiff stalks), thereby initiating development of the successful BSSS maize population continued by Russell. BSSS is widely used at Iowa State University and has been the origin of a large number of successful inbred lines, many of which are widely used in various countries throughout the world. The practices Sprague began in the 1940s are still prominent in generating successful breeding material (see the Hallauer entry in Part B, Chapter 24). What was his rationale for such an undertaking? The systematic assembly of genotypes from the original 16 lines in the origination of the BSSS series was a unique opportunistic event. From any plant, only two alleles could operate. This Sprague knew; but little did he know that approximately five decades later, two of the lines in the original 16 would be found to contain two active transposons—the *Uq* (*Ubiquitous*) and the *Mrh* (*Mutator*) Rhoades and Dempsey, 1982; Peterson and Salamini, 1986—that could have contributed to the variability pervasive in the BSSS series.

Sprague has delved into many genetic and plant-breeding areas during his long career and initiated many kinds of experiments. His efforts with the barley stripe mosaic virus in order to induce mutations inspired a number of subsequent researchers. Even 20 and 30 years beyond his "retirement," he continued to produce papers of significance. His extensive cadre of students has been successful at carrying out his strategies, illustrating and demonstrating his foresight, and developing of superior maize-breeding material (Fig. 24).

Sprague's impact on plant-breeding initiatives and policy is significant and is a testimony to his conviction that traditional plant-breeding efforts by the USDA should not be altered. When molecular biology emerged in the 1960s as a potential tool for plant breeding—for example, with the then-emerging isozyme technology—strong consideration was being given by the USDA (Washington, D.C.), to a reallocation of resources within the agricultural community. This question was often raised: Shouldn't some of these classical programs be phased out and replaced with more modern areas of study? Sprague was then a USDA administrator in Washington D.C. Though sympathetic to the new initiatives, he strongly defended the retention of traditional programs. He often reminded the USDA corn breeders not to discard their pollinating aprons. Today, the maize community is still benefitting from "traditional" programs in genetics and plant breeding. Strategy development for the breeding of maize remains largely unaffected by molecular biology, despite the enormous output of new molecular findings. Maize-breeding programs have prospered in the development of superior germplasm, leading to improved hybrids, but have been minimally affected by the findings and technology of molecular biology, despite the extensive RFLP effort. Sprague showed considerable foresight and courage in defending the retention of the traditional programs, which now appear viable well into the next century.

Arnel Hallauer: As previously stated, the Iowa State University corn-breeding program has a long history of success. The program began with the USDA's effort in Ames initiated by Dr. M. T. Jenkins, then continued with his successor George Sprague. Sprague's successor was Arnel Hallauer. Further evidence of the success of the ISU corn breeding program is the attraction to the environs of the Ames' campus of numerous companies that have located in the central Iowa, Story County area with more than 50 plant breeders now located in the area. A product of the Sprague tradition at ISU, Hallauer was in a position to continue long-range plant breeding programs and thereby foster those initiatives for the next 30 years. Maize breeding programs can require years, or even decades, to develop products, as well as the patience to ignore the demands of the publication process. It takes confidence to pursue a strategy developed from intuitive and rational thinking, taking clues from numerous experiments generated by graduate students, technicians, and a principal investigator. Only at a university or in the USDA could scientists be sufficiently buffered to pursue such objectives. Hallauer's efforts toward population improvement have been the basis for the supreme success of the Iowa Corn Breeding Program. This effort to improve populations has been a continuation of the Sprague tradition; it includes widely dispersed, statewide yield plots, long days, cold, late-fall harvests, driving rains, sleet, and fingers frozen getting in the last harvest of the fall. With today's machines and computer-assisted harvesting, field work is somewhat easier; but those who survived training in the Sprague "field tradition" ultimately were successful in their own careers. Hallauer is a true field-oriented maize breeder. A maize breeder has to know his breeding materials, and this knowledge comes from field observations.

Hallauer has conducted many kinds of studies, many demonstrating the vast volatility of the maize genome. This trait was most noticeable in the long-ear/ short-ear experiment (Salazar and Hallauer, 1986), which was achieved quickly by the selection of the two extreme types with mass-selection procedures (Figs. 4 and 5). He has long been active in manipulating the corn genome in the development of new heterotic groups coming from subtropical regions. He is a prolific developer of students for the maize community and has held together a productive maize-breeding group in the ISU Agronomy Department. Hallauer's program attracts an international clientele, and he is frequently called upon to advise and consult with the international community, including Europe, Asia, and South America. His textbook written with Fo Miranda (Hallauer and Miranda, 1988) is highly regarded worldwide as a resource for teaching maize breeders and plant breeders. Hallauer's successes are well outlined in the tribute written by Drs. Lamkey and Sprague (Part B, Chapter 33). In 1991, he was voted a member of the U.S. National Academy of Sciences, a testimony to his contribution to the science of plant breeding.

William L. Brown: Brown is another kind of success story in plant breeding. He studied at Washington University with Professor Edgar Anderson (an East student, see Fig. 24) with a broad plant-science background; he subsequently joined the research staff of Pioneer Hi-Bred International. Despite a commercial orientation, he maintained a dedication to basic science during all his years in administration. He was a devoted scientist who excelled in cytogenetics and consistently pursued his research, publishing a number of significant papers related to the diversity and range of germplasm in maize. Genome diversity and its conservation and utilization were his main focus. Brown's career was one of the most diverse of the people in this collection. From his early studies of improving maize at the parental level, Brown went on to do basic cytogenetics work, including a postgraduate period with Barbara McClintock at Cold Spring Harbor. His effort with field crosses coupled with lab-bench analysis was short-lived because his research and leadership qualities were recognized by Pioneer Hi-Bred International, where he was propelled to administrative duties. He rose rapidly through the ranks and eventually the company's president, then chairman of the board during a period in which biotechnology was ready to complement classical breeding techniques. During his tenure with the company, Pioneer gained prominence and has become one of the most successful distributors of hybrid seed in the world. This company has locations worldwide and its success is due largely to the efforts of its breeders and to Brown, who provided strong support and leadership to the research effort. Following his retirement from Pioneer, he was associated with a number of initiatives of the National Academy of Sciences (He was also a member) and other groups. He was a significant proponent of germplasm conservation and usages.

Brown's most effective efforts were to encourage the improvement of maize in the lowland tropics of Latin America and other areas of the world. He had a strong interest in maize improvement in Latin America, particularly because of his contacts with the Centro Internacional De Mejoramiento De Maiz Y Trigo (CIMMYT) group, which included Norman Borlaug and E. Wellhausen. Brown was one of the first maize geneticists to promote the use of tropical germplasm in the improvement of U.S. corn-belt maize by incorporating tropical germplasm into temperate maize races. This he called "tropicalization" of temperate maize for Latin America. While a member of a task force with the National Academy of Sciences, he promoted global awareness of what was considered the erosion of genetic materials. Thus, he encouraged the collection and conservation of this irreplaceable genetic diversity in germplasm banks throughout the world, and founded the noted germplasm-related journal, *Diversity*. This global germplasm effort was continued by the National Academy of Sciences National Research Board, and stands as a landmark study in effecting global germplasm conservation. William Brown passed away in 1991.

Charles Gardner: Continuing in the area of plant breeding and quantitative inheritance, we come to the long-term Nebraska corn breeding programs led by Charles Gardner. Charles Gardner emphasized the study of quantitative genetics of maize populations and the application of quantitative genetic information to corn breeding. His years at the University of Nebraska brought him to great prominence in both theoretical and applied study of quantitative inheritance. In response to the apparent ineffectiveness of mass selection techniques for improvement of maize populations, Gardner introduced the grid method, which reduces the confounding effects of the environment. He conducted long-term mass selection for yield improvement in Hays Golden and showed that the modifications for conducting mass selection were effective. [Note: this author uncovered an active *Ac* (not a *Uq*) in the Nebraska-originated Hays Golden]. The grid system of mass selection has been used in many other crop species. Gardner, his students, and his colleagues also developed models and analyses that extended our understanding of the importance of epistatic effects in the inheritance of quantitative traits; partitioning the variability among parents and their crosses for average, general, and specific heterosis; the effects of selection on changing frequency of alleles; and developing general formulae for predicting genetic advance for different methods of selection. Of course, the concept of epistasis, though as old as genetics itself, has not become more understandable, even with the bountiful revelations flowing out of molecular biology and biochemistry. That it occurs is not deniable, but its role in plant breeding is difficult to measure. That different loci collaborate in interactions is well-documented, such as, for example, the anthocyanin pathway requiring the *C* and *R* loci (Hattori *et al.,* 1992; McCarty, 1992) or the analysis of the heterotetrameric adenosine diphosphoglcose pyrophosphorylase that is a lead player in starch synthesis (Giroux *et al.*, 1996). Even the most significant plant-breeding concept, *heterosis*, has defied explanation. Although the corn industry is dependent on it, and although an enormous assemblage of molecular geneticists is allied with the industrial research programs, we are no nearer an insight into this concept than when it was first recognized (Shull, 1909).

Sources of germplasm have been of interest to Gardner during his career, and he determined the breeding values of adapted and exotic sources of germplasm using diallel and topcross data. Gardner investigated the potential contribution of exotic germplasm to determine the appropriate combination of adapted and exotic germplasm for maximum genetic advance. Recently, he and his students have used molecular genetic markers in the study of mating systems. Gardner attracted a large number of graduate students and postdoctoral researchers during his career at the University of Nebraska between the mid-1940s to 1988.

9   The Maize Evolution Group

Mangelsdorf, Iltis

Paul Mangelsdorf (Fig. 11) was another student from the East group at the Bussey Institute (along with D. F. Jones, R. A. Brink, and Edgar Anderson) to carry on maize genetic research. After finding opportunities in a maize area, he continued with a breeding position at the Texas Agricultural Experiment Station at College Station. Mangelsdorf was most influential in his efforts to generate thinking and discussion on the evolution of maize after he left Texas and returned to Harvard University. His research into the origin of corn led to his appointments at Harvard as Professor of Economic Botany and Director of the Botanical Museum. More of his career is discussed by Thimann and Galinat (1991). Even after retirement in North Carolina, his interest in maize evolution continued. His strong views on the unknown other parent in corn's origin did not enjoy universal agreement among his peers; however, his views on maize evolution did provoke considerable thought. This debate generated further discussion, which led to numerous studies designed to counter his arguments (Beadle, 1972). Though Mangelsdorf proposed and supported the "tripartite theory" of the origin of maize, suggesting that *teosinte* was just a hybrid between maize and *Tripsacum*, he did modify this view in his book *Corn, its Origin, Evolution and Development* (1974). He had a number of noted students; some pursued his ideas on germplasm (G. Wilkes), and some went on to commercial ventures (Surinder M. Sehgal). Maize origin and evolution was, and still is, a major area of research. With the advent of molecular methodology, the role of *teosinte* has taken center stage, especially in the work of Doebley (1993), who has provided support for Beadle's teosinte-origin postulate in the Beadle-Mangelsdorf debate.

Mangelsdorf's extensive involvement in world-wide agriculture began in 1961, when the Rockefeller Foundation began efforts to aid Mexico in modernizing its agriculture. This venture eventually led to the International

Mexican Corn and Wheat Improvement Center (CIMMYT) and then to the International Rice Research Institute (IRRI). Mangelsdorf passed away in 1989.

Professor Hugh Iltis at the University of Wisconsin has continued this line of scientific inquiry into maize's origin. He, however, received his training at Washington University with Edgar Anderson and colleagues—training similar to that of W. L. Brown and Don Duvick, although oriented more toward botany and taxonomy. Iltis brought botanical thought into the field of maize research. His basic mode of thinking was botanical and, as such, enriched discussion on basic aspects of maize evolution. Bruce Benz, one of his students, describes many of Iltis' activities and the expansion of his thought on the maize genome (see Part B, Chapter 29). Iltis has provoked much thought on the origin of maize (was maize's origin really cataclysmic?). This line of inquiry has been pursued by several of his students [including John Doebley with his mitochondria research and more recently with his uncovering of five quantitative trait loci (QTL) segments related to morphological traits of *teosinte* (Doebley *et al.*, 1990)].

Iltis has made several ground-breaking contributions to our understanding of the evolution of maize. Specifically, his taxonomic studies were the first to recognize formally that maize and some forms of *teosinte* are members of the same biological species. Consequently, he changed the name of these *teosintes* from *Zea mexicana* to *Zea mays ssp mexicana* (Iltis, 1972). Subsequently, Iltis and his student Doebley fully revised the taxonomy of *Zea*, building upon the pioneering work of Garrison Wilkes. This was a significant departure and created a basis for molecular research into the problem.

Iltis and his students demonstrated that there were two primary groups of *teosintes*: (1) the more primitive *teosintes*, *Z. perennis*, *Z. diploperennis*, and *Z. luxurians*, those relatively distantly related to maize; and (2) the more advanced *teosintes*, *Z. mays* subsp. races *mexicana* and *parviglumis*, which show a very close relationship to maize. Based on these findings, a strong taxonomic argument can be made in favor of the hypothesis that *teosinte* is the direct ancestor of maize, something Beadle (1939) proposed earlier on genetic grounds.

Iltis' ideas have provoked many challenges. His students have carried on his efforts and have made great progress in uncovering the mysteries of the origin of maize. Iltis is a highly enthusiastic and determined researcher, out collecting in the field on horseback through swamps in Central America, and represents, more than other geneticists honored here, both a taxonomic and evolutionary outlook. His students have under-taken many types of investigations. Doebley's work began with taxonomic studies under Iltis, continued with isozyme studies at North Carolina, and have moved into RFLP and gene sequence analysis at the University of Minnesota. Bruce Benz will explore botanical aspects in his endeavors at the conservancy at Manatlan, Mexico, where *Zea diploperennis* originated. Future students of maize evolution, especially molecular geneticists will benefit from Iltis' earlier studies. They will enjoy especially the fruits of his effort to create the Sierra de Manatlan Biosphere Reserve, southwest of Guadalajara, Mexico, which will be a center for the study of ecological diversity.

10 A Later Genetics-Cytogenetics Group

Nelson, Schwartz, Peterson, Robertson

Along with the Cornell group under Emerson, the Jones group at the Department of Genetics at the Connecticut Experiment Station was also prominent in maize genetic research, especially as applied to breeding. Oliver E. Nelson, who studied at Yale, completed his doctorate under D.F. Jones (Fig. 11). Nelson's early years were associated with seed companies and then with popcorn genetics at Purdue University. Later, in the early 1950s, he developed an interest in physiological traits during a sabbatical year at the University of Stockholm, Sweden. Nelson returned to Purdue and then moved to the University of Wisconsin, in both places pursuing this interest in plant physiological genetics. He has maintained that line of investigation following his retirement.

His concentration on physiological mutants, especially those associated with starch, was a forerunner of molecular studies which followed when the "proper" molecular tools were developed in the late 1970s and early 1980s. Curt Hannah, one of Nelson's students, expounds on Nelson's career (see Part B, Chapter 30). If the *MAYDICA* commemorative volumes (see the biographical chapters in Part B) had originated earlier, Nelson's mentor D. F. Jones certainly would have been a prominent figure in them, as would John Laughnan's mentor, L. J. Stadler, memorialized by the annual Stadler symposia held annually at Columbia, Missouri.

Returning from his sabbatical leave in Stockholm, Nelson explored many problems in the physiology of maize mutants, and then, in the early 1980s embarked on research into various features of transposons. But questions of starch biosynthesis were a primary focus, especially evident in his studies of starch debranching enzymes. Despite the significance of starch to seed corn companies ethanol production, it is not altogether clear how the starch amylose-amylopectin balance proceeds.

Early in the 1960s, studies indicated that ADPglucose was more effective than UDPglucose for starch synthesis; however, it was not clear which nucleotide sugar was the more important substrate physiologically. If ADPglucose was the substrate for starch biosynthesis, what was the pathway involved in the formation of ADPglucose? Here again, the repertoire of previously discovered maize mutants came into play. Investigations of starch-deficient maize mutants *shrunken-2* and *brittle-2* from the Nelson laboratory clearly demonstrated that starch biosynthesis in maize endosperm is catalyzed by enzyme systems that utilize ADPglucose as the primary substrate, and that the latter is formed chiefly by ADPglucose pyrophosphorylase. Nelson retired in 1991, but he, like others in maize genetics, continues investigations into his interesting mutants.

Drew Schwartz: A Nelson contemporary, Drew Schwartz covered many areas, starting with classical cytogenetics under M. M. Rhoades at Columbia University's experimental fields at "Hastings on the Hudson" and continuing to isozyme studies and molecular genetics. Schwartz was honored on his 70th birthday in 1991 at a gathering of most of his students in Bloomington, Indiana. It was a heart-warming occasion for which a generation of students spanning 25 years traveled to this college town. Many students came to study with him at Indiana University (Bloomington) following his move from Case Western Reserve University (Cleveland) in 1964. This move began an important era in maize genetics as numerous students sought new ways (other than classical) to study plants.

Plant physiological studies during the 1960s had reached a plateau with regard to approaches to gene function. The anthocyanin biochemical products were readily measured (Harborne, 1967), but direct gene products were not readily accessible. At this time, researchers did not have access to today's network of computer technology. Thus, though the Smithies' initiative with electrophoretic movement of proteins on a gel was apparent in the 1950s (Smithies, 1955), plant geneticists did not assimilate this technology until the 1970s. Now, gel electrophoresis could provide access to gene products and maize genetics, with the prospect of probing the enormous diversity of alleles at numerous loci. Schwartz initiated isozyme studies; this focus attracted numerous students to Bloomington to pursue a botanical thrust in gene products and to exploit the extensive array of mutants and the diversity of alleles. They began with the study of locus polymorphism, beginning with the esterase loci.

Many of the students coming out of the Schwartz lab continue today with active maize programs. In fact, his "progeny" are the largest group that continued into active maize research. Schwartz's move to Bloomington came at a time when plant genetics, with its abundance of genetic resources, was ready to burst out of its classical sphere. The isozyme thrust with the esterases that began with *Drosophila* following Smithies' (1955) development of zone electrophoresis, coupled with genetic studies, gave plant geneticists a window into gene products. Maize displays infinitely available polymorphism that can be genetically, and therefore readily, manipulated in maize crosses, so considerable progress was made. This exciting plant genetics research in Bloomington attracted Tony Pryor, who arrived from Australia in the mid-1960s as Schwartz's first student. He was soon followed by Michael Freeling and many others.

Thus this early beginning in plant genetics, in a pre-molecular era, spawned other focal centers. Professor Seymour Fogel at Berkeley (University of California) was looking for a plant geneticist in 1972 and attracted Michael Freeling to the Department of Genetics to continue the plant genetics tradition earlier begun by R. E. Clausen and his students (ploidy studies in *Nicotiana*). The Freeling laboratory in Berkeley is quite prominent in attracting plant-oriented researchers to the West Coast and has been prolific in training graduate and postdoctoral students, who in turn are conducting laboratories with students and "postdocs" in maize. Note that Freeling began with isozymes in the Schwartz laboratory and a quarter of a century later is in the molecular area in studies on development.

Another Schwartz "satellite" is found in the current molecular laboratories at the University of Missouri with Kathy Newton and Jim Birchler. Further evidence of the extension of the Schwartz laboratory in continuing the maize-genetics tradition is found in the group at Gainesville, Florida (Chourey and Ferl). This group of Schwartz's students extends the scientific pedigree started four or five generations earlier in the Emerson laboratory at Cornell (Fig. 24).

Peter A. Peterson: One of the next volumes in this series commemorates P.A. Peterson.* Peterson began as a student of Marcus M. Rhoades at the

* The Peterson section was authored by Professor A. Hallauer.

University of Illinois, following his research year at the Carnegie Institute of Washington at Cold Spring Harbor (post-World War II research on carcinogens with *Drosophila* with M. Demerec). His studies of the genetics and cytogenetics of maize, begun in the late 1940s, have emphasized the role of transposable elements of maize. In his postgraduate period, Peterson moved to Riverside, California, to study and improve peppers including *Capsicum* linkage groups (first gene linkage in *Capsicum*), cytoplasmic male sterility (first male sterile in *Capsicum*), and avocados (diurnal periodicity of the avocado flower). But given an opportunity to pursue the cytogenetics of maize at Iowa State College (later Iowa State University), Peterson moved back to the Midwest in 1956 to the Department of Genetics, which was then headed by John Gowen and included Joseph O'Mara (a Sax student) in Plant Genetics. With this opportunity, Peterson began nearly five decades of transposon studies and continues today.

In the early 1950s, three laboratories in the United States independently were emphasizing the study of transposable elements of maize: Barbara McClintock at the Carnegie Institution of Washington Laboratories, Cold Spring Harbor, Long Island; R. A. Brink and his students at the University of Wisconsin, Madison; and P. A. Peterson and his students at Iowa State University. Independent research from the three laboratories established the universality of transposable elements in maize, and the information derived on them from the maize research became the "benchmark" for the study of transposable elements in other organisms.

Peterson and his students continue the study of the transposons in maize. Peterson's research on mobile elements over the past 40 years has been recognized, and he has been asked by several review journals to summarize his research findings on the importance, role, and mode of action of mobile elements in maize genetics. Peterson, colleagues, and students also have expanded the study of transposable elements to include the significant question of why transposons exist; with F. Salamini they have studied the presence and frequency of occurrence of transposable elements within important maize breeding lines and the presence of active transposable elements in maize breeding populations. How do transposons survive and what is their benefit to maize populations? Further, transposable elements have been used successfully for tagging desirable genes within the maize genome.

In fact, most of the genes isolated in maize have come from transposon tagging. Questions also arise on their possible role in long-term selection experiments in maize breeding. The information and genetic materials derived from Peterson's research have become an integral part of other research laboratories throughout the United States and the world.

Peterson started in the Department of Genetics at ISU in 1956, then moved in 1960 to the Department of Agronomy, to explore problems of agronomic interest. It was some time before a critical mass of students developed. By the 1970s, students trained in this laboratory moved in several directions. Because some students were in the Department of Agronomy, they majored in plant breeding and went into plant breeding positions (Hsu, sugar cane breeding; Huang, agronomy). Other students studied various aspects of molecular biology, including transposons, bacteria (Fowler), potato (Pereira), gene expression (Schnable, Scheffler, Muszynski), gene transformation (Secor-Sukapinda), and others. A number of students joined biotechnology groups (A. Pereira at Wageningen) and others worked for commercial companies (e.g., Secor-Sukapinda at Dow-Elanco and M. Muszynski at Pioneer Hi-Bred International). Again, as with the Schwartz group, these students represent the fourth generation of Emerson origin.

Peterson's effort to move transposons into molecular biology started with an attempt to posit the maize transposons as analogous to the mobile *P2* phage in bacteria while on an National Institute of Health (NIH)-sponsored study-leave with Professor Joe Bertani at the Karolinska Institute in Stockholm, Sweden. Later, to further explore the molecular biology of transposons, he joined Professor H. Saedler in 1977 in Freiburg, Germany, in *Is2* studies and asked how they were related to maize transposons. These were the first attempts to explore the universality of transposable elements, and later studies of transposable elements exploded in laboratories among the biological community (Berg and Howe, 1989). During Peterson's three-year interrupted stay in Freiburg the maize studies became an attractive diversion and focal point for Saedler. This led to an 18-year collaboration (Ames-Cologne) which uncovered many aspects of transposons (Fig. 22). Cologne, Germany, with Starlinger, Saedler, and Salamini, became a world center for advances in maize molecular biology during the 1980s.

Peterson also has made contributions in several other aspects of maize and plant genetics. He has studied the mode of diurnal flowering,

Fig. 22. P.A. Peterson with H. Saedler in 1991 on the occasion of the Peterson Commemorative Volume at Ames, Iowa. Collaboration between the two laboratories at ISU and Cologne, Germany, continued for 18 years in the pursuit of the En/Spm transposon and the cloning of several genes.

flower pollination, and fruit set in the avocado; inheritance of male sterility and gene linkages in peppers; influence of B-chromosomes on pollen size in maize; an unstable locus in soybeans; properties of activities that cause cytoplasmic male sterility in maize; isozymes to predict performance of lines in maize hybrids; and factors that affect callus initiation in maize endosperm, as well as an incompatibility system in maize. Peterson, his colleagues, and his graduate students have maintained an active current-research program in the research-oriented Agronomy Department at Iowa State University. The information and genetic materials derived from his research efforts will continue to have an impact in the future as different aspects of molecular genetics are developed. With his genetic insight and agronomic surroundings, he has attempted to wed current advances in molecular biology to basic problems in maize breeding (Peterson and Salamini, 1986). He perceives a need to augment problem-solving in maize breeding with new insights from the molecular biology of maize.

11 Stock Center, Genetics, and Pathology

Patterson, Ottaviano, Coe, Neuffer, Hooker, Walden, Smith, Robertson

At the time of this writing, the commemorative issues of *MAYDICA* honoring E. Patterson, E. Ottaviano, E. H. Coe, M. G. Neuffer, A. L. Hooker, D. Walden, J. Smith, and D. Robertson were being developed. Earl Patterson's devoted and sincere dedication to maize genetics and breeding progress has been recognized by the authors of the Patterson volume (see Part B, Chapter 34) and indicate the appreciation of the maize community for Earl's dedicated and long stewardship of the Maize Stock Center at Urbana, Illinois. Based on the sound foundation that Patterson established over nearly four decades, the stock center will remain valuable for the future of maize research. Today, following Patterson's initiative, the Maize Stock Center is an exceptional resource for the maize community now under the leadership of Marty Sachs and Phil Stinard.

E. H. Coe: Ed Coe began his genetic studies with Charlie Burnham at the University of Minnesota for his early graduate degree but was attracted to physiological studies on anthocyanins being pursued by John Laughnan. As noted previously, anthocyanins were the most prominent access to gene products in the early 1950s, and because of that, Coe joined Laughnan in this effort. In 1954, he assumed a USDA position at the University of Missouri, where he still resides. Advances in computer technology found Coe ready and waiting to augment and adapt this tool to the collection and recording of the explosion of maize genetics information. The maize community is indebted to his efforts to funnel this extensive and updated data base into the *Maize Genetics Cooperation Newsletter*, which has approximately 1200 subscribers. In 1995, Ed Coe was awarded the Morgan Medal for his efforts to coordinate the maize database. The reader may want to examine Sheila McCormick's tribute to E. Coe in Part B, Chapter 39, and further discussion of Coe with reference to anthocyanin in Part A, Chapter 13.

M. G. Neuffer: Gerry Neuffer spent his graduate days among the rich maize collection of L. J. Stadler at the University of Missouri. In his graduate work, Neuffer investigated transposons with his study of *Dt* in natural populations. But in the 1960s, Neuffer, with the aid of his long-time colleague Ed Coe at Missouri, began the chemical mutagenesis of maize silks and then embarked on a mutagenesis program that uncovered a large number of mutants of immense value to the maize community. He patiently studied their inheritance and pursued detailed linkage studies. His treatment methods are being used by numerous researchers as ever-increasing number of maize molecular biologists chase an ever-decreasing number of options. Along with Coe, Neuffer formed the base which continues the Stadler maize tradition at Missouri and has attracted a number of cytogeneticists, geneticists, and molecular biologists to Columbia to provide a critical mass of genetically oriented researchers.

A. L. Hooker: The pathology of maize diseases received the most intense study by Arthur Lee Hooker and his students. He was most prominent in uncovering genetic resistance to plant diseases by pursuing one disease-resistant gene after another. Most gene resistance studies in maize came from his laboratory, as Hooker was one of the few pathologists to couple pathology investigations with genetic experiments. These discoveries are well-developed in Don White's essay on Hooker (see Part B, Chapter 35). Hooker passed away prior to the dedication of the *MAYDICA* volume honoring him, which was accepted by Mrs. Hooker in December, 1993.

D. S. Robertson: Don Robertson was discussed by the author in a commemorative volume; this laudation is found in Part B, Chapter 38. The presentation to Don was made during the 1997 Iowa State University Plant Science Lecture Series. This event came after his retirement from the newly created Department of Zoology/Genetics. This hybrid department development from the dismemberment of the Genetics Department that had a history dating from 1932 following the arrival of E. W. Lindstrom to the then Iowa State College. Don's arrival followed a series of plant geneticists beginning with E. W. Lindstrom's arrival (see Part B, Chapter 6). Between Don's teaching and other duties, Don laboriously carried on his genetic research with geneticist's watchful eye for that ever-recurring mutant in a geneticist's experimental cultures. It was this type of alertness that allowed him to uncover the Mu element that has been a benefit to so many. The maize community is grateful to these geneticists such as Don who have left a legacy as illustrated in the maize geneticists and breeders pedigree in Chapter 16.

J. D. Smith: The retirement of JD was highlighted by the presentation of the commemorative volume at a luncheon held on the 11th floor of the Memorial Union at Texas A & M University overlooking the expanse of this campus. Many of his colleagues and former students were there to honor their colleague and mentor. One can readily see from the laudation organized by David Hole (see Part B, Chapter 42) that there was expressed a deep admiration for JD. Greetings were also received from his former students in Korea and elsewhere. From his coterie of former students who ventured into very different areas, it is readily apparent that JD was a magnet for many students from the Texas bush country looking for a "home." JD provided that refuge and lead them through degree programs that were far removed from their initial entry to Texas A and M. He was able to generate interest in new faculty who found many interesting experiments to pursue in his laboratory. From his field programs and experiences beginning with his help on his father's fields, he possibly wondered what generated such bountiful growth in a corn plant. Thus, when he looked at possible research programs in his beginning days in Texas, he considered the hormone aspects of growth that continued with his students and colleagues for many years.

Ercole Ottaviano: Ercole Ottaviano's thriving career brought a new dimension to the Italian genetic scene. His contributions were abruptly ended by his sudden death as noted in Part B, Chapter 37. His training under Professor Angelo Bianchi, who himself had some post-graduate training with another of the *MAYDICA* lauded geneticists (Professor Mangelsdorf) and in some elite laboratories in England (with K. Mather in Birmingham), provided Ercole to explore interesting areas of quantitative genetic studies. Yet, scientists always leave a legacy of well-trained students and associates. And, this was no less so with the successors in his laboratory as expressed by Professor Mirella Sari-Gorla in her write-up of Ercole, providing a fitting tribute to his scientific career.

David Walden: The banquet held at the 1997 Annual Meeting of the Canadian Genetics Society provided the scene for the expression of gratitude by the friends, former students, associates and peers of David Walden. What was quite evident was the sincere appreciation of David's contribution to genetics and to the Canadian Society of Genetics. This accolade is portrayed by his former students V. R. Bommineni and C. L. Baszczynski in their story in Part B, Chapter 41. These tributes and retirement gatherings are always final career-ending summation of one's scientific accomplishments and legacy. David's contribution to Canadian genetics development and progress is certainly noted, and he should be proud of this accomplishment that began in his "internship" as an undergraduate at Wesleyan and his summer program with D. F. Jones (also noted in the Pedigree Tree in Chapter 16.) Again, as with others, he has left a legacy of thorough training and scientific attitude that will long endure by his followers.

12 The Italian Maize Genetics Scientists

The reader may find that this essay on maize genetics, cytogenetics, and breeding is biased toward North American research. Note that the commemorative volumes focus on individuals who have contributed to the growth and development of the genetics and breeding of maize, and are generated by students and colleagues of the honoree.

In the previous pages, recognition has been given to Correns (1902) and Kuwada (1911) for their contributions. However, the focus of this book centers on the commemorative issues and how the individuals noted are professionally related.

In Italy, students were (and still are) encouraged to seek postgraduate training outside Italy; many students choose to come to the United States. One of the first maize researchers to seek such training was Professor Angelo Bianchi, who traveled to Cambridge, Mass., to study with Paul Mangelsdorf (Fig. 11) at Harvard University between 1954 and 1956. On returning to Italy, he began the "Bianchi School" in Bologna, moving to Pavia University and then to Milan University, and introduced maize genetic methodology to his students. One of his first students was Francesco Salamini (now one of the directors at the Max Planck Intitüt in Cologne, Germany). Salamini spent a period with O. Nelson at Purdue in 1969 and then began studies at Piacencia (Italy) and later at a breeding-genetic-molecular biology Institute in Bergamo (Italy). There, Salamini began and advanced the study of the genes controlling the zein proteins and finally cloned the *opaque* gene. He later moved to Cologne (Max Planck Intitüt), where he continued research into the opaque gene and broadened his approach to include other grasses.

Other Italian researchers who received postgraduate training in maize included A. Ghidoni, who traveled to Bloomington, Indiana, to study with M. M. Rhoades; G. Gavazzi, who studied with R. A. Brink in Madison, Wisconsin, and with Professor Harborne in England; and E. Ottaviano, who studied with Kenneth Mather at the University of Birmingham. Each brought his acquired expertise back to Italy to continue these studies. Ottaviano's biography (see Part B, Chapter 37) can give the reader further insight into genetics development and progress in Italy.

13 The Development of an Understanding of the Anthocyanin Pathway

Most of the basic components of the anthocyanin pathway were genetically isolated in the early decades of the 20th century (Table 2). Included in this collection were the *a1*, *c1*, *r1*, *a2*, *a3*, *in*, and numerous other genes. In place and genetically understood when the tools of molecular biology became available, these genes became the targets of gene tagging (Table 2) (Peterson, 1995).

In 1950, the genetic world was as yet unaffected by the DNA explosion to come in the latter part of the decade. In studies of maize and *Drosophila*, the main access to genetic units was to study loci with alleles that could be separated (pseudoalleles). What, then, was the limit of the gene? Didn't the separation of a pseudo-allelic series suggest *"beads on a string?"* The DNA orientation that would develop from the Hershey and Chase (1952) experiment did not appear until the Watson-Crick revelation a short time later. Yet, the DNA-discovery impact did not have an immediate wide acceptance among non-specialists though George Beadle (for example, in 1954) was one of the first to spread the "new gospel" among scientists in general. Thus in the early 1950s, any aspiring plant geneticist with a physiological interest might look at how genes make a product and what that product might be. A number of discoveries became apparent at this time. It was not until the 1980s that transposons could be put to use to clone and sequence maize genes.

A significant development was the anthocyanin pathway that Harborne had assembled in his seminal volume *Comparative Biochemistry of the Flavonoids* (Harborne, 1967). Earlier research had revealed much about the structural components of the anthocyanins. The structural integrity of the phenyl propinal pathway was being investigated. Now, it was only a matter of more investigation to combine the genetics, their associated genes, and their structural components. What genes began the pathway? What genes were related to the first coupling of the A and B ring of the anthocyanin molecule? It was not until 1980 that the *C2* gene was associated with flavone synthase (later identified as chalcone synthase) (Dooner, 1983) and only a short time later, if not at the same time, that the parsley chalcone clone was isolated (Kreuzaler *et al.*, 1983). But first, the *C2* gene had to be identified genetically.

And so in 1950, Ed Coe, with a master of science degree earned at the University of Minnesota with Charlie Burnham came to Illinois to join John Laughnan in dissection of the anthocyanin pathway. Note the interlinking. Laughnan, trained with Stadler at Missouri, was focusing early investigations on gene mutation. He was sending out his student disciples to further study the pathway, though Laughnan was one of the few to remain with the maize plant. This became a central focus of Coe's laboratory at the University of Missouri. A series of students from India arrived, including K. R. Sarkar and G. M. Reddy, to pursue these studies.

During this period Coe recognized a need to homogenize the genetic background. Was this because of the plant-breeding setting in the Agronomy Department at Missouri, or because of a growing awareness of epistasis in gene interactions? Certainly, genetic modifiers were apparent in all reactions. It was obvious that to study genes, one must control the background. Genes do not act in a vacuum, they interact with other genes; so it was necessary to introduce each gene in a recurrent backcrossing program to a common background. This is a painstaking effort—not difficult, but costly when there are limited resources. It was this recurrent crossing program that led to the uncovering of the "duplicate genes" *C2* and *Whp* (Coe, *et al.*, 1981) and finally to the *intensifier* (*in*) gene. The genetic relationship of these three genes had important implications when molecular investigations began (Franken *et al.*, 1994).

It was only with a long-term effort to incorporate *C2* into a Missouri white dent corn that the *Whp* gene could be uncovered. Here was a white line (*C2 C2*) harboring a recessive *Whp* (*whp whp*). Only when the *c2* gene was crossed into the line could the *Whp* allele be uncovered. The white line was (*C2 C2 whp whp*). Obviously, the investigator had to be astute and observant. This plant segregating in the row (*c2 c2 whp whp*) obviously did not set seed upon selfing. This would not be realized during the usual summer crossing program because the white pollen (*c2 whp*) appeared normal in a standard crossing protocol; but when the harvest followed, no kernels on the ear could be seen. Was this simply a poor set? But, this occurred with more than one plant. Yet at harvest, it was too late to check the pollen; therefore it was necessary to try again at the next appropriate season. At this time the color of the white pollen became obvious, but its lack of function was yet to be detected.

This is the geneticist's role: to dissect and to frame windows of opportunity which will guide the molecular biologist. As the *C2* gene was finally cloned, the second unexpected band in a southern probed with the *c2* probe became meaningful, especially when coupled with the RFLP location of the second gene. So it was similar, as was found with the *a1* probe to find a second band on chromosome 8, as well as with the *in* and *a3* duplication—again, coupling the background genetics with the disturbing second band (such findings of duplicate genes are no longer disturbing because of the precedent established by the *c2* and *whp* example, especially with the prevailing notion of the duplication of maize genes). The combination of these complicated genetics, RFLP analysis, and molecular probes all coalesced to provide a framework to make order out of chaos. As Coe's students matured into active genetic scientists in new positions, their training led them to additional ventures. Sheila McCormick at the USDA Albany lab in California has pursued the floral activities in the tomato; G. M. Reddy has developed a large center of plant genetic activity in Hyderabad (India). Much of this gene duplication would not have been detected without the development of the recombinant lines by B. Burr *et al.,* (1983).

The Robertson saga also began with one of the graduates of the Emerson laboratory. The author has discussed Robertson's training and successful career in Part B, Chapter 38 (Peterson, 1995). Robertson also was trained when the prime focus, as for Coe, was to analyze genes genetically. So many new interesting mutants had to be located, mapped, and extensively analyzed to investigate their genetic bases. Thus, Robertson is one of that group of geneticists trained during the 1940s and 1950s, who has been commemorated in these issues and contributed heavily to molecular investigations. More is considered on this point in the Robertson biography (Part B, Chapter 38). The contributions of these dedicated "bench-tied" geneticists cannot be emphasized enough. This is convincingly evident in the current use of the "gene-machine" (Meeley and Briggs, 1995) with the *Mu* element and now with the *En* element in *Arabidopsis* (Spellman *et al.*, 1997).

14 The Origin of the Allerton Maize Genetics Meetings

"That Old-Time Religion"

When Marcus M. Rhoades settled at the University of Illinois in 1948 at Oswald Tippo's (Chair, Botany Department) invitation, maize cytogenetics finally exploded in the Midwest. Programs at Missouri, Minnesota, and Purdue were just then emerging into the "genetic era." From the Botany Department at Columbia University in New York, Rhoades brought with him Ellen Dempsey, Drew Schwartz, and E. Dollinger. Also coming from the Carnegie Institute at Cold Spring Harbor to join this group were Sally Rohrer (later Peterson) and Peter A. Peterson. They were joined by several "walk-ons," and soon enough were assembled to form a "critical mass" to start a cytogenetics seminar and finally a full class in cytogenetics. Here Rhoades presented a survey of genes and chromosomes of the first half century of genetics, emphasizing basic concepts. Note that this was 1948; the double-helix of Watson and Crick (1953) was still five years away. Later, in the fall of 1948, John Laughnan arrived from Princeton University, and now a "critical mass" had developed in the Botany Department at the University of Illinois. With such a concentration of maize genetics workers in Urbana, the stocks of the Maize Cooperative were moved there from Cornell University in New York; Earl Patterson assumed able leadership of that enterprise in 1954 and has continued there for four decades. Some researchers from the University of Illinois Department of Agronomy, D. Alexander and P. Bauman, soon joined in seminars and joint research. Purdue University was relatively close, and H. Kramer and O.E. Nelson drove over from W. Lafayette, Indiana, every Tuesday morning in the fall of 1948 to listen to the 10 o'clock cytogenetics lectures. Rhoades brought to this group current genetic thinking developed at Columbia University, the Cold Spring Harbor Symposia, and the Rockefeller Institute (later, Rockefeller University).

In the mid-late 1950s, universities in the Midwest "corn belt" were beginning to hire maize geneticists. P. A. Peterson left the University of California at Riverside and joined the Genetics Department at Iowa State College (now Iowa State University) in 1956; D. Robertson from California followed in 1957. The vacuum in genetics research at the University of Missouri created by the unexpected passing of L. J. Stadler was soon filled by M. G. Neuffer and E.H. Coe. At the University of Wisconsin, R. A. Brink had resurrected maize research with the re-investigation of variegated pericarp. At the University of Minnesota, C. Burnham was working with a small group, mainly on chromosome translocations. Oliver Nelson and H. Kramer worked at Purdue University. With all these labs across mid-America actively engaged in maize genetics, most within a medium-long day's driving distance from Urbana, Illinois, it seemed a fruitful idea to gather researchers together at a two day meeting. Here was an opportunity to resurrect the Emerson "Corn Fab" (Rhoades, 1984).

The Illinois group of Laughnan, D. Alexander, and Patterson (see the Patterson biography in Part B, Chapter 34, by John Laughnan, for more insight on these beginnings) initiated the maize meeting (meanwhile, M. M. Rhoades moved to Indiana University). And so on a cold, snowy weekend in January 1959, a meeting was arranged in the country setting near Allerton, Illinois. It was held at an English country house containing formal gardens and sculptures (now under a month of a winter snow in the quiet Illinois countryside in Monticello, approximately 25 miles southwest of the Illinois campus, built by the cattle baron, Mr. Allerton. (This is the same Mr. Allerton that is associated with the Allerton Hotel in Chicago and who had a son who made a donation of land that became the Pacific Tropical Gardens in Kauai, Hawaii). For most laboratories, such as those in Iowa, Wisconsin, Minnesota, and Missouri, Allerton House at Monticello was one-half to a full day's drive. About 25 maize geneticists attended, including the Indiana group of Rhoades and Schwartz, the Brink group from Wisconsin, the growing Minnesota group led by Charlie Burnham, the young Coe and Neuffer groups from Missouri, and others. The faculty and students arrived on a Wednesday night and left on Friday, after the last talk. It was exciting for students and new faculty to meet researchers previously known only through the literature. They presented talks and exchanged research notes "elbow-to-elbow" at a long table in the elongated library room off the main hallway at Allerton House. Especially stimulating were the "open" meetings—reports of material before publication. When these maize geneticists share six meals, several coffee breaks, and late night sessions, it is relatively easy to meet and exchange ideas. Here in a short two- to three-day meeting, everyone met all in attendance, not once, but many times. The ambiance of the rural Allerton estate with its large central atrium was most suitable for this type of meeting. The excitement generated among students must have had a lasting effect, because many have become the maize research leaders of today.

There was no organized agenda; John Laughnan writes, the meetings were "delightfully informal." Earl Patterson gathered volunteers' names on cards, and the talks started. Patterson might have an agenda for the two morning sessions; the afternoon sessions were open. No time clock was used; if a speaker could hold the audience's interest, more power to the speaker. *Ad-hoc* night sessions were common. This was 1959, only a few years after McClintock's transposons became known, so they became a major topic (*P-vv, Ac, En*, and later *Fcu*). Paramutation (Brink) was just being analyzed. And a vast number of talks focused on cytogenetics; abnormal 10, the elongate gene, trisomics, preferential pairing, inversions, knob investigations, neocentromeres, and a number of other studies presented mostly by Rhoades' Indiana group. The meeting room at that first meeting was, by today's standards, very small; everyone sat around one wooden conference-table or on seats and benches along the wall. The chalkboard was the mobile type on a stand, and one had to hold it upright with one hand as one wrote. A speaker was as close as four feet from a listener across the table, and could readily see whether the presented data was understood, or needed clarification. It was more like a series of spontaneous *ad-hoc* sessions.

This intimacy did not last long. By the next year, this small alcove was too small, and the group was moved to the library of the Allerton house a two-story, balconied, book-lined room with a Pickwickian flavor. This was the meeting room for a number of years, through studies of the isozyme (*esterase* and *Adh*), genes controlling the zein proteins, and the beginnings of molecular studies and genetics of transposons; however, molecular studies on transposons did not appear on the Allerton scene. The study of the genes controlling the zein proteins brought in the molecular period and an influx of new maize researchers. Many students were trained in maize genetics and cytogenetics during the 1960–1980 period; some are pictured in the group photograph in the

Laughnan commemorative issue (see Part B, Chapter 28). Interested maize researchers were coming from diverse areas: London, Ontario (D. B. Walden and his students brought cases of Lablatt "blue"), Yale University (Poethig), Waltham, Mass. (Galinat), and Austin, Texas (McGuire). This was before the number of maize workers exploded on the West Coast.

During the 1970s and early 1980s, the assembled maize geneticists at Allerton numbered approximately 100. Most knew one another (See Fig. 23 for one of the early "Allerton" groups.)

Fig. 23. Participants at the 1960 Maize Genetics Conference, Allerton House, Monticello, Illinois. (Photo contributed by Susan Gabay-Laughnan.): 1. J.R. Laughnan; 2. I. Mizukami; 3. D.L. Shaver; 4. G.Y. Kikudome; 5. A. Taylor; 6. E.G. Anderson; 7. D.E. Alexander; 8. M.H. Emmerling; 9. D. Schwartz; 10. H. Peterson; 11. R. Runge; 12. B.C. Mikula; 13. K. McWhirter; 14. E.R. Leng; 15. I.M. Greenblatt; 16. H.H. Kramer; 17. B. Ashman; 18. L.F. Bauman; 19. J. Arison; 20. E. Dempsey; 21. E.B. Patterson; 22. R.J. Lambert; 23. R.A. Brink; 24. M.M. Rhoades; 25. P.A. Peterson; 26. D.M. Steffensen; 27. G.M. Reddy; 28. E.H. Coe; 29. O.L. Miller; 30. M.G. Neuffer; 31. G.G. Doyle; 32. S.K. Sinha; 33. D.F. Brown; 34. D.S. Robertson; 35. J.L. Kermicle; 36. G.P. Hanson; 37. H.B. Cooper, Jr.

By the early 1980s, molecular-biology studies began with the improvement of the technology associated with gene handling. Zein proteins were the first focus of study. Increasing numbers of researchers pursued maize genetics as laboratories became familiar with a focus on molecular technology. The students who followed their mentors became leaders in maize genetics in their own labs and appear as authors in a number of the *MAYDICA* commemorative volumes. As molecular investigations began, it became obvious that a move from Allerton was imminent. An explosion of interest in molecular studies of maize in the mid-1980s added the few laboratories in the Midwest to laboratories on each coast, a number of European laboratories [for example, Cologne and Freiburg (Feix) in Germany]; numerous seed corn companies were establishing their own laboratories. Maize was now a viable molecular tool. Molecular biologists felt this enormous genetics was "ripe" for the taking (Veit *et al.*, 1993). There weren't enough dorm rooms or meeting space at Allerton. In 1984, researchers met at a motel in Champaign-Urbana, Illinois, and then moved meetings to Lake Delavan in Wisconsin (with an intervening attempt in Madison). Finally, for a change of pace and to equalize access to the western labs, Asilomar in California became the meeting site in 1992 and 1995. The meeting was held in the Chicago area (Pheasant Run) in 1993, 1994, and 1996, with a move to Florida scheduled for 1997. Of course, the budgets that prevailed in the 1960s could not sustain the travel to these new meeting sites—California and Florida.

From those early Allerton meetings with the no-agenda, give-and-take of earlier times, the maize conference has evolved into a formal program of competitive funding, producing a bound volume of typed abstracts, and includes invited principal speakers, increased budget grant funding, and larger attendance. Most presentations are made after their publication and largely represent a summation or review. Unlike the earlier period of single-author papers, presentations now represent multiple authors (Figs. 25 and 26). Surprises are rare (like the first report on paramutation, or on the analysis and isolation of the *C1* locus and its homology to *myb* proteins, etc.). Discussion is minimal or nonexistent. The "meatiest" and most meaningful parts of maize research gatherings are the posters; one can interrogate and clarify questions with the presenter. And as each year passes, the posters become more meaningful and more presentable.

So, what has happened in the 38 years of the Annual Maize Meetings? We have learned more about maize—much more. A large number of genes have been cloned and their interactions deciphered. Finally, we know that regulatory genes exist (critical negative comments on the theory of regulatory genes on proposal reviews of the 1970s make good reading), and that they function as transcription activators. Those who were critical of regulatory genes on proposals finally were convinced that some genes could be regulatory. In 1992, the *Maize Genetics Cooperation Newsletter* listed 365 *Zea mays* nucleic acid sequences (there are now more), as the *c-myc* and *b-myb* regulator genes have also brought the maize researchers to the attention of a generalized genetic community. The computerized data bank has made genetic phenomena universally accessible. The grass family can be treated as a single molecular family. These findings of interactions are meaningful to a broad group of plant and animal workers, and this is recognized by their assemblage at general plant meetings.

What of the future? Because sequences are finding their homolog in other plants, a change in outlook is inevitable. The homology of genes among the *Gramineae* is making the grass family a unitary area of study (Bennetzen and Freeling, 1993). Many wish to have a foot in two camps—*Arabidopsis* and maize—to take advantage of fruitful opportunities. This futuristic outlook is discussed by Francesco Salamini, Mario Motto, and Angelo Bianchi in the final section of this book (Part C).

When one examines the maize maps today, it is evident that they are filling at different rates. The genetic map is generally static compared to the RFLP map (Fig. 18), but cytological maps have not changed in the last 10 years—possibly because of a lack of emphasis on this portion of the program. With no continuation of the thrust by Jack Beckett at the University of Missouri to fully exploit the *B*-translocations and therefore provide greater access to the physical map, it is not likely that this portion of the map will grow very rapidly, if at all.

Look what has happened to the molecular thrust. The protein domains had been determined for a number of genes. For example, in the *C1* gene, a prominent regulatory gene in the maize anthocyanin pathway has been determined to demonstrate functional aspects of regulation. Beyond this, protein engineering has been demonstrated by a natural process of transposon mutagenesis. Amino acids have been substituted, added, and altered leading to a rapid evolution. An abundance of RFLP opportunities exists in maize, especially in contrast to many other plants (Helentjaris, 1989; Shattuck-Eidens *et al.*, 1990). From the availability of computer-assisted protein elucidation, amino acids changed by a mutation can be envisioned (Franken *et al.*, 1994). Transposon mutagenesis has initiated new proteins in important genes (Giroux *et al.*, 1996). On another front, Ben and Francis Burr and their colleagues at Brookhaven have advanced maize research with recombinant inbreds. This has facilitated gene location and the location of sequences of genes with unknown functions. So the maize community has grown from its beginning in that smoke-filled room in a New York City hotel at the 1928 annual Genetics Society meeting in late December under the leadership of R. A. Emerson (Rhoades, 1984), through seven decades to the formal maize meetings of today. The *Maize Genetics Cooperative Newsletter* has evolved from typed and stapled sheets to the computer-enhanced publication of the 1990s produced by E. H. Coe.

15 The Maize Community

Will the maize community continue to be "colleague-friendly?" From the early period, characterized by minimal costs (bags, field aprons, hand-planters, and an annual budget of $500-$1000), to the current high-cost enterprises (labeled isotopes at $500 per week, centrifuges, isotope counters, freezer compartments, and half-million-dollar budgets), a new dimension emerges: competition. Priorities are important and domains are defended. For example, though McClintock recognized *En* as the same transposon as *Spm*, it was most convenient for her to retain the *Spm* nomenclature, because a decade had passed since her *Spm* identification before the two were found to be genetically homologous (Peterson, 1965). In a compromising gesture, the Cologne (Max-Planck-Institüt) group identifies it as *En/Spm*, which provides the reader with a handle on the subject—a feature that other labs have not reciprocated.

This kind of conflicting literature will become more troublesome as sequences are exhibited. Note the recent case of "tourist" (Bureau and Wessler, 1992). As that sequence became known, similarities to past discoveries became apparent (M1SD-1 by Zack *et al.*, 1986) as did similarities to simultaneous studies (heartbreak by Johal *et al.*, 1993)? How will the interested reader reconcile the findings?

It is hoped that the features in this book will provide budding maize breeders and geneticists a background on the maize legacy that has been handed to them. These maize breeders' and geneticists' careers provide an exposure of the kind of effort and pursuit that has led to the vast assemblage of maize information and individual tutoring that is the maize legacy.

Of paramount importance is the expression of esteem by former students and colleagues of the commemorative laudations seen in Part B. This warmth and esteem, mostly written in the mellowing period after leaving the laboratories of those written about will be obvious to the reader.

This author has spanned more than a half century of observing plant breeders and general geneticists. Some contacts do stand out. Obviously, the first contacts in the field of genetics in the 1940s and 1950s are most prominent. Starting with a year in the Demerec *Drosophila* laboratory in Cold Spring

Harbor, there were many interesting interactions. Most prominent were the discussions over meals with a number of the Cold Spring Harbor group and especially the meals and evening "small talk" with Barbara McClintock. Also, the patient genetic tutelage with Bruce Wallace who guided both Sally and the author through the various mutants and description of the chemically induced changes in *Drosophila*. Not obvious at the time, but the soon-to-be-famous visitors were so prominent during the Cold Spring Harbor summers (see Part B, Chapter 32). The casualness of that summer scene was quite disarming and very significant in this author's future perspective.

What became clear over this half century is how these scientists differ in their approach to fellow workers and individual students. Certainly what comes to mind is the critical and forthright thinking and ideals of H. Muller in genetic policy. In the 1950s he was critical of the lack of support from the United States Genetics Society for the treatment by the Stalinist regime in Russia against fellow Russian geneticists. This author can still remember his plea and cry "where are our principles?"

Other individuals were singularly prominent. Most that had contact or were tutored by the Drosophilist, Curt Stern, will remember the warmth extended to them during his noon lunch sessions or at Genetics Society meetings. John Laughnan, like Bruce Wallace, was always available for a thoughtful explanation of things genetic. Some of these same expressions of feeling are sprinkled throughout the various laudations, and these will be evident to the reader.

16 The Maize Pedigree Tree

The pedigree tree of maize geneticists commemorated in these issues is illustrated in Fig. 24. The roots of the tree nourished the growth of genetics with significant botanical discoveries, some of which are shown. Without a basic understanding of botany, many of the observations would have mystified the observing scientist.

The Bussey Institution played a significant role in "funneling" these early discoveries as genetics was acknowledged as a legitimate scientific endeavor in the early part of the 20th century (Weir, 1944). It was fortunate that E. M. East established a base at the Bussey and eventually emerged as a leader in plant genetics and who could act as a magnet for those interested in pursuing plant genetics. Where else in the United States could a genetics-oriented student go? East provided the hub from which most of 20th century leaders in maize genetics emerged. This is readily apparent as we see the Stadler, Emerson, Mangelsdorf, E. Anderson, Jones, and Brink branches extending from that trunk.

The main trunk of the tree extends to R. A. Emerson and the branches beyond him, where we see a proliferation of many who went on to make numerous contributions and to develop and train students who are active today in maize research. This represents a massive increase in maize workers, and the extended branches would be too numerous if all participants were included. But the tree represents those commemorated in the *MAYDICA* volumes and is confined to maize genetics.

Though maize genetics and breeding have been the focus of these pages, it is appropriate to look at the other contributions these maize workers have made to other aspects of plant science. Several did not pursue a career in maize, and among those are two of Professor East's students: E. R. Sears and Karl Sax.

Ernie Sears spent a long and productive period in wheat genetics and cytogenetics as a USDA employee at the University of Missouri. Through his steadfast and devoted work ethic, he and his colleagues uncovered many of the principles significant to the evolution and composition of wheat. He was a devoted hands-on worker, often seen making crosses and attending to his plants in the greenhouse. Many of Sears' discoveries currently are finding their way into molecular investigations of the wheat genome.

Karl Sax, a leading plant geneticist and cytologist during the early decades of the 20th century, held a pivotal position at Harvard University. He attracted a large number of students, and they, in turn, trained a number of students who are leaders in genetics today. Possibly because Sax investigated X-ray radiation of plants, an interesting and attractive field in the 1930s following the Muller and Stadler discoveries of X-ray-induced mutations, a number of students joined the Sax group. Charlie Rick was one of the early Sax-trained students who, after getting his doctoral degree, went to the University of California-Davis to join the Vegetable Crops Department. There he devoted (and still is active) more than four decades to the investigation of the genetics of the tomato, making that plant a compelling genetic tool. Currently one of his students, Steve Tanksley, has undertaken the enormous task of relating the *Solanaceae* to evolving molecular science.

Another Sax student was Norman A. Giles. Giles, though beginning with plant studies (*Tradescantia*), turned his attention to *Neurospora* and proceeded to uncover the molecular understanding of the quinic acid pathway. Among his students at Yale University was Gerry Fink, who along with yeast studies has undertaken studies in *Arabidopsis*. Another of Giles' students was Dow Woodward, who went to Stanford University to study mitochondria. At Palo Alto, he attracted a number of postdoctoral students, including Heinz Saedler (Cologne, Germany) and Richard Flavell (now at Norwich, England), who have gone on to make significant contributions to maize genetics and to plant science in general. It is readily apparent that the trunk of the pedigree tree (East) still supports current efforts in plant science.

This departure from our focus on maize workers of the 20th century is not intended to be exhaustive but only to illustrate the roots of some current thrusts in plant science. Of course, some such as Elliot Meyerowitz, who has made pioneering studies in floral development at the California Institute of Technology, originally investigated other areas (RNA and *Drosophila*) before embarking on plants. The same can be said of Enrico Coen at the John Innes in Norwich, England, who worked with *Drosophila* before his efforts with plants, although his mentor's mentor, Ralph Riley, was a highly successful plant cytogeneticist. Jonathan Jones, now a successful molecular worker at

Norwich, began his career at the Flavell lab in Trumpington, England, though he enriched his genetic understanding under the tutelage of Hugo Dooner.

Fig. 24. The pedigree tree of the professional training of these maize plant breeders and geneticists commemorated in this book in Part B. The beginnings in the Bussey Institute under Professor East leads to branches of 4th and 5th generation "progeny."

17 Legacy of the Past

The genetic findings in the first five decades of the 20th century provided a broad base of accumulated genetic information (Table 1 and the *Maize Genetics Cooperation Newsletter*, 70, for example). These genetic findings included the accumulation of a large number of genes, their inheritance, and location in the genome. The early development of the idiogram map of maize chromosomes, coupled with the extensive documentation of maize genes and linkage maps, has provided the new generation of maize geneticists a firm foothold for further investigation. Progress on these genes has included the linkage relationships to other genes and their possible association. The *A1* gene, for instance, was utilized in early X-ray studies by Stadler, and the accumulation of X-ray-induced deficiencies studied byRoman and Stadler provided a glimpse of the effects induced by gamma rays. Further, the *A1* gene became the focus of Laughnan's duplicate gene study with the α and ß locus. This provided an extensive insight into possible gene duplication and intrachromosomal crossing-over. It further developed that these units of the *A1* gene were duplicates of an original gene maintained in some South American populations. Now the *A1* gene is a focus in gene physical distance studies (Civardi *et al.*, 1994).

The original *a1* allele contained a *dt* insert; this was the forerunner of many studies with the *Dt* transposon. Thus, Emerson's accumulation of mutants such as the *a1* allele in the early decades of the 20th century led to many subsequent investigations.

The *A2* gene originally described by Jenkins (1932) lay dormant for many years. It was used in linkage studies with the *bt* gene (*bt1*) and others such as Stadler's investigations of *A2* and gamete factor linkage. It wasn't until 1990 (Menssen *et al.*, 1990), however, that *A2* was cloned and described and found to be an intronless gene, possibly one of the first found in maize. Further, experiments were initiated to clarify whether the insert in the *a2mII* allele was the basis for the origin of introns in genes (Thathiparthi *et al.*, 1995). The *A2* investigations also provided an insight into the gene product developed by the *A2* gene.

83

Other genes came into focus. The *A3* gene, a recessive (*a3a3*) that produces color in plants, was identified originally by Lindstrom in 1935. How could a recessive gene produce a colored plant phenotype that is absent with its dominant allele? Many years later, in 1994, the gene was found to be homologous to the *In* gene in maize, also a recessive (*in in*)—which produces a phenotype by facilitating the translation of the *Whp* transcript in the kernel. The *In* gene when in the recessive condition intensifies aleurone color; this was first described by Fraser (1924). Further, in the presence of the recessive *c2*, pale coloration is formed, and it was finally understood that *in* was enhancing the expression of *Whp*. With probes of the *C2* and *Whp* genes, Northern blots could reveal RNA transcripts, but Western blots indicated that the *c2 In Whp* genotype did not show an expressed protein (Burr *et al.*, 1996)

Brn1 was originally described only recently, but the genetics of this gene and its interaction with other genes was characterized more fully by Stinard (1994). The *Bt1* gene, however, was originally found by Mangelsdorf (1926), and was not cloned and identified until 1990. This would have been a readily observable gene to find in view of its extreme nature on the kernel. (It was simultaneously described by Wentz in 1926.)

How the *B-Peru* gene was found is a classic model in genetic research. The first report of this allele appeared in the Cornell memoirs in a report by Emerson (1921). A researcher had to be quite astute to follow up the plant versus kernel characteristics of this gene. Very likely, Emerson was scanning the South American accessions for new genes. Here was an SS allele that differed from others in the allelic series by the purple coloration in the kernel.

The studies with the *B-Peru* allele lay dormant until Mangelsdorf (*Ac1497*) sent this colored kernel to R. A. Brink, who was gathering different-colored alleles to incorporate into his colored inbred lines. To extend the study of *B-Peru*, Brink asked Derek Styles, then a graduate student at Wisconsin [and now retired from the University of Victoria (Canada)], to locate it and make further studies. Using the laborious *wx*-translocation method, Styles located this allele on chromosome 2 and, in consultation with fellow graduate student Jerry Kermicle, concluded that this allele, unlike other *b* alleles, expressed a colored aleurone in addition to the colored plant. These early *B-Peru* studies by Styles evolved into molecular studies by the Chandler group (Chandler *et al.*, 1989; Patterson and Chandler, 1995).

One of the most exciting genes to be described molecularly was the *c1* gene. Sequenced by Paz-Ares *et al.,* (1986) and Cone *et al.,* (1986), it was reported originally by East and Hayes in 1911. This finding was not unusual, in that most of the commercial corn lines had the *c1* gene. East and Hayes, in crosses with these materials with the color lines, easily uncovered the *c1* gene along with the *R1* gene. What made this sequencing of the *C1* gene exciting was its homology to the *myb* proto-oncogene products in chickens, which spurred an early awareness of the relation of genes between the plant and animal kingdoms; similarly, with the *r1* gene, finding the *myc* oncogene homology was of note. Further, *C1* was one of the first regulatory genes uncovered, and it gave plant geneticists a handle on regulatory control of genes.

18 Science Progress

All scientific endeavor is not as it appears. The discovery of transposons is a case in point. Current admirers of the McClintock period (Fedoroff and Botstein, 1992), fail to acknowledge her past treatment. At the 1951 Cold Spring Harbor meeting, few questioned McClintock's acute transposon observations (Comfort, 1995); the detailed crosses were clearly shown. Her claims, however, were questioned. McClintock was dismayed by the lack of interest in her later paper on the *A1* dotted relation (McClintock, 1953), evidenced by a lack of reprint requests. McClintock vowed that when colleagues eventually discovered the nature of the transposable elements themselves, they would begin to accept her findings (personal communication).

McClintock's transposon discovery could be considered a "punctuated" discovery. This revelation of elements moving in the genome did change biologists view of the genome and can be allied to other discoveries that Kuhn (1962) describes as "episodic" in scientific revolution. Yes, maize genetic research, prior to the transposon studies was in a quiet interlude that was followed by a "violent eruption" following the assimilation of transposons in later studies.

Thus, in the early 1950s McClintock was unwilling to submit papers for publication. After that last refereed paper in 1953, she confined her presentations to invited papers or to the Cold Spring Harbor Carnegie Annual Reports. McClintock in the 1950s and 1960s was fully confident in her observations but aware that the scientific public was years away from acceptance of the phenomena. The 1960s and early 1970s saw little funding or support for transposons. Discoveries of transposons in bacteria (Jordan *et al.*, 1968; Shapiro, 1969) at last established their universality. But in scientific progress, projections are made to be challenged and modified, and to provoke further experimentation.

The great discoveries of science are the result of a range of earlier discoveries in which an initial notion was suggested but final understanding requires diligent effort. This was especially true with the initial input by McClintock and the resulting effort required over the next half century to render her ideas acceptable. And, that of the 1997 Nobel Prize (Medicine) winner, Stanley Prusiner, whose 1972 discovery (prions) is also a case in point. Going against the tide, he was finally vindicated with his revelation.

Changes in science progress also appear in a cursory examination of published papers from the 1930s to the 1990s. In examining the authorship of papers accumulated as part of the *Maize Genetics Cooperation Newsletters*, a striking change becomes evident (Fig. 25): single-author papers have been replaced by those with dual and multiple authorships. Even more striking are the number of authors in papers published in *Genetics*. In the 1930s most papers have a single author, however, in the 1990s, most or almost all have multiple authorship (Fig. 26). This authorship record mirrors the trend with the maize group and science in general. Research now requires teams of individuals, especially when it is driven by competition for funding.

It is inescapable that multiple authorships will continue. What will be the consequence? Would a specific site breakage be recognized and distinguished from a random-site breakage by an assistant assigned to kernel examination, as with the observation that originated the transposon study explosion? Or, in the case of the discovery of the *Mutator* transposon (Robertson, 1977), would the extraordinary abundance of newly discovered mutant seedlings in a seedling bench be acknowledged today? Then, one could examine the sequence of discoveries (Part A, Chapter 13) that elucidated the various components of the anthocyanin pathway. Or, in the case of the discovery of restriction enzymes that led to the RFLP consequences (Arber and Dussoiz, 1962), would such a phenotypic change be identified? Large-group science will alter the research climate of the future.

Most recipients of the Nobel Peace Prize, including E. B. Lewis, B. McClintock, G. Beadle, and Ch. Nousard, showed a common pattern. Unmistakable in these studies is "hands-on" investigation of a problem, whereby clear observation leads to ultimate confidence in the material gestating in their minds. The value of inner conviction is that the investigator is undeterred by those who criticize the material because it does not fit a norm. Thousands of current grant proposals are rejected because they lack clarity of purpose (Eaves, 1984; Zimmerman and Lehman, 1993; Morrison and Russell, 1996), or possibly because hurried reviewers are anxious about their own projects. Impatience and curtness are evident in the review process.

Is there a better means of judging science quality? The intrusion of computerized analyses of scientific publications was discussed recently in *Nature* (Motta, 1995; Metcalfe, 1995). These citation frequency values have been translated into journal impact factors (IF) and citation index (CI). The increasing use of these values to make judgment calls during promotion and hiring deliberations will minimize the human element and, as Motta and Metcalfe argue, become too simplistic and misleading.

Gina Kolata (1993) looked into the growth of "big science." The change from "organismal" studies to "machine" genetics requires more workers, each contributing a segment to a final result. And this change in the scientific climate demands other skills, such as managing people. The lone investigator, oblivious to those around him or her, is no more. The current laboratory principal scientific investigator must motivate people (psychology), write grants (English), and manage and administer a laboratory (business administration). When genome studies uncover sequences of whole chromosomes, as was the recent case of chromosome 3 in yeast, multiple authorship will be especially necessary because of the massive endeavor for such an effort. As molecular biologist Dr. Shirley Tilghman relates in an interview with the New York Times, "Taking a sabbatical was necessary in order to visit my own laboratory" (Kolata, 1993).

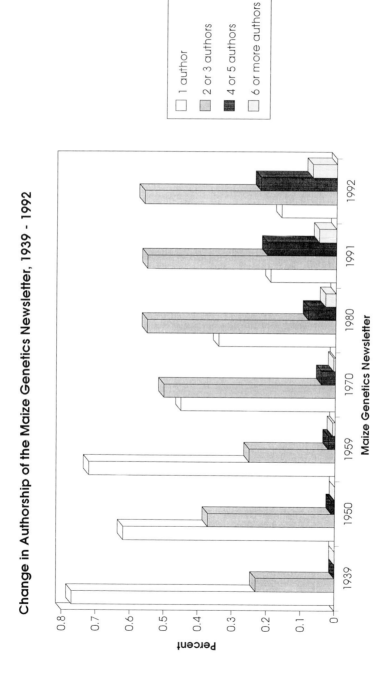

Fig. 25. Change in frequency of authorship of maize workers of individual papers as accumulated in several volumes of the *Maize Genetics Cooperation Newsletters* 1939–1992.

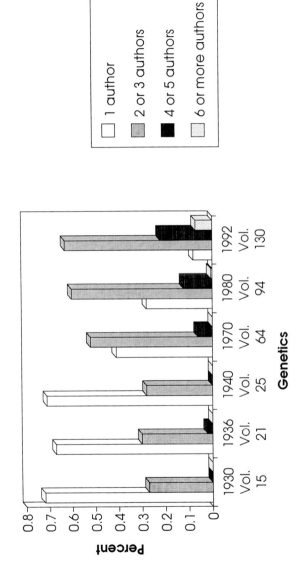

Fig. 26. Change in frequency of authorship of author number in individual maize papers published in several volumes of *Genetics* from 1930–1992.

References

Anderson, E. G., A. E. Longley, C. H. Li, and K. L. Retherford, 1949. Hereditary effects produced in maize by radiations from the bikini atomic bomb. I. Studies on seedlings and pollen of the exposed generation. Genetics 34: 639-646.

Anonymous, 1873. Bussey Institution Bulletin I: 1-7.

Arber, W., and D. Dussoix, 1962. Host specificity of DNA produced by Escherichia coli. I. Host controlled modification of bacteriophage. J. Mol. Biol. 5: 18-36.

Avery, O. T., C. M. MacLeod, and M. McCarty, 1944. Studies on the chemical nature of the substance-inducing transformation of pneumococcal types; induction of transformation by a deoxyribonucleic acid fraction isolated from pneumococcus type. III. J. Exp. Medicine 79: 137-158.

Bateson, W., 1916. The mechanism of Mendelian heredity, by T. H. Morgan, A. H. Sturtevant, H. J. Muller and C. B. Bridges (a review). Science 44: 536-543.

Bateson, W., 1930. Mendel's principle of heredity. 4th Imp., Cambridge Univ. Press.

Bateson, W., and R. C. Punnett, 1911. On the interrelation of genetic factors. Proc. Roy. Soc. of London, B. 84: 3-8.

Baum, J. A., and J. G. Scandalios, 1982. Multiple genes controlling superoxide dismutase expression in maize. J. Hered. 73: 95-100.

Beadle, G. W., 1939. Teosinte and the origin of maize. J. Hered. 30: 245-247.

Beadle, G. W., 1950. Rollins Adams Emerson. Genetics 35: 1-3.

Beadle, G. W., 1972. The mystery of maize. Field Museum of Natural History Bulletin 43: 2-11.

Belling, J., 1927. The attachments of chromosomes at the reduction division in flowering plants. J. Genet. 18: 177-205.

Bennetzen, J. L., and M. Freeling, 1993. Grasses as a single genetic system: Genome composition, collinearity, and compatibility. TIG 9: 259-261.

Benzer, S., 1955. Fine structure of a genetic region in bacteriophage. Proc. Natl. Acad. Sci. USA 41(6): 344-354.

Berg, D. E., and M. M. Howe (eds.), 1989. Mobile DNA. Am. Soc. Microbiol., Washington, DC.

Bridges, C. B., 1916. Non-disjunction as a proof of the chromosome theory of heredity. Genetics 1: 1-52.

Brink, R. A., and D. C. Cooper, 1935. A proof that crossing over involves an exchange of segments between homologous chromosomes. Genetics 20: 22-35.

Brink, R. A., and I. M. Greenblatt, 1954. Diffuse, a pattern gene in Zea mays. J. Hered. 45: 47-50.

Brown, J. J., M. G. Mattes, C. O'Reilly, and N. S. Shepherd, 1989. Molecular characterization of rdt, a maize transposon of the "dotted" controlling element system. Mol. Gen. Genet. 215: 239-244.

Bryan, A. A., and J. E. Sass, 1941. Heritable characters in maize 51-"knotted Leaf." J. Hered. 32: 342-346.

Bureau, T. E., and S. R. Wessler, 1992. Tourist—a large family of small inverted repeat elements frequently associated with maize genes. Plant Cell 4: 1283-1294.

Burr, B., S. V. Evola, F. A. Burr, and J. S. Beckmann, 1983. The application of restriction fragment length polymorphism to plant breeding. In Genetic Engineering: Principles and Methods, edited by J. K. Setlow, and A. Hollaender. New York: Plenum Publ. Corp., pp. 45-59.

Burr, F. A., B. Burr, B. E. Scheffler, M. Blewitt, U. Wienand, and E. C. Matz. 1996. The maize repressor-like gene *intensifier1* shares homology with the *r1/b1* multigene family of transcription factors and exhibits missplicing. The Plant Cell 8:1249-1259.

Chandler, V. L., J. P. Radicella, T. P. Robbins, J. Chen, and D. Turks, 1989. Two regulatory genes of the maize anthocyanin pathway are homologous: Isolation of *b* utilizing *r* genomic sequences. Plant Cell 1: 1175-1183.

Chomet, P., D. Lisch, K. J. Hardeman, V. L. Chandler, and M. Freeling, 1991. Identification of a regulatory transposon that controls the mutator transposable element system in maize. Genetics 129: 261-270.

Civardi, L., Y. J. Xia, K. J. Edwards, P. S. Schnable, and B. J. Nikolau, 1994. The relationship between genetic and physical distances in the cloned *a1-sh2* interval of the *Zea mays* L. genome. Proc. Natl. Acad. Sci. 91: 8268-8272.

Coe, E. H., S. M. McCormick, and S. A. Modena, 1981. White pollen in maize. J. Hered. 72: 318-320.

Collins, G. N., 1909. A new type of Indian corn from China. USDA Plant Industry Bureau Bull. 161: 1-30.

Collins, G. N., 1917. Hybrids of *Zea tunicata* and *Zea ramosa*. Proc. Natl. Acad. Sci. 3: 345-349.

Collins, G. N., and J. H. Kempton, 1920. Heritable characters of maize. I. Lineate leaves. J. Hered. 11: 3-6.

Collins, G. N., and J. H. Kempton, 1927. Variability in the linkage of two seed characters of maize. USDA Bull. 1468: 1-64.

Comfort, N. C., 1995. Two genes, no enzyme: A second look at Barbara McClintock and the 1951 Cold Spring Harbor Symposium. Genetics 140: 1161-1166.

Cone, K. C., F. A. Burr, and B. Burr, 1986. Molecular analysis of the maize anthocyanin regulatory locus *Cl*. Proc. Natl. Acad. Sci. USA 83: 9631-9635.

Correns, C., 1899. Untersuchungen über die Xenien bei *Zea mays*. Ber. deuts. Bot. Ges. 17: 401-417.

Correns, C., 1901. Bastarde zwischen Maisrassen mit besonderer Berucksichtigung der Xenien. Bibliotheca Bot. 53: 401-417.

Correns, C., 1902. Scheinbare Ausnehmen von der Mendels'schen Spaltungsregel für Bastarde. Ber. Deutsch Bot. Gesells. 20: 159.

Crabb, R. A., 1947. The hybrid-corn makers. New Brunswick, NJ: Rutgers University Press, p. 331.

Crabb, R. A., 1992. The hybrid corn-makers. 2nd Ed. West Chicago, IL: West Chicago Publishing Co., p. 331.

Creighton, H. B., and B. McClintock, 1931. A correlation of cytological and genetical crossing over in *Zea mays*. Proc. Natl. Acad. Sci. USA 17: 492-97.

Dellalporta, S. L., I. M. Greenblatt, J. Kermicle, J. B. Hicks, and S. Wessler, 1988. Molecular cloning of the maize *R-nj* allele by transposon tagging with *Ac*. Ed. by J. P. Gustafson and R. Appels. 18th Stadler Genetics Symposium, Chromosome Structure and Function. New York: Plenum Press, pp. 263-282.

Demerec, M., and Z. Hartman, 1956. Tryptophane mutants in salmonella typhimurium. Carnegie Inst. Wash. Publ. 612. Genetic studies with bacteria, p. 5.

Dempsey, E., 1994. Marcus Morton Rhoades. Proc. Am. Phil. Soc. 138: 561-567.

deVries, H., 1901. *Die Mutations Theory*. Leipzig: Verlag Weit and Co., p. 63.

Doebley, J., 1993. Genetics and the morphological evolution of maize. In The Maize Handbook, ed. by M. Freeling and V. Walbot. New York: Springer-Verlag, pp. 66-67.

Doebley, J., A. Stec, J. Wendel, and M. Edwards, 1990. Genetic and morphological analysis of a maize-teosinte *F2* population: Implications for the origin of maize. Proc. Natl. Acad. Sci. USA 87: 9888-9892.

Doebley, J., A. Stec, and C. Gustus, 1995. Teosinte branched-1 and the original of maize: Evidence for epistasis and the evolution of dominance. Genetics 141: 333-346.

Dole, R., and E. D. Porteus, 1990. *The Story of James Dole*. Aiea, Hawaii: Island Heritage Publishing.

Dooner, H. K., 1983. Coordinate genetic regulation of flavonoid biosynthesis enzymes in maize. Mol. Gen. Genet. 189: 136-141.

Dunn, L. C., 1965. *A Short History of Genetics*. New York: McGraw-Hill, Inc., pp. 261.

Duvick, D. N., 1956. Allelism and comparative genetics of fertility restoration of cytoplasmically pollen sterile maize. Genetics 41: 544-565.

Dyer, D., 1994. Heterosis and pollen transmission issues for producing hybrid soybeans; novel production strategies for corn. Proceedings of the 1994 Keystone Meetings, Winter Park, CO.

East, E. M., 1910. A Mendelian interpretation of variation that is apparently continuous. Am. Nat. 44: 65-82.

East, E. M., and H. K. Hayes, 1911. Inheritance in maize. Connecticut Agric. Exp. Stn. Bull. 167: 1-142.

East, E. M., and A. J. Mangelsdorf, 1925. A new interpretation of the hereditary behavior of self-sterile plants. Proc. Natl. Acad. Sci. 2: 166-171.

Eaves, G. N., 1984. Preparation of the research-grant application: opportunities and pitfalls. Grants Magazine 7: 151-157.

Emerson, R. A., 1911. Genetic correlation and spurious allelomorphism in maize. Nebraska Agr. Exp. Stn. Ann. Rep. 24: 59-90.

Emerson, R. A., 1914. The inheritance of a recurring somatic variation in variegated ears of maize. Am. Naturalist 47: 87-115.

Emerson, R. A., 1917. Genetical studies of variegated pericarp in maize. Genetics 2: 1-35.

Emerson, R. A., 1918. A fifth pair of factors, *Aa*, for aleurone colour in maize and its relation to the *Cc* and *Rr* pairs. Cornell Agric. Exp. Stn. Memoir 16: 225-289.

Emerson, R. A., 1921. The genetic relations of plant colors in maize. Cornell Univ. Agric. Exp. Stn. Memoir 39: 1-156.

Emerson, R. A., 1929. The frequency of somatic mutation in variegated pericarp of maize. Genetics 14: 488-511.

Emerson, R. A., and E. M. East, 1913. The inheritance of quantitative characters in maize. Nebraska Agric. Exp. Stn. Bull. 2: 1-120.

Emerson, R. A., 1934. Relation of a differential fertilization genes, *Ga ga*, to certain other genes of the *Su-Tu* linkage group of maize. Genetics 19: 137-156.

Emerson, R. A., G. W. Beadle, and A. C. Fraser, 1935. A summary of linkage studies in maize. Cornell Univ. Agric. Exp. Stn. Memoir 180: 1.

Eyster, W. H., 1924. A second factor for primitive sporophyte in maize. Am. Nat. 58: 436-439.

Eyster, W. H., 1931. Vivipary in maize. Genetics 16: 574-590.

Fant, K., 1993. *Alfred Nobel: A Biography.* New York: Arcade Publications, p. 342.

Faure, J.-E., C. Digonnet, and C. Dumas, 1994. An *in vitro* system for adhesion and fusion of maize gametes. Science 263: 1598-1600.

Fedoroff, N., S. Wessler, and M. Shure, 1983. Isolation of the transposable maize controlling elements *Ac* and *Ds*. Cell 35: 235-242.

Fedoroff, N., and D. Botstein, 1992. *The Dynamic Genome: Barbara McClintock's Ideas in the Century of Genetics.* Cold Spring Harbor: Cold Spring Harbor Lab Press, p. 422.

Fedoroff, N., D. Furtec, and O. Nelson, 1984. Cloning of the *bronze* locus in maize by a simple and generalizable procedure using the transposable controlling element *Ac*. Proc. Natl. Acad. Sci. USA 81: 3825-3829.

Franken, P., S. Schrell, P. A. Peterson, H. Saedler, and U. Wienand, 1994. Molecular analysis of protein domain function encoded by the myb-homologous maize genes *C1*, *Zm1*, and *Zm38*. Plant J. 6: 21-30.

Franklin-Tong, V. E., and F. C. H. Franklin, 1993. Gametophytic self-incompatibility: Contrasting mechanisms for *Nicotiana* and *Papaver*. Trends in Cell Biology 3: 340-345.

Fraser, A. C., 1924. Heritable characters of maize. XVII. Intensified red and purple aleurone color. J. Hered. 15: 119-123.

Freeling, W., and V. Walbot, 1994. *The Maize Handbook.* New York: Springer Verlag, p. 759.

Friedmann, P., and P. A. Peterson, 1982. The *Uq* controlling-element system in maize. Mol. Gen. Genet. 187: 19-29.

Giroux, M. J., J. Shaw, G. Barry, B. Greg Cobb, T. Greene, T. Okita, and L. C. Hannah, 1996. A single gene mutation that increases maize seed weight. Proc. Natl. Acad. Sci. 93: 5824-5829.

Guignard, L., 1899. Sur les antherozoides et la double copulation sexuelle chez les vegetaux angiospermes. C. R. Acad. Sci. Paris 128: 864-871.

Hallauer, A., 1984. George Sprague: 60 years of corn research. *MAYDICA* XVIX: 351-356.

Hallauer, A., and F. Miranda, 1988. Quantitative genetics in maize breeding. Rev. ed. Ames, Iowa: Iowa State University Press, p. 468.

Harborne, J. B., 1967. *Comparative Biochemistry of the Flavonoids*. New York: Academic Press, p. 383.

Haring, V., J. E. Gray, B. A. McClure, M. A. Anderson, and A. E. Clarke, 1990. Self-incompatibility: A self-recognition system in plants. Science 250: 937-941.

Hattori, T., V. Vasil, L. Rosenkrans, L. C. Hannah, D. R. McCarty, and I. K. Vasil, 1992. The *viviparous-1* gene and abscisic acid activate the *C1* regulatory gene for anthocyanin biosynthesis during seed maturation in maize. Gene Develop. 6: 609-618.

Helentjaris, T., 1989. Future directions for plant RFLP technology and its applications. In *Development and Application of Molecular Markers to Problems in Plant Genetics*, ed. by T. Helentjaris and B. Burr. Cold Spring Harbor, NY: Cold Spring Harbor Lab, pp. 159-161.

Hershey, A. D., and M. Chase, 1952. Independent functions of viral protein and nucleic acid in growth of bacteriophage. J. Gen. Physiol. 36: 39-56.

Iltis, H. H., 1972. The taxonomy of *Zea* (*Gramineae*). Phytologia 23: 248-249.

Jackson, W. H., 1902. "The Man with the Hoe" (photograph), Moki Indian from Arizona. Copyright 1902, Detroit Photographic Company. Photographer: Vroman.

Jansen, R. C., 1993. Interval mapping of multiple quantitative trait loci. Genetics 135: 205-211.

Jansen, R. C., and P. Stam, 1994. High resolution of quantitative traits into multiple loci via interval mapping. Genetics 136: 1447-1455.

Jenkins, M. T., 1932. An additional pair of factors affecting anthocyanin pigment in maize. J. Agric. Res. 44: 495-502.

Johan, G., P. Close, and S. Briggs, 1993. Is our "heartbreaker" a "tourist?" Maize Genet. Coop. Newsl. 67: 53.

Jones, D. F., 1918. The effects of inbreeding and cross-breeding upon development. Connecticut Agric. Exp. Stn. Bull. 207: 5-100.

Jordan, E., H. Saedler, and P. Starlinger, 1968. O° and strong-polar mutations in the *gal* operon are insertions. Mol. Gen. Genet. 102: 353-363.

Kahmann, R., 1992. Barbara McClintock—A personal view. Trends in Genetics 8: 407.

Keller, E. F., 1983. *A Feeling for the Organism: The Life and Work of Barbara McClintock*. San Francisco, CA: W. H. Freeman and Co., p. 235.

Kolata, G., 1993. Labs, a scientist's foreign country. Education Life Supplement, *New York Times*, 4 April 1993, p. 19.

Kreuzaler, F., H. Ragg, E. Fautz, D. N. Kuhn, and K. Hahlbrock, 1983. UV-induction of chalcone synthase mRNA in cell suspension cultures of *Petrosilinum hortense*. Proc. Natl. Acad. Sci. 80: 2591-2593.

Kuhn, T. S., 1962. The structure of scientific revolutions. Chicago: University of Chicago Press, p. 172.

Kuwada, Y., 1911. Meiosis in the pollen mother cells of *Zea mays* L. Botan. Magazine (Tokyo) 25: 163-181.

Kuwada, Y., 1915. Über die chromosomenzahl von *Zea mays*. Bot. Mag. 29: 83-89.

Lechelt, C., A. Laird, and P. Starlinger, 1986. The *P*-locus of *Zea mays*. Abstract presented at the 1986 meeting on Eukaryotic Transposable Elements. Cold Spring Harbor, NY, May 7-11. Abstracts arranged by G.R. Fink, G. M. Rubins, S. L. Dellaporta, and J. Hicks. p. 50.

Lechelt, C., T. Peterson, A. Laird, J. Chen, S. L. Dellaporta, E. Dennis, W. J. Peacock, and P. Starlinger, 1989. Isolation and molecular analysis of the maize *P* locus. Mol. Gen. Genet. 219: 225-234.

Lederberg, J., E. M. Lederberg, N. D. Zinder and E. R. Lively, 1951. Recombination analysis of bacterial heredity. Cold Spring Harbor Symposium Quant. Biol. 16: 413.

Lee, H., 1984. *Roswell Garst: A Biography*. The Henry A. Wallace Series. Ames, Iowa: Iowa State University Press, p. 310.

Lindstrom, E. W., 1923. Heritable characters of maize. XIII. Endosperm defects-sweet defective and flint defective. J. Hered. 14: 126-135.

Lindstrom, E. W., 1935. Some new mutants in maize. Iowa State Coll. J. Sci. 9: 451-459.

Lock, R. H., 1906. Studies in plant breeding in the tropics. III. Experiments with maize. Roy. Bot. Gard. (Peradeniya, Ceylon) Annals 3: 95-184.

Mains, E. B., 1949. Heritable characters in maize. Linkage of a factor for shrunken endosperm with the a_1 factor for aleurone color. J. Hered. 40: 21-24.

Mangelsdorf, P. C., 1926. The genetics and morphology of some endosperm characters in maize. Connecticut Agric. Exp. Stn. Bull. 279: 509-614.

Mangelsdorf, P. C., 1955. George Harrison Shull. Genetics 40: 1-4.

Mangelsldorf, P. C., 1970. Donald F. Jones (1890-1963). Genetics 65: 1-7.

Mangelsdorf, P. C., 1974. Corn: its origin, evolution, and improvement. Cambridge, Mass.: Belknap Press of Harvard University Press. p. 262.

Mangelsdorf, P. C., 1975. Donald Forsha Jones (1890-1963). National Academy of Sciences, Biographical Memoirs 46: 135-156.

May, M. 1992. Hidden glory in the Hole. Maize Genetics Coop. Newsl. 6: 217-220.

McCarty, D. R., T. Hattori, C. B. Carson, V. Vasil, M. Lazar, and I. K. Vasil, 1991. The viviparous-1 developmental gene of maize encodes a novel transcriptional activator. Cell 66: 895-905.

McCarty, D. R., 1992. The role of *Vp-1* in regulation of seed maturation in maize. Biochem. Soc. Trans. 20: 89-92.

McClintock, B., 1929. Chromosome morphology in *Zea mays*. Science 69: 629.

McClintock, B., and H. E. Hill, 1931. The cytological identification of the chromosome associated with the *R-g* linkage group in *Zea mays*. Genetics 16: 175-190.

McClintock, B., 1931. Cytological observations of deficiencies involving known genes, translocations and an inversion in *Zea mays*. Missouri Agric. Exp. Stn. Bull. 163: 1-30.

McClintock, B., 1933. The association of non-homologous parts of chromosomes in the mid-prophase of meiosis in *Zea mays*. Zeitschrift Zellforschung u. Mikrosopische Anatomie 19: 191-237.

McClintock, B., 1944. The relation of homozygous deficiencies to mutations and allelic series in maize. Genetics 29: 478-502.

McClintock, B., 1953. Induction of instability at selected loci in maize. Genetics 36: 579-599.

McClung, C. E., 1902. The accessory chromosome-sex determinant. Biol. Bull. III, 1 and 2.

McFadden, E. S., and E. R. Sears, 1947. The genome approach in radical wheat breeding. J. Am. Soc. Agron. 39: 1011-1026.

Meeley, B. and Briggs, S. Briggs, 1995. Reverse genetics for maize. Maize-Genetics-Cooperation Newsletter. 69, 67, 82.

Mensson, A., S. Hohmann, W. Martin, P. S. Schnable, P. A. Peterson, H. Saedler, and A. Gierl, 1990. The *En/Spm* transposable element of *Zea mays* contains splice sites at the termini generating a novel intron from a *dSpm* element in the *A2* gene. EMBO J. 9: 3051-3057.

Metcalfle, N. B., 1995. Journal impact factors. Nature "Letters" 376: 720.

Moose, S. P., and P. H. Sisco, 1994. Glossy15 controls the epidermal juvenile-to-adult phase transition in maize. Plant Cell 6: 1343-1355.

Morgan, T. H., 1910. Sex-limited inheritance in *Drosophila*. Science 32: 120-122.

Morrison, D. C., and S. W. Russell, 1996. Handout for grant-writing seminar, "Write winning grants." Grant Writers' Seminars and Workshops, Taos, New Mexico.

Motta, G., 1995. Journal impact factors. Nature "Letters" 376: 720.

Nawaschin, S., 1899. Neue beobachtungen über befruchtung bei *Friatillaria* und *Lilium*. Bot. Centrabl. 77: 62.

Nelson, O. E., 1993. A notable triumvirate of maize geneticists. In *Anecdotal, Historical and Critical Commentaries on Genetics*, ed. by J. F. Crow and W. F. Dove. Genetics 135: 937-941.

Neuffer, M. G., 1954. Activator of *bz-m2* mutability. MNL 28: 63-65.

Neuffer, M. G., L. Jones, and M. S. Zuber, 1968. *The Mutants of Maize*. Madison, Wis.: Crop Science Society of America, p. 74.

Neuffer, M. G., E. H. Coe, and S. R. Wessler. 1996. *Mutants of Maize*. Cold Spring Harbor Press (Forthcoming, December 1996).

Neilsen, K. J., R. L. Heath, M. A. Anderson, and D. J. Craik, 1995. Structures of a series of 6-kDA trypsin inhibitors isolated from the stigma of *Nictotiana alata*. Biochemistry 34: 14304-14311.

O'Reilly, C., N. S. Shepherd, A. Pereira, Z. S. Schwarz-Sommer, I. Bertram, D. S. Robertson, P. A. Peterson, and H. Saedler, 1985. Molecular cloning of the *a1* locus of *Zea mays* using the transposable elements *En* and *Mu1*. EMBO J. 4(4): 877-882.

Patterson, G. I., and V. L. Chandler, 1995. Timing of *b* locus paramutation. *MAYDICA* 40: 35-41.

Paz-Ares, J., U. Wienand, P. A. Peterson, and H. Saedler, 1986. Molecular cloning of the *c* locus of *Zea mays*: A locus regulating the anthocyanin pathway. EMBO J. 5: 829-833.

Pereira, A., ZS. Schwarz-Sommer, A. Gierl, I. Bertram, P. A. Peterson, and H. Saedler, 1985. Genetic and molecular analysis of the *Enhancer* (*En*) transposable element system of *Zea mays*. EMBO J. 4: 17-23.

Peterson, P. A., 1953. A mutable pale green locus in maize. Genetics 38: 682-683.

Peterson, P. A., 1965. A relationship between the *Spm* and *En* control systems in maize. Am. Nat. 99: 391-398.

Peterson, P. A., 1992. Quantitative inheritance in the era of molecular biology. *MAYDICA* 37: 7-18.

Peterson, P. A., 1993a. Transposable elements in maize: Their role in creating plant genetic variability. Adv. Agron. 51: 79-124.

Peterson, P. A., 1993b. Transposons in maize (*Zea mays* L.) and their role in creating variability. In *International Crop Science I*, ed. by D. R. Buxton *et al.*, Madison, Wis.: Crop Science Society of America, pp. 641-645.

Peterson, P. A., 1994. Transposable elements in plants. In *Encyclopedia of Agricultural Science*, ed. by C. J. Arntzen. Vol. 4. New York: Academic Press, pp. 363-373.

Peterson, P. A., 1995. The development of transposon biology: From variegation to molecular confirmation. *MAYDICA* 40: 117-124.

Peterson, P. A., and F. Salamini, 1986. A search for active mobile elements in the Iowa Stiff-Stalk Synthetic Maize Population and some derivatives. *MAYDICA* XXXI: 163-172.

Pisabarro, A. G., W. F. Martin, P. A. Peterson, H. Saedler. and A. Gierl, 1991. Molecular analysis of the ubiquitous (*Uq*) transposable element system of *Zea mays*. MGG 230: 201-208.

Ralston, E. J., J. J. English, and H. K. Dooner, 1988. Sequence of three *bronze* alleles of maize and correlation with the genetic fine structure. Genetics 119: 185-197.

Rhoades, M. M., 1935. A new aleurone color in maize. Am. Nat. 69: 74-75.

Rhoades, M. M., 1936. The effect of varying gene dosage on aleurone colour in maize. J. Genet. XXXIII: 347-354.

Rhoades, M. M., 1952. The effect of the *bronze* locus on anthocyanin formation in maize. Am. Nat. 86: 105-108.

Rhoades, M. M., 1984. The early years of maize genetics. Annu. Rev. Genet. 18: 1-29.

Rhoades, M. M., and E. Dempsey, 1953. Cytogenetic studies of deficient-duplicate chromosomes derived from inversion heterozygotes in maize. Amer. Jour. Bot. 40: 405-424.

Rhoades, M. M., and E. Dempsey, 1982. The induction of mutable systems in plants with the high-loss mechanism. Maize Genet. Coop. Newsl. 56: 21-26.

Robertson, D. S., 1977. Consideration of some of apparent non-mendelian-*Mu* aspects of a *mutator*-system in maize. Genetics 86(2): S52-2715.

Robertson, D. S., 1984. A recessive brown aleurone mutant (*brn*). MNL 58: 18.

Robinett, D., E. Coe, and K. Cone, 1995. Map location of anthocyanin3. Maize Genet. Coop. Newsl. 69: 46.

Salamini, F., 1980. Genetic instability at the *opaque-2* locus of maize. Mol. Gen. Genet. 179: 497-507.

Salazar, A. M., and A. R. Hallauer, 1986. Divergent mass selection for ear length in maize. Rev. Brasil. Genet. IX: 281-294.

Sax, K., and H. J. Sax, 1924. Chromosome behavior in a genus cross. Genetics 9: 454-464.

Schmidt, R. J., F. A. Burr, M. J. Auekerman, and B. Burr, 1990. Maize regulatory gene *opaque-2* encodes a protein with a "leucine-zipper" motif that binds to zein DNA. Proc. Natl. Acad. Sci. USA 87: 46-50.

Schnable, P. S., and R. P. Wise, 1994. Recovery of heritable, transposon-induced, mutant alleles of the *rf2* nuclear restorer of T-cytoplasm maize. Genetics 136: 1171-1185.

Schnable, P. S., P. S. Stinard, J.-J. Wen, S. Heinen, D. Weber, M. Schneerman, L. Zhang, J. D. Hansen, and B. J. Nikolau, 1994. The genetics of cuticular wax biosynthesis. *MAYDICA* 39: 279-287.

Shapiro, J. A., 1969. Mutations caused by the insertion of genetic material into the *galactose* operon of *Escherichia coli*. J. Mol. Biol. 40: 93-105.

Shattuck-Eidens, D. M., R. N. Bell, S. L. Neuhausen, and T. Helentjaris, 1990. DNA sequence variation within maize and melon: Observations from polymerase chain reaction amplification and direct sequencing. Genetics 126: 207-217.

Shull, G. H., 1909. A pure-line method of corn breeding. Rep. Am. Breeders' Assn. 5: 51-59.

Singer, M., and R. Berg, 1991. *Genes and Genomes*. Mill Valley, Calif.: University Science Books.

Singleton, W. R., and D. F. Jones, 1935. Unpublished. p. 17.

Smithies, O., 1955. Zone electrophoresis in starch gels: Group variations in the serum proteins of normal human adults. Biochem. J. 61: 629-641.

Speulman, E., L.J. Metz, G.M. Aarts, B. Lintel Hekkert, W. J. Stiekema, and A. Pereira, 1996. Functional analysis of the Arabidopsis genome using the *En-1* transposon mutagenesis system. Plant Molecular Biology Book of Abstracts.

Stern, C., 1931. Zytologische-genetische untersuchungen als beweise für morganische theorie des factorenaustauches. Biol. Zentralb. 51: 547-587.

Stinard, P. S., 1994. The brown1 (*brn1*) locus of maize (*Zea mays* L.). *MAYDICA* 39: 273-278.

Stinard, P. S., D. S. Robertson, and P. S. Schnable, 1993. Genetic isolation, cloning, and analysis of a *Mutator*-induced, dominant antimorph of the maize *amylose extender1* locus. Plant Cell 5: 1555-1566.

Sturtevant, A. H., 1913. The linear arrangement of six sex-linked factors in *drosophila* as shown by their mode of association. J. Exp. Zool. 14: 43-59.

Sullivan, T. D., L. I. Strelow, C. A. Illingworth, R. L. Phillips, and O. E. Nelson, Jr., 1991. Analysis of maize *brittle-1* alleles and a defective *suppressor-mutator*-induced mutable allele. Plant Cell 3: 1337-1348.

Sutton, W. S., 1900. The spermatogonial divisions in Brachystola magna. Kansas Univ. Quart. Vol. 4: 135-160.

Sutton, W. S., 1902. On the morphology of the chromosome-group in brachystola magna. Biol. Bull. 4: 24-45.

Thatiparthi, V. R., S. P. Dinesh-Kumar, and P. A. Peterson, 1995. Permanent fixation of a transposable element insert in the *A2* gene of maize (*Zea mays* L.). J. Hered. 86: 167-171.

Thimann, K. V., and W. C. Galinat, 1991. Paul Christoph Mangelsdorf (20-22 July 1989). Proc. Am. Phil. Soc. 135: 469-472.

Townsend, R. F., 1992. The ancient Americas: Art from sacred landscapes. Chicago, Ill: The Art Institute of Chicago.

van Ooijent, J., and R. C. Jansen (eds.), 1994. Biometrics in plant breeding: Applications of molecular markers. Proc. 9th Meeting Eucarpia Section Biometrics. CPRO-DLO, Wageningen.

Veit, B., R. J. Schmidt, S. Hake, and M. F. Yanofsky, 1993. Maize floral development: New genes and old mutants. Plant Cell 5: 1205-1215.

Veit, B., E. Vollbrecht, J. Mathern, and S. Hake, 1990. A tandem duplication causes the *kn1-0* allele of knotted, a dominant morphological mutant of maize. Genetics 125: 623-631.

Vineyard, M. L., and R. P. Bear, 1952. Amylose content. Maize Genet. Coop. Newsl. 26: 5.

Wallace, H. A., and W. L. Brown, 1988. *Corn and Its Early Fathers*, ed. by R. S. Kirkendall. Ames, Iowa: Iowa State University Press, p. 141.

Watson, E. L., 1991. Houses for Science. New York: Cold Spring Harbor Laboratory Press, p. 352.

Watson, J. D., 1980. *Double helix: A personal account of the discovery of the structure of DNA*. New York: Norton, p. 298.

Watson, J. D., and F. H. C. CRICK, 1953. Molecular structure of nucleic acids. Nature 171: 737-738.

Weir, J. A., 1994. Harvard, agriculture and the Bussey Institute. Genetics 136: 1227-1231.

Wentz, J. B., 1926. Heritable characters of maize. XXVI-Concave. J. Hered. 17: 327-329.

Wickelgren, I., 1995. Protein sculptors that help turn on genes. Science 270: 1587-1588.

Wienand, U., U. Weydemann, U. Niesbach-Klösgen, P. A. Peterson, and H. Saedler, 1986. Molecular cloning of the *c2* locus of *Zea mays* – The gene coding for chalcone synthase. Mol. Gen. Genet. 203: 202-207.

Wilson, E. B., 1925. *The Cell in Development and Heredity*. New York: MacMillan Company, p. 1232.

Zack, C. D., R. J. Ferl, and L. C. Hannah, 1986. DNA sequence of a shrunken allele of maize: Evidence of visitation by insertional sequences. *MAYDICA* XXXI: 5-16.

Zimmerman, R. M., and A. W. Lehman, 1993. Grantwriting and Fundraising Fundamentals. Zimmerman, Lehman & Associates. San Francisco, CA.

PART B

THE COMMEMORATIVE
ISSUES OF *MAYDICA*

Biographies of 24 Plant Breeders
and Geneticists

19 William A. Russell

A Major Contributor to Maize Breeding*

By A. R. Hallauer and G. F. Sprague

Dr. W.A. Russell has completed twenty-five years of maize breeding research at Iowa State University. On July 16, 1952, he became a member of the cooperative state-federal maize breeding program to handle the hybrid testing program and has continued to be an important member of the project since that time. During his 25 years on the project, he has played a prominent role in the release of inbred lines that have made a significant contribution to the hybrid maize seed industry of the United States. In addition to the selection and testing of new inbred lines, Russell has maintained a comprehensive basic research program to complement the final phases of inbred line development. He has made significant contributions for the conversion of lines to T-cytoplasm, inheritance of rust resistance and conversion of lines that possess resistance, selection studies for stalk rot resistance, inheritance and selection for resistance to the European corn borer, selection for tolerance to corn root worms, and selection studies to determine the optimum regime for inbred line development. Thoroughness in planning, attention to detail, accuracy of analysis, and clarity in writing are characteristics of his research efforts. Most of the information from his basic research studies has direct use in applied breeding programs. Hence, Russell has been able to effectively combine basic research with the applied testing phase of the cooperative maize breeding program; both have a significant impact on inbred line and hybrid development in the United States.

Wilbert Ambrick Russell was born August 3, 1922, in Lenore, Manitoba, Canada, of parents from England. His early life was spent on the family farm.

* *MAYDICA* 1977, Vol. 22, No. 4

His primary and secondary education was obtained in the Manitoba Provincial schools, and he was graduated from Lenore High School in 1938. He obtained a fellowship in agriculture and attended the University of Manitoba from 1938 to 1942 and obtained his B.S.A. there in 1942. While attending the University of Manitoba, he distinguished himself as a student and received the Lieutenant Governor's Gold Medal for highest standing for the class during his undergraduate years and the University of Manitoba Gold Medal for highest standing for the class in his senior year.

Russell's education was interrupted in 1942 by World War II. He entered the service in November 1942 and served until September 1945 as a flying officer of the Royal Canadian Air Force. He was an able navigator and spent most of his tour of duty on the western coast of Canada.

After his military service, Russell resumed his education. He was awarded an Agricultural Institute of Canada fellowship for post-graduate training in 1946 and 1947 at the University of Minnesota. He obtained his Master of Science (M.S.) degree at the University of Minnesota in 1947. He then returned to Canada and was appointed an Agriculture Research Officer at the Experiment Station in Morden, Manitoba. He served there until July 1952, conducting plant breeding research in sunflowers and maize. While at the Morden Experiment Station, Russell played an important role in the development of the first two maize hybrids licensed by the Canadian Department of Agriculture for production in Manitoba. One of those hybrids remained on the recommended list until about 10 years ago. His main contribution to the sunflower program was the identification of good sources of rust resistance. One of the sources of rust resistance that he identified has since been widely used in sunflower programs in many countries. In addition to his research responsibilities, Russell continued his graduate training on an intermittent basis at the University of Minnesota and received his Ph.D. in plant breeding from the University of Minnesota in 1952.

In July 1952, W.A. Russell was appointed assistant professor of Agronomy at Iowa State University. He, L. H. Penny, and G. F. Sprague formed a team of maize scientists who conducted a vigorous maize breeding program during the 1950s. Russell's productivity resulted in rapid promotion through the academic ranks. He was promoted to associate professor, July 1, 1955, and to full professor, July 1, 1962, his present academic rank. He also is a member of the Graduate College faculty. In addition to his research activities, he teaches an advanced plant breeding course for graduate students and directs graduate student programs in plant breeding.

Russell's research is adequately documented. He has been an author or co-author of about 90 scientific publications. Several of the publications are the result of thesis or dissertation problems of his graduate students. Most of these problems involved some aspect of maize breeding. He has supervised about 25 graduate student programs for either the Master of Science or Ph.D. degree. In addition to the formal graduate programs, he has acted as the advisor of special students who have come to Iowa State University to obtain practical experience in maize breeding and to take some special courses. An important part of his research program has been the inbred and hybrid development program. He plans and conducts the topcross and hybrid yield tests for the State of Iowa, and the information from these trials is eagerly sought out by other maize breeders, both public and private. They know that the data are reliable and are used extensively in the selection of lines to be used in hybrid seed production. During his 25 years at Iowa State University, Russell has had the responsibility for numbering and releasing for public use several inbred lines (e.g., B37, B49, B52, B57, B64, B68, and B73) that have been the cornerstones in many maize breeding programs and pedigrees of hybrids.

Russell's achievements in maize breeding research are recognized by those who wish to know about the organization, planning, and conducting of maize breeding programs. He has been a consultant and has presented seminars in Romania, Yugoslavia, Brazil, and Argentina. The demand for his time is also great in the United States because maize breeders frequently visit his research plots and laboratories to learn of the latest information on maize breeding and materials available for release to other breeders. His achievements in maize breeding were recognized by Northrup King Seeds through dedication of an advertisement to his 22 years of maize breeding in Iowa in 1975. He was program chairman for the American Seed Trade Association, Corn and Sorghum Division, in 1975, and program chairman for Crop Breeding, Genetics, and Cytology Division of the Crop Science Society in 1971. He is a member of Gamma Sigma Delta, Sigma Xi, American Society of Agronomy, and Crop Science Society of America.

W.A. Russell married Dorothy Taylor of Winnipeg, Canada, in 1943. Mrs. Russell actively follows her husband's career and has provided a warm home for her family. They have three sons, Bernard L., Barry A., and

W. Kenyon, who also have distinguished themselves in their educational programs. He is totally involved with his research professionally, but he is an active member of the Ames Lions Club and pursues his hobbies of gardening and reading of history in his spare moments. In the limited time that he has available, he is a fierce competitor in golfing and bowling. Although he sets high standards for his professional and personal life, he is affectionately called "Russ" by his colleagues and is well recognized for his accomplishments during the past 25 years. When the history of the cooperative state-federal maize breeding program at Ames, Iowa, is recorded, Russ will be found to have played a vital role in its contribution to maize breeding research.

20 Charles R. Burnham

Long-Time Contributor to Maize Breeding*

By R. L. Phillips,[1] E. D. Garber,[2] O. L. Miller, Jr.,[3] H.H. Kramer[4]

Dedicated, hard-working, intelligent, inquisitive, perceptive, warm, inspiring and knowledgable—words that describe Charles Russell Burnham. Yet one cannot capture in words the appreciative feeling students, associates, and friends have for Dr. Burnham. This issue of *MAYDICA* commemorating his 75th birthday, January 13, 1979, is a means of expressing gratitude to Dr. Burnham—noted teacher, author and scientist. Although many former students contributed papers for this commemorative issue, it is intended to represent a thank you, a congratulations for outstanding career accomplishments and a Happy Birthday from Dr. Burnham's friends around the world.

Professor C. R. Burnham retired from the Department of Agronomy and Plant Genetics, University of Minnesota, St. Paul, in 1972. He continues research on corn and Datura cytogenetics, manages a 1-1/2 acre summer corn nursery, writes, and does cytology in the laboratory. He occasionally substitute teaches in the University of Minnesota Cytogenetics course and enjoys contacts with students. Dr. Burnham has two daughters, Barbara, living in northern Minnesota, and Sally, living in Chicago. His beloved wife, Lucille, passed away November, 1977, several years after an incapacitating stroke. His dedication to his wife and family has been unfailing.

* *MAYDICA* 1979, Vol. 24, No. 1
[1] University of Minnesota
[2] University of Chicago
[3] University of Virginia
[4] Purdue University

Professor Burnham was born January 13, 1904, in Hebron, Wisconsin. He grew up on a farm near Fort Atkinson, Wisconsin, and was graduated from Fort Atkinson High School in 1920. He attended the University of Minnesota for two years, and then the University of Wisconsin, where he received the B.A. degree in 1924, the M.S. degree in 1925 and the Ph.D. degree in 1929, majoring in genetics and minoring in plant pathology. At Wisconsin he was a graduate assistant under Dr. R. A. Brink and was employed full-time for one year to carry on the research of Dr. Brink during his leave of absence. He then was a National Research Council Fellow working on corn cytogenetics at Cornell University with Dr. R. A. Emerson. Others working with Emerson at that time included Drs. George W. Beadle, Barbara McClintock and Marcus M. Rhoades as shown in the 1929 photograph (Part A, Fig. 14). All of these individuals have been pioneers in genetics.

While Dr. Burnham was a National Research Council Fellow, he also studied at the Bussey Institution at Harvard with Dr. E. M. East and at the California Institute of Technology with Dr. E. G. Anderson. He then was a teaching fellow at the California Institute of Technology where he taught a course in cytological technique and continued corn cytogenetics research. In 1932 to 1933 Dr. Burnham was at the University of Missouri working with Dr. L. J. Stadler doing cytogenetics research on wheat, sorghum and corn. In 1933 to 1934 he was a Sterling Fellow in Botany at Yale University and also worked on corn cytogenetics in Dr. D. F. Jones' laboratory.

He was appointed Assistant Professor of Genetics at West Virginia University in 1934 and attained the rank of Associate Professor. His responsibilities included breeding field corn, sweet corn and watermelons as well as maize cytogenetics.

In 1938 Dr. Burnham joined the staff of the Department of Agronomy and Plant Genetics at the University of Minnesota as an Associate Professor and attained the rank of Professor in 1943. Later he was a Gosney Fellow at the California Institute of Technology during a sabbatical leave in 1947 to 1948 and was a visiting professor at Purdue University for one semester in 1961 to 1962.

Dr. Burnham has published over 60 scientific articles in refereed journals. He has worked on many plant species and has published on corn, beans, flax, barley, wheat and tomatoes. In addition, he has worked on Datura, watermelon, sorghum, onions and sweet corn and published notes or bulletins on these species.

The scientific topics on which he has published include: description, behavior and uses of chromosomal interchanges; chromosome pairing; genetics of pollen-tube growth; pigment patterns; disease resistance; male sterility; gene mapping; mutations; breeding; statistical genetic methods; linkage; crossing over; inversions; trisomy and polyploid inheritance. In addition to these publications, Dr. Burnham has published over 80 articles in the *Maize Genetics Cooperative Newsletter* and approximately 15 articles in the *Barley Genetics Newsletter*. He is the author of a book entitled "Discussions in Cytogenetics" which is unique in presenting an in-depth treatment of the genetic aspects of cytogenetics. He has contributed chapters to two books and recently published with R. L. Phillips a book entitled "Cytogenetics" in the *Benchmark Series in Genetics* collection.

Dr. Burnham taught courses in cytological techniques, advanced genetics and on methods in plant genetics. He advised almost 50 graduate students—14 M.S. and the remainder Ph.D.s. His students have held many responsible positions in basic and applied plant breeding and genetics research and in administration; former students include professors, commercial breeders, dean of a college of agriculture, four heads of biology departments, a member of the National Academy of Sciences, and an experiment station director. He has former students on the faculties of 18 universities throughout the United States alone. Approximately 350 of his graduate students were from other countries and are now carrying on work in at least nine foreign countries.

Honors include election to Fellow of the American Society of Agronomy, Distinguished Service Award of Sigma Xi, and the Gamma Sigma Delta Award of Merit. Dr. Burnham is an avid fisherman and over the years has enjoyed instructing the art and science of fishing to graduate students.

For many of us, Charlie Burnham had more influence on our careers and development than any other person. Our thanks and admiration for him as a person and scientist continue to grow with each year.

21 R. Alexander Brink

Sixty Years of Contributions to Genetics*

By R. A. Nilan[1]

This commemorative issue of *MAYDICA* constitutes a tribute to R. Alexander Brink—eminent geneticist, innovative investigator, productive scholar, dedicated teacher. It is a grateful acknowledgment and sincere appreciation by his former graduate students and his colleagues of 60 years of fundamental and applied contributions to the science of genetics—since it was 1921 when R. A. Brink published his first scientific article. Throughout these years, which spanned the early post-Mendelian classical period and the era of molecular genetics, his numerous and often pioneering basic investigations initiated and/or developed several important areas of genetics. The fact that all of these investigations utilized plants, especially maize and alfalfa, was and has been a boon to plant science, especially plant physiology and plant breeding, and to agriculture culture generally.

Professor Brink's major research contributions are widely known and easily identified. They include the role of the pollen tube and the endosperm in plant reproduction; an early demonstration of gene action on a biochemical basis; the concept of semisterility and reciprocal translocation as it cause in plants; confirmation that homologous chromosomes are involved in crossing over; and the nature and action of controlling elements. This latter research in maize led to one of the most intriguing and intensely studied new fields in genetics—mobile or transposable genetic elements. A consistent and unique aspect of Brink's experimental thrust has been the selection and analysis of problems that often were assumed to have been solved, but in fact had received

* *MAYDICA* 1981, Vol 26, No. 3
[1] Program in Genetics and Cell Biology, Washington State University

little scientific attention. His thorough knowledge of the literature enabled his highly ordered mind to create decisive experiments which usually led to the correct solution. His curiosity and unbounded scientific vigor which explored all facets of a problem and dedication to quality research are hallmarks of R. A. Brink's professional career and are evident in all of his major investigations.

In his several foremost discoveries Brink was not aided by sophisticated equipment and facilities—his tools and facilities consisted of a light microscope, a small laboratory, a greenhouse, a field and the simplest of physiological and biochemical facilities. It is his vast and detailed knowledge of botany combined with his scientific genius and skills that permitted him to develop and pursue plant genetic problems beyond most of his peers.

R. Alexander Brink was born September 16, 1897, in Oxford County, Ontario, Canada. He received his B.S.A. degree in Chemistry and Physics from Ontario Agricultural College in 1919 and his M.S. in Agronomy and Botany from the University of Illinois in 1921. He then went to the famous Bussey Institute of Genetics of Harvard University and within two years received a Doctor of Science in Genetics under the guidance of Professor E. M. East. He joined the faculty of the University of Wisconsin in 1922.

Between 1922 and 1930, he and Professor Leon Cole comprised the Genetics Department of the College of Agriculture. They were later joined by M. R. Irwin, L. E. Casida, N. P. Neal, G. H. Rieman and W. K. Smith. As Chairman from 1939–1955, he built the department into one of the foremost in the USA. Through this period, A. B. Chapman, D.C. Cooper, R. D. Owen, R. M. Shackelford, J. F. Crow, W. H. Stone and Sewell Wright were added to the faculty. He was instrumental in moving the department into the molecular era with the hiring of Joshua Lederberg in 1947. In spite of a heavy teaching and administrative load, the productivity and quality of Brink's research increased. Throughout his chairmanship, as in his entire career, the departmental research and teaching activities in genetics were basic with always the view of relating genetical concepts to agricultural practice. Although he became Professor Emeritus in 1968, he continued to train graduate students for several years and remains active in research. In recognition of his scientific stature as well as his valuable contributions to many facets of the development of the University, in an unprecedented action the University of Wisconsin recently established the "Brink Professor of Genetics." This Professorship is currently occupied by Oliver Nelson.

In his early years at Wisconsin, Brink's overriding basic research interest was in plant reproduction, and he chose to investigate areas in which considerable deficiencies of knowledge were evident, such as pollen tube growth, the role of genes in differential pollen tube growth, and embryo-endosperm relationships. His initial experiments, which continued from his doctoral research, concentrated on pollen physiology as he realized that "there is still wanting a comprehensive interpretation based upon a sufficient body of well-ascertained facts as to the fundamental processes involved in the growth of the pollen tube with its cargo of nuclear material from stigma to ovules." In a remarkable series of papers, he developed methods for determining the rate and amount of pollen tube growth in artificial media and in turn ascertained the effects of a variety of chemical and physiological factors.

It was during these analyses that Brink wondered about the role of differential pollen tube growth as a cause of differential seed set of specific genotypes, and he selected the waxy trait to help solve this problem. Using iodine staining, he provided independently and at the same time as Demerec the first illustration in maize that waxy and nonwaxy pollen segregated 1:1 in the anther. However, in some genotypes, waxy (*wx*) and nonwaxy (*Wx*) seeds did not occur in this 1:1 relationship. His experiments led to the clear conclusion that the genetic and chemical constituents of pollen tubes were important factors in differential pollen tube growth and differential fertilization—in this case a bias against male gametes carrying the waxy allele and its chemical product.

While engaged in the waxy analyses in the middle 1920s, Brink, ahead of his time and with considerable insight, realized that chemists could not provide a direct answer to the structure of the gene. He concluded that such knowledge could only be acquired by analyzing gene products in the haploid or gametophytic stage and he developed a biochemical genetic analysis of the action of the waxy and nonwaxy alleles.

He demonstrated that the two alleles produced different forms of starch and that probably different enzyme products were responsible. It was another 15 years and with greater knowledge of starch chemistry before a precise determination of the chemical composition of waxy starch was made and the differing proportions of amylopectin and amylose determined in waxy and nonwaxy endosperm and pollen.

As was the research style of this geneticist, one investigation often led to another. In analyzing waxy pollen, Brink observed that defective pollen grains in some strains reached about 50% and that the basis of the sterility appeared to be genetic. This was the first record of semisterility in maize, and he, D. C. Cooper and a graduate student, Charlie Burnham, soon demonstrated the basis of this phenomenon to be reciprocal translocations.

These studies led to Brink's cytogenetical period which included determining pairing relationships in reciprocal translocation heterozygotes, elegant confirmation that translocations were the basis of semisterility in maize, and a monumental study with E. G. Anderson at California Institute of Technology at Pasadena in 1938 of 21 translocations involving chromosome 3 in which the break points were used for confirming linkage data on this chromosome. This maize cytogenetic research culminated in the substantive and convincing paper of Brink and Cooper proving, along with Stern, Creighton and McClintock, that genetic crossing over involves exchange of segments between homologous chromosomes.

In accord with the mission of the Department of Genetics and his own interests, Brink often turned his attention to the solution, with the aid of genetics, of practical farming problems. In his research and graduate training activities he never lost sight of the applied benefits of science and particularly genetics. One year after arriving at Madison, he initiated the Wisconsin maize breeding program to demonstrate, as he says, "the potential of genetics for Wisconsin agriculture." This action was followed by his long and intensive research on alfalfa which included pollination mechanisms, improving seed set, and heterosis, and the release in 1954 of the bacterial wilt-resistant, winter-hardy alfalfa cultivar, Vernal, one of the most widely grown cultivars in Wisconsin and other alfalfa-producing states. He also helped found the Wisconsin forest-tree and potato breeding programs, made several contributions to pea production in the state, and he and his colleagues detected, among numerous collections of sweet clover, a nonbitter strain that was free of the toxin coumarin and nontoxic to cattle.

It was the investigation of differential seed set and embryo mortality in alfalfa that led Brink and Cooper to a prodigious series of studies (1938–1947) with numerous plants of different reproductive processes that produced new and remarkable discoveries about the major role of the endosperm in seed development. Until these studies, seed failure was attributed entirely to embryo failure. Brink and Cooper provided unassailable proof that the behavior of the endosperm, the tissue arising from the unique secondary fertilization process in plants, affords the principal clue in understanding embryo and seed collapse (somatoplastic sterility) in interspecific matings and in incompatibility. Brink asserted that by acting as a barrier to interspecific hybridization, the endosperm is an important factor in evolution. He provided further proof of the vital role of the endosperm by removing the embryos from intergeneric hybrids, such *as Hordeum jubatum x Secale cereale*, and producing viable plants.

Throughout his career Brink made numerous contributions to understanding the genetic composition of maize by analyzing the inheritance and linkage relations of numerous mutants. These include defective seed, ragged leaf, brown midrib-2, liguleless-2, pale midrib, and diffuse.

As a capstone to an already distinguished career, Brink initiated a monumental series of investigations in maize on Modulator in the late 1940s and on Paramutation in the middle 1950s. Through the analyses of Modulator, an element repressing the action of the *P* locus, he independently and coincidentally with McClintock initiated the concept of controlling elements in maize. He and his students advanced considerably an understanding of the nature and action of these elements. One of the more notable achievements was the proof that such elements as Modulator could move or transpose away from the affected locus and may subsequently move back to it. Brink recognized this transposition phenomenon as a form of somatic segregation and a possible basis for differentiation. Another type of chromosome repression element is involved in Paramutation, an interaction between alleles that leads to directed heritable change with high frequency. Although not a new phenomenon as revealed by the literature, Brink and his colleagues and students developed much new knowledge about the nature and mechanism of action of this repression element.

The nature and basic mechanism of action of the maize elements remain obscure. But Brink's prodigious investigations have provided a basis for further understanding as more information is obtained about the eukaryotic chromosome organization and function.

Throughout his career, Brink has turned to many colleagues for counsel and advice in his research and in turn has enriched and inspired these colleagues with his knowledge and technical skills. He often extended these interactions outside the USA. He was an International Education Board Fellow (1926) in Berlin where he worked with Erwin Baur and in Birmingham, where he learned about starch chemistry; a Haight Traveling Fellow, at the Universities of London and Oxford, 1960, where he prepared an excellent review on plant gene repression; and a National Science Foundation Senior Postdoctoral Fellow in Australia in 1967.

Brink's contributions have brought him numerous honors and awards. Among the most prestigious are his election to the National Academy of Science (1957), as President of the Genetics Society of America (1957), and as President of the American Society of Naturalists (1967). He was Chairman of the Mendel Centennial Committee of the Genetics Society of America during 1964–65, and Managing Editor of Genetics (1952–57). As one colleague has stated, Brink made significant contributions to the formation of the Genetics Society of America and the development of its policies.

The results of Brink's numerous research investigations have been presented to the world in approximately 150 published articles. Equally important, his scientific philosophy has been transmitted to future generations through his numerous graduate students who have been trained by collaboration in his research, informed by his knowledge and inspired by his example. Moreover, Brink imparted much knowledge to his students through his formal lectures which were always excellently organized and well presented. He has exercised a profound effect on their careers and attitudes towards science and life. Some of his knowledge and philosophy would be imparted in the summer time during lunch breaks under a shady tree next to the maize field. Students were impressed by the fact that Professor Brink always brought two flasks with his field lunch—one milk and one coffee. As he would say, "one for the body and one for the soul." He gave students a sense of modesty as well as a sense of history of genetics and science and the encouragement of his warmth and his humor. In addition, he was always hospitable and entertained the students frequently in his home. Indeed, the highlights of the social calendar of most of his students were the Brink's Thanksgiving or other dinners where they were royally dined and would learn bridge, play charades, or become involved in some other novel form of entertainment.

Even in retirement, Professor Brink explored the secrets of maize genes and their inheritance with a focus on sugary. It is the sincere and heartfelt wish of his graduate students and colleagues that this exploration will continue for years to come and remain a challenge to all engaged in the study of genetics.

22 William L. Brown

A Contribution to the Genetics and Commercial Improvement of Corn*

By Donald N. Duvick[1]

This commemorative issue is a tribute to William L. Brown in acknowledgment of his scientific achievements, his accomplishments in commercial plant breeding and his contributions to the cause of germplasm conservation.

Dr. Brown has earned a prominent place in science and private industry in both national and international agricultural sectors. His dedication to maize evolution, breeding and cytogenetics and his leadership in germplasm conservation and utilization are known and respected worldwide. His research and leadership contributions have spanned more than forty years in academia, private industry and scientific societies. He has stimulated plant breeders and research programs around the world.

William Lacey Brown was born in Arbovale, a farm community in the mountains of West Virginia, on July 16, 1913. At nineteen he entered Bridgewater College, earning his B.A. in biology in 1936. For a short time during his senior year at Bridgewater he taught at a small country school in the hills of western Virginia. After graduation he moved to the southern United States and spent one year in graduate school at Texas A & M University.

In 1937 he became a Research Fellow at the Missouri Botanical Garden at St. Louis, Missouri. During the next four years he earned the M.S. and Ph.D. degrees in botany and cytogenetics, studying under Edgar Anderson. He received his degrees from Washington University in 1940 and 1941, respectively.

* *MAYDICA* 1983, Vol. 28, No. 2
[1] Plant Breeding Division, Pioneer Hi-Bred International, Inc.

Even in his early years Brown's energy and leadership abilities were evident. He served as class president in all four years at Bridgewater College, and excelled in a variety of sports activities, serving as captain of both football and basketball teams. In Texas he played on a professional basketball team to earn college funds.

In 1941 he worked in Washington, D.C., as a cytogeneticist for the United States Department of Agriculture. In 1942 he joined the Rogers Brothers Seed Co. in Minnesota. He then joined the Corn Breeding Department of the Pioneer Hi-Bred Corn Company, as it was then called, at Johnston, Iowa, in 1945. His primary responsibility was to do basic research in corn breeding. In his early work with Pioneer he used cytogenetic analyses of knob numbers in maize as a basis for study of the evolution of maize. Particular emphasis was placed on investigation of ways to maximize heterosis in maize hybrids, utilizing knowledge about racial affinities of Corn Belt inbred lines. He also began to consider ways to adapt exotic germplasm for inclusion in Corn Belt varieties. As time went on, major emphasis was placed on morphological measurements and descriptions as additional aids in classifying races of maize.

He was a Fulbright Advanced Scholar in 1952 at the Imperial College of Tropical Agriculture in Trinidad. His studies involved research on races of maize collected from the West Indies. The results of these studies, along with his other work on exotic germplasm and its subsequent inclusion into domestic germplasm pools, have had major impact on our ideas about the diversity and ancestry of North American hybrid maize.

Upon his return to the United States he became more involved in applied aspects of the corn breeding research conducted at Pioneer under the direction of Raymond Baker. At the same time he continued his interest in and study of the races of maize.

Three distinct careers can be described for Dr. Brown. His most easily recognized contribution outside the field of science has been as an administrator. In 1958 he was named Assistant Director of Research for Pioneer. He then advanced through the positions of Director of Corporate Research, Vice President and President of the company. He retired in 1981 and at present is chairman of the company's Board of Directors. The financial success of his company is evidence of his success as an administrator.

Brown's second career has been that of a successful research director and director of research directors. The research program now in place at his company, largely set up according to his plans, is looked upon as a model for commercial breeding research programs. He has worked consistently to build competent, well-staffed and well-financed research teams for development of better hybrids and varieties of the seven crops—corn, sorghum, soybeans, wheat, alfalfa, cotton and sunflowers—that are bred, produced and sold by his company. The breeding techniques and attitudes that Dr. Brown has fostered show extreme practicality, based on continuing analysis of successes and failures, and a readiness to incorporate useful new breeding ideas and techniques as they are developed in the profession and within the corporation's research staff. In common with other successful administrators, Dr. Brown has always had a high regard for his associates. His genuine interest in his fellow workers and the admiration and affection that he has given to his colleagues have made him a popular administrator, very human and approachable.

Brown's third career, that of plant scientist, is likely to have the greatest worldwide impact on agriculture and long-term importance to the agricultural sciences. His personal research and publications in classification and use of exotic maize germplasm, his long-standing and effective efforts as member and chairman of committees for collection, preservation and classification of exotic germplasm, and his efforts—by example and by persuasion—to foster cooperation between public and private agricultural scientists have been of great value. He has saved from extinction numerous germplasm collections, is fostering the study of those collections that have been made, and is encouraging attempts to broaden the base of maize germplasm in the United States and other countries by the introduction and use of materials from the exotic collections.

Brown published his first paper in 1937 on modified root tip smear techniques. Since then he has published over 30 scientific papers, was co-author of the book "Corn and Its Early Fathers" with H. A. Wallace, wrote with Major Goodman the chapter "Races of Corn" for the book "Corn and Corn Improvement," and also wrote the chapter "Major Germplasm Banks in the Western Hemisphere" for the book "Crop Genetic Resources for Today and Tomorrow." He was a member of a select panel of plant breeders charged with writing the book "Plant Breeding Perspectives." A series of papers by Brown and Anderson on the northern flint corns and the southern dent corns of the United States have had major impact on ideas about the origins of the dent corns of the American Corn Belt. Traces of the scientific ideas of Edgar Anderson and Raymond Baker, and of the philosophy of H. A. Wallace, founder of Pioneer and a close personal friend, can be seen in Brown's scientific accomplishments, but in the final analysis, the body of work is his own unique product.

Dr. Brown's service on committees concerned with collection, preservation and classification of exotic germplasm has been of special importance in national and international efforts to collect, classify, maintain and utilize exotic maize germplasm. In this area he has presented numerous invitational papers on subjects such as maize germplasm banks, races of maize and their importance in maize breeding, genetics and evolution. He has been active in the conservation and utilization of plant genetic resources through membership on national and international boards and committees such as the National Academy of Sciences Committee for Preservation of Indigenous Strains of Maize, the National Academy of Sciences Committee on Vulnerability of Major Food Crops, The President's Science Advisory Committee of World Food Supply, National Plant Germplasm Committee, Rockefeller Foundation Maize Germplasm Committee, Maize Committee of International Board of Plant Genetic Resources (Chairman), National Plant Genetic Resources Board (Vice Chairman), Governing Board of the Agricultural Research Institute, and the Joint USDA-SAES Task Force on Corn and Grain Sorghum (Advisor). In addition, he has served on the Board of Directors of the Iowa Academy of Science and currently is president of the Crop Science Society of America.

His concern for world agriculture has taken him to numerous underdeveloped and developed countries to help them learn the advantages of hybrid maize. Tours and working visits have sent him to Central and South America, Mexico, Southeast Asia, eastern and western Europe and, recently, New Zealand and Australia. In all of these places Brown's name and reputation as corn breeder and teacher are widely known and respected. His broad knowledge of maize origins and maize breeding, the scientific and cultural insights he has given to students of maize breeding and maize classification, and the moral and financial support he has brought to struggling researchers have all contributed to his reputation as elder statesman of maize breeding.

One of his tours, in the service of hybrid maize, caught him in the middle of the Hungarian rebellion of 1956, unable to contact anyone for nearly ten days. He escaped via a Danube riverboat and eventually made his way to

Holland. Most of his maize breeding trips, however, were not as perilous. They have brought to him close personal friendships as well as valuable professional relationships that have helped him in his efforts to improve world agriculture.

His continued interest in the evolution of maize has recently led him to renew study of a research problem that he had set aside some 30 years ago. He has obtained a grant to study the cytology and evolutionary history of maize grown by the Eastern Cherokee tribes of the Qualla Reservation in North Carolina. The study will add to knowledge of the origins and diversity of maize in much the same way his earlier studies have contributed to plant breeding knowledge.

William Brown has recently embarked on yet a fourth career by accepting a position with the National Academy of Sciences as chairman of a newly established Board on Agriculture. The Board will function as a part of the Academy's National Research Council with findings reported directly to the Council's Governing Board. The twelve-member Board is intended to foster the development of closer liaison and working relationships between the Academy and other public and private organizations engaged in agricultural research. As chairman for the Board on Agriculture, Brown also assumes responsibilities on the Governing Board of the National Research Council.

Dr. Brown has received many honors for his research and scholarship. Most prominent was his election to membership in the National Academy of Sciences in 1980. For many years he was an Extramural Professor of Botany at Washington University and in 1981–82 was a University Fellow at Drake University. He has been elected a Fellow of the American Society of Agronomy as well as of the Iowa Academy of Sciences. The American Society of Agronomy awarded him its Agronomic Service Award in 1979. In 1981 he was named Distinguished Alumnus at Bridgewater College where he is also a Trustee, and he was honored again in 1981 as Distinguished Economic Botanist by the Society for Economic Botany.

Dr. Brown's life has not totally been absorbed in his research. During his final year of graduate school he married Alice Hevener Hannah. He and Alice attended school together in rural West Virginia and later carried on a long-distance correspondence between St. Louis, Missouri and the Belgian Congo, Africa, where Alice spent three years teaching children of American missionaries. Bill and Alice recently celebrated their 41st wedding anniversary. They have a daughter, Alicia, a son, William and two granddaughters.

Travel has been an integral part of their lives together. Brown's early research took them to a variety of places, some of which did not have the best accommodations. Some of their earlier experiences included campfire cooking and river bathing in rural Missouri, and sandbar camping on maize collecting trips to the American southwest. Recent trips, however, have been more elegant. Whether their trips together were business or personal, they have always come away with warm personal feeling for the friends they've gained. Many friends from abroad have visited the Browns in Iowa. They have opened their home and hearts to a great many people over the years. On behalf of all of their friends, I am pleased to have this opportunity to thank Bill and Alice for their generosity, encouragement and friendship.

23 Marcus M. Rhoades

Over 50 Years of Major Contributions to Genetics*

By Ellen Dempsey[1]

Ten years ago, on the occasion of his 70th birthday, Marcus Rhoades was presented with a commemorative issue of Theoretical and Applied Genetics edited by Professor W. Seyffert of the University of Tubingen and Professor Peter A. Peterson of Iowa State University. Included in the Festschrift volume were articles written by former students and by geneticists who had worked in his laboratory, as well as two biographical sketches, one emphasizing the scholarly contributions of Professor Rhoades and the other describing his personal characteristics. The presentation ceremony and his official retirement the same year did not mark the end of his scientific career. Indeed, the ten years from 1973 to 1983 proved to be a productive period and in some ways
the most rewarding decade of his career. Freed from the tedious duties of administration and committee work and from the demands of teaching, Rhoades pursued his investigations with renewed vigor and enthusiasm. It is no exaggeration to say that his dedication to maize genetics has increased over the years and his intuitive grasp of complex hereditary patterns devised by the plant has become more profound. The laudation presented here will summarize some of his recent activities and accomplishments.

Marcus M. Rhoades, the Later Years

At the time of his formal retirement in 1973, Rhoades was engaged in an exhaustive analysis of chromatin loss. First described by Rhoades, Dempsey and Ghidoni in 1967, the high-loss phenomenon involves an interaction of B chromosomes and heterochromatic knobs on the A chromosomes leading to

* *MAYDICA* 1983, Vol. 28, No. 3
[1] Indiana University

segmental loss of A-chromatin at the second microspore division. When plants of the high-loss stock are used as pollen donors in crosses with recessive female parents, the ears reveal a variable number of kernels with the maternal phenotype, a result that could easily be dismissed as self contamination. Rhoades refused to accept the possibility that faulty pollinating techniques were responsible for the exceptional kernels and he decided to continue the investigation. A series of papers followed in which the circumstances necessary for loss were delineated and the data obtained with chromosome 3 markers were supported by tests with chromosomes 4, 5, 9, and 10. In 1973, a mechanism was proposed to account for the observations; according to the hypothesis, microspores with two or more B chromosomes experience a delayed replication of heterochromatic knobs on the A chromosomes at the second microspore mitosis, leading to formation of bridges at anaphase, breakage of bridge chromatin, and loss of genetic material in the acentric fragment. Although the mechanism was not susceptible to direct-testing because of the difficulty of viewing chromosome bridges in the starchfilled pollen grain, subsequent studies strongly supported the hypothesis. In retrospect, the high-loss investigation was of considerable significance; not only did it reveal an unexpected interaction between chromosomes possessing heterochromatic segments, but the instability in the genome appears to have been responsible for activation of mutable systems identified in this stock. Moreover, deficient chromosomes produced by high-loss breakage were later utilized to great advantage in dissecting the complex organization of abnormal chromosome 10.

In 1975, certain aspects of the high-loss problem were given to a Brazilian student, Luiz Saraiva, as a subject for his doctoral research. With the advice and encouragement of Professor Rhoades, Saraiva searched for and identified a number of half-translocation and dicentric chromosomes. Their occurrence and the absence of full translocations were predicted by the proposed mechanism of high-loss and Saraiva's findings offered independent confirmation of its validity.

The cornerstone of the hypothesis explaining high-loss is the assumption of late replication of heterochromatic knobs. Heterochromatin had been demonstrated to undergo late replication in many organisms and the work of Abraham and Smith indicated that B chromosomes of maize likewise show delayed replication. In order to provide definitive evidence of the replication times of various kinds of heterochromatin and euchromatin in maize, Dr. Rhoades, in cooperation with his Australian colleagues, A. Pryor, K. Faulkner, and W. J. Peacock, undertook a detailed investigation of DNA synthesis in plants possessing varying numbers of knobs and B chromosomes. His visit to Canberra in the winter of 1978 led to a fruitful collaboration resulting in an elegant demonstration of the asynchronous replication of the four classes of heterochromatin found in the maize genome. Knobs proved to be the last component of the chromosome to undergo DNA replication. Again, the findings were consistent with the postulated high-loss mechanism. In addition, this study showed that heterochromatin is not uniformly late in replication time; some heterochromatin (namely, that situated in the nucleolus organizer region) replicates with the euchromatin, while the proximal heterochromatin of the B chromosome has a later replication time than the distal B heterochromatin.

Over the years, Professor Rhoades has studied the behavior of chromosomes in normal and abnormal situations in an effort to understand such fundamental processes as meiotic pairing, recombination and segregation. The genomes of maize plants vary in their endowment of heterochromatin and it seemed possible that chromosomal maneuvers might be affected by the amount, the specificity, and the location of this type of chromatin. When invited to address the International Symposium on Maize Genetics and Breeding held at Urbana in 1975, Dr. Rhoades took this opportunity to summarize a number of experiments revealing diverse effects of heterochromatin. While heterochromatin has often been described as genetically inert, he demonstrated and documented several intranuclear activities influenced by heterochromatin. These include enhanced recombination due to B chromosomes, preferential segregation and neocentromere formation in plants containing abnormal chromosome 10, the differential effect of knobs on recombination in male versus female flowers and the above-mentioned high-loss phenomenon. In some of these cases, the heterochromatic element has a localized influence over neighboring chromosomal regions; in other cases, heterologous chromosomes of the complement may be affected. Cellular events, such as recombination, have been investigated at a genetic level by analysis of progeny ratios and at a molecular level by identification of enzymes causing chromosome breakage and repair. Rhoades recognized that a full understanding of these events requires a close examination at the chromosome level, where a variety of interactions between parts of the genome may alter the genetically observable outcome.

During his 1979 visit at the CSIRO in Canberra, Marcus Rhoades participated with W. J. Peacock, E. S. Dennis, and A. J. Pryor in collaborative research on the molecular composition of maize heterochromatin. Their studies led to an important and interesting paper published in the Proceedings of the National Academy of Sciences in 1981. A 185 base pair repeated DNA segment isolated from the maize genome showed in situ hybridization with the late replicating heterochromatic knobs including the satellite of chromosome 6 and the proximal heterochromatin in the long arm of the B chromosome. This sequence was not detected in other types of heterochromatin or in the euchromatin. The nucleotide order was determined and estimates of the number of copies per knob of this highly reiterated sequence revealed a correlation with knob size. Nonspecific associations of maize knobs during pachynema as well as in mitotic interphase have often been observed and were studied in detail by Sally Peterson, J. A. Gurgel, and D. T. Morgan, Jr. These knob fusions become understandable on the basis of their possession of multiple copies of a common DNA sequence. Previous studies on heterochromatin of maize by Rhoades had indicated that the different classes of heterochromatin have specific cellular effects. The 1981 paper represents the first attempt to identify the chemical differences responsible for these diverse activities, hopefully leading eventually to a better understanding of the role of heterochromatin in cellular processes.

Current research by Rhoades centers on two studies; one dealing with new systems of transposable elements and the other involving the structure and origin of abnormal chromosome 10. Both problems owe their inception to prior investigations of the high-loss phenomenon. During the analysis of crosses involving high-loss stocks, three unstable phenotypes appeared. Genetic studies revealed the presence of two-unit systems similar to the original *a-Dt* case described by Rhoades in 1938. With characteristic thoroughness, Rhoades expanded his study of these systems and has uncovered some significant deviations from the classical *Ac-Ds* mutable system studied by McClintock. Receptors at the *a1*, *bz1*, and *bz2* loci were found to be controlled by independent regulator elements. The *Ac2* controller of the *bz2-m* system shows interesting dosage effects and unexpected interactions with the controlling elements of McClintock's *Ac-Ds* system. Some aspects of the new systems are not yet fully understood; their complete elucidation may require a molecular analysis.

In the past few years, the abnormal chromosome 10 of maize, first described by Longley in 1937 and 1938 and subsequently analyzed by Rhoades in 1942, 1952, and 1966, has received renewed attention. The origin of this unusual chromosome, with its large heterochromatic knob and its ability to induce neocentromeres, preferential segregation, and enhanced recombination, has long puzzled maize geneticists. Some have speculated on its possible relationship to the heterochromatic B chromosome. Rhoades has shown that the abnormal chromosome 10 differs from normal 10 not only in the acquisition of the heterochromatic knob and a differential segment containing three prominent chromomeres, but also in the disposition of the euchromatin homologous to the tip of the long arm of the normal 10. This finding has significance for any hypothesis on the origin of this chromosome. Multiple chromosome breaks, insertion of alien segments of chromatin at two points, and inversion of the linear order of genes residing in the terminal segment of the long arm of normal 10 must have occurred. No simple rearrangement involving components of the B chromosome could give rise to the complex structure described above and the derivation of abnormal 10 from a B chromosome appears to be highly unlikely. This conclusion is in agreement with Carlson's finding that the knob of abnormal chromosome 10 lacks the inducer of B chromosome nondisjunction present in the distal B chromatin and the results of Peacock, Dennis, Rhoades, and Pryor showing that the abnormal 10 knob possesses a DNA sequence not found in the distal B chromosome segments. In the detailed analysis of the architecture of abnormal chromosome 10, Rhoades made use of a series of deficient abnormal chromosomes 10, each lacking the large knob and variable segments of the tip of the long arm. These deficiencies were products of the high-loss system and proved to be particularly valuable since they represent simple deficiencies. Their origin at the second microspore mitosis rules out the possibility of repeated breakage-fusion-bridge cycles in the gametophyte, which would generate complex structural changes.

Throughout his career, Marcus Rhoades has maintained a high level of interest and enthusiasm in his scientific inquiries; his attention cannot long be diverted from his ongoing experiments. Even after a major heart attack in 1968 and subsequent health problems requiring brief hospitalization, he rebounded to face new challenges. His mind is wide-ranging, receptive to new ideas and stimuli, and retentive of details. He became acquainted with some of the techniques of molecular genetics while in Australia and although he makes no claim to have mastered this new field of genetics, he has an intuitive grasp of what is significant. He characterized his 1978 and 1979 sojourns in Australia as rejuvenating experiences; he greatly enjoyed his participation in the exciting and competitive atmosphere of a modern molecular genetics laboratory.

After more than 50 years of research on the genetics and cytology of the maize plant, Marcus Rhoades retains his painstaking and rigorous style of investigation. He is never satisfied with a single experiment, but is impelled to repeat and check the original result, sometimes even after publication. He is a perfectionist, insisting that novel or bizarre findings be confirmed. He takes pride in the soundness of his work and was pleased to learn that some of his experiments have been used as reliable and instructive classroom exercises. An interesting comment on Rhoades's approach to research is preserved in the oral history of Th. Dobzhansky: "Dr. Rhoades is inclined to be, as I said, a Puritan in science... To Rhoades, any sort of theorizing is a little bit sinful. Speculation is a rather bad word. And since I am addicted to this sin, we were very good friends but did not publish together." Rhoades may have mellowed since the early days of his interaction with Dobzhansky. Certainly, he welcomes discussions with his more imaginative colleagues and willingly listens to controversial ideas and speculations, calling it "good, clean fun"; but in the end, his Puritanical instincts emerge and he asks "but, how are you going to test it?"

Although Rhoades has labored long hours in the field and laboratory and accumulated a vast array of data, many of his experiments remain unreported in his files. He has published only the truly significant findings and has no interest in producing "pot-boilers." He has an uncommon gift for words and his papers are notable examples of clarity and style. His scientific contributions include more than 70 articles in refereed journals; in addition he has been a regular supporter of the *Maize Genetics Cooperation Newsletter*, with nearly 160 research reports to his credit. A considerable portion of Rhoades's research findings served as the basis for student theses; in every case, the professor freely gave research notes, suggested breeding strategies, and often assisted with the thesis preparation. He relinquished all rights to joint publication; all of his students are sole authors of the papers based on doctoral thesis research.

The significance of Rhoades' research endeavors is evidenced by numerous citations in widely used textbooks and monographs. He is unquestionably one of the world's leading cytogeneticists.

Although officially retired at age 70, Distinguished Professor Rhoades was generously provided with laboratory and office space by Indiana University. This made it possible for him to continue his scientific research and his participation in department affairs. His office remains open to visitors and his advice is sought by colleagues and administrators. He continued to serve on the Guggenheim Selection Committee until 1976, reviewing hundreds of applications each year from scholars in the arts as well as science. In 1976, he traveled to Denmark to join a panel of geneticists in screening candidates for the Chair in Genetics at the University of Copenhagen. From 1974 to the present he has served as Director of the Genetics Training Grant awarded by the National Institutes of Health to Indiana University, a position which keeps him in touch with a select group of graduate students in genetics. Rhoades continues to respond to invitations to talk about his research. Although a senior member of the academic community, he presents his lectures with undiminished and infectious enthusiasm. He is frequently asked to review manuscripts or read student theses and, when he agrees to do so, the author finds his opus has received a rigorous and critical scrutiny. His interest in recent developments in all areas of genetics is evidenced by his regular attendance at student seminars and lectures by visiting scientists; he can generally be found near the front of the room where he has an unobstructed view of the slides and can best appreciate the nuances of the speaker's presentation.

In the past decade, special honors have come to Marcus Rhoades. He was elected a Foreign Fellow of the Royal Danish Academy of Sciences and Letters in 1977. In 1981, he and Barbara McClintock were the first recipients of the Thomas Hunt Morgan Medal awarded by the Genetics Society of America. The Medal is presented annually to senior members of the Society in recognition of a lifetime's contribution to genetics. In 1982, Indiana University awarded him an honorary Doctor of Science degree for his contributions to genetic research and for his outstanding service to the University as Chairman of the Botany Department, teacher, and investigator. Indiana University takes pride in the accomplishments of its faculty. On display in the lobby of the Indiana Memorial Union Building are photographs of Indiana professors who have been elected to one or more of the most prestigious honorary societies, namely the National Academy of Sciences, the American Philosophical Society, and the American Academy of Arts and Sciences. Rhoades is the only living member of this select group to have been elected to all three societies.

Marcus Rhoades was one of the early maize workers honored at a symposium on "The Golden Age of Corn Genetics" held at Cornell University in the summer of 1982. The program included short talks by several of the former students of R. A. Emerson, the leader of the famous Cornell school of maize genetics. In his presentation, Rhoades described a research study begun with Emerson in the 1930s. Following his talk, Barbara McClintock, the illustrious cytogeneticist and a member of the original group, asked for the microphone and proceeded to express her admiration for the work of her colleague, Marcus Rhoades. The mutual regard of Barbara McClintock and Marcus Rhoades has not diminished with the years. They have been friends since the 1920s and Rhoades was among the first to recognize her unusual intelligence and ability. In the early days of her work on mutable systems, McClintock kept him informed of the progress of her studies and her startling conclusions about the transposability of controlling elements. He followed her experiments closely and was one of the few geneticists at that time who understood and appreciated her work.

I have dwelt at some length on the research findings of Professor Rhoades in the years from 1973–1983. These contributions, together with his earlier work on recombination, mutation, cytoplasmic inheritance, genetic control of meiosis, structural aberrations and preferential segregation, constitute the scientific record on which his reputation as a geneticist rests. Perhaps of equal importance, although less tangible, is the influence of Marcus Rhoades on his own students, as well as on a host of graduate students, postdoctoral fellows, and colleagues with whom he has been closely associated. Many attributes of Rhoades's scientific style are reflected in his Ph. D. students; some have adopted his opportunistic approach, others emulate his passion for detail and thoroughness of analysis. Some have chosen a different organism for their studies but were stimulated by Rhoades' experiments to pursue a specific direction in their research. The affection and admiration shown by Rhoades' students for their professor are recorded in the dedication statements in the Festschrift volume. Over the years, they have kept in touch with the professor and his attentive interest and approval of their accomplishments remain important to them. Rhoades' impact on students of genetics is not restricted to his own doctoral students. Sally and Peter Peterson have described his kindness to young researchers at scientific meetings; he would single out promising young people and, with obvious sincerity and without condescension, indicate his appreciation of the importance of their contributions. The Petersons conclude "His legacy is surely the endowment of a competent and enthusiastic generation of students who continue to make significant contributions in the field of Genetics." Although now in his 80th year, Professor Rhoades's enthusiasm and dedication for research remain unabated. He retains the intellectual curiosity of his youth and any new fact uncovered brings him deep satisfaction. His devotion to research is legendary. He spends seven days a week in the laboratory or in the experimental field. The obvious pleasure he derives from his research efforts is both a challenge and a stimulation to younger colleagues. Professor Rhoades is one of the fortunate few who are able to live a full and zestful life after official retirement.

This volume is dedicated to Marcus Rhoades on his 80th birthday as an expression of the respect, admiration, and affection of his fellow maize geneticists. I am pleased to join the many friends and associates of Marcus Rhoades in wishing him many more years of satisfying research activities and continuing success in unraveling the mysteries of maize genetics.

24 George F. Sprague

60 Years of Contributions to Genetics and Breeding

by Arnel R. Hallauer

The name of George F. Sprague is and will be prominent in the recorded history of corn genetics and breeding research conducted during the 20th century. His research career has spanned seven decades: from the initial development of the hybridization techniques for hybrid corn to the present-day acceptance of hybrid-corn breeding techniques. He has had a wide range of interests relative to corn genetics and breeding, and his contributions range from classical genetics to quantitative genetics, from line and hybrid development to recurrent selection, and from practical and applied aspects of corn genetics and breeding to theoretical and basic studies of gene action. Dr. Sprague has had substantial influence impact on the development of the breeding and selection methods that have contributed directly to improving the effectiveness and efficiency of identifying superior lines and hybrids. His career has paralleled the development and expansion of the hybrid seed corn industry in the United States.

Dr. Sprague's career began as a junior agronomist with the U.S. Department of Agriculture in 1924 and has continued uninterrupted; he is still active today at the University of Illinois. During these 60 years, Dr. Sprague has been a productive and visionary scientist who has gained the respect of his peers for his planning of research studies, attention to detail, thorough analysis of the data, and lucid interpretations of the results. Dr. Sprague's first publication appeared in the Journal of Heredity in 1927,

* *MAYDICA* 1984, Vol. 29, No. 4

and he recently had an article in a 1983 issue of *MAYDICA*, a period of 56 years and including more than 130 research reports published.

George Frederick Sprague was born at Crete, Nebraska, in 1902. His academic education and training first were at the University of Nebraska, where he earned the B.S. degree in agriculture in 1924 and the M.S. degree in agronomy in 1926. He continued his graduate studies at Cornell University in 1928, where he earned the Ph. D. degree in genetics in 1930. After beginning his career with the U.S. Department of Agriculture at North Platte, Nebraska, from 1924 to 1926 and then attending Cornell, he resumed employment with the U.S. Department of Agriculture in 1929 at Arlington Farms, Virginia. His research career continued with the Department until his retirement in 1972. During more than 40 years of service with the Department, Dr. Sprague was a very productive scientist, particularly while stationed at Columbia, Missouri, and Ames, Iowa. In 1958, he returned to the Washington, D.C., area and assumed the responsibilities of Investigations Leader, Corn and Sorghum Investigations, Agricultural Research Service, U.S. Department of Agriculture. Although the position of Investigations Leader was primarily administrative, he continued his genetic studies on corn. Upon his retirement from the Department in 1972, he accepted a position at the University of Illinois, Urbana, where he is still actively engaged in corn genetics research.

Dr. G. F. Sprague's research career has included nearly all facets of research related to corn genetics and breeding. During his career, there have been great advances in the study of classical Mendelian genetics, quantitative genetics, and corn breeding methods. He has been a keen student of all phases of corn research and has kept abreast of the latest developments within each. He regularly attends research conferences and workshops to maintain his knowledge of the most recent developments in biotechnology, genetics, and breeding.

Although Dr. Sprague has made contributions to a broad area of corn research, his greatest contributions have been made in the application of the principles of quantitative genetics to corn breeding methods. He has always maintained a strong active interest in basic research. But he also was a firm believer in relating the information derived from basic research studies to applied breeding programs. It was only after the information derived from basic research studies could be demonstrated to be applicable and effective in applied research programs that the basic information would be accepted; one was a corollary of the other. This relation between basic and applied research was instrumental in his influence for developing corn breeding programs throughout the world. His counsel was widely sought by others in organizing breeding programs that included basic and applied aspects of research.

Dr. Sprague's contributions to corn research are pervasive and well documented, ranging from the genetics of scutellum color (1927) and heterofertilization (1932), effects of mutagenic agents (1936), and aberrant ratios caused by virus infection (1971) to estimates of number of plants required to sample a corn variety (1939), relative importance of general and specific combining ability (1942), early testing of inbred lines (1946), inheritance of oil and protein (1949), estimates of rates of mutation (1955), effectiveness of recurrent selection (1952, 1961), and cytoplasmic-genic interaction (1983). Of his more than 130 research articles, several have been classics in corn breeding. Probably the most frequently cited paper is the one by Sprague and Tatum in 1942, in which the terms "general" and "specific" combining ability were introduced and defined. Combining ability was a term used by corn breeders in referring to the potency of inbred lines of corn in crosses. The Sprague and Tatum paper partitioned combining ability among diallel crosses of inbred lines into a) average performance of a line in crosses with other lines (general combining ability, GCA) and b) performance of a specific pair of lines relative to their average performance (specific combining ability, SCA). This paper stimulated theoretical research on the type of information that can be obtained from use of diallel crosses, types of gene action expressed by lines in crosses, and also provided a method for evaluating lines for use in hybrids. Today, the terms GCA and SCA are widely used in plant breeding and are commonly used in the description of lines included in breeding programs.

Early testing of lines and recurrent selection for the improvement of corn populations are two other aspects of corn breeding in which Dr. Sprague made important contributions. On the basis of studies by M. T. Jenkins in 1935 and 1940, Sprague recognized early the potential importance of early testing and recurrent selection for increasing the effectiveness of corn breeding methods. Although early testing and recurrent selection usually are considered two separate aspects of corn breeding, they are important corollaries to increasing the efficiency of line development and germplasm improvement. Sprague's studies (1946 and 1952) substantiated Jenkins' 1935 results and laid the foundation for use of recurrent selection methods because nearly all recurrent selection methods are based on early generation test results. Dr. Sprague was one of the first to initiate formal recurrent selection studies, which were designed to gradually increase the frequency of favorable alleles in corn populations in a systematic manner. He initiated half-sib recurrent selection in Iowa Stiff Stalk Synthetic in 1939, reciprocal recurrent selection in Iowa Stiff Stalk Synthetic and Iowa Corn Borer Synthetic No. 1 in 1948, half-sib and S_1 recurrent selection in a strain of Krug in 1953, and half-sib selection with an inbred tester in the open-pollinated variety 'Alph' in 1949. All these recurrent selection programs are still in progress at Ames, Iowa, and have contributed significantly to the information that is available on the relative importance of gene action involved in selection within and between populations of corn and the relative effectiveness of the different methods of recurrent selection. Dr. Sprague's foresight was instrumental for initiating the studies to provide information on the potential and usefulness of early-testing methods to increase the effectiveness of recurrent selection.

Although Dr. Sprague usually is recognized for his accomplishments in basic research, he also has had a major impact on applied breeding in the U.S. Corn Belt. The formation of the yellow, dent synthetic variety, designated as Stiff Stalk Synthetic, in the early 1930s was one of the greatest corn populations ever developed. Stiff Stalk Synthetic was the foundation material for many of his basic research studies conducted at Ames (and it still is today), but it also has become one the most important sources of lines used in the U.S. Corn Belt for the past 30 years. Because Dr. Sprague insisted that basic research should be integrated with applied research, Iowa Stiff Stalk Synthetic has become one of the premier populations as a source for lines that have above average combining ability and stalk quality. B14 (released in 1953) and B37 (released in 1958) were two of the most important lines used in hybrids produced and grown during the 1950s and 1960s; both lines were derived from his studies on early testing and recurrent selection. Both lines also had extensive use in pedigree selection programs, and recoveries of B14 (B14A, B64, B68, A632, A634, A635, CM105, etc.) and B37 (B76, H84, H93, H94, ND481, Oh561, etc.) are prominent in present-day hybrids. The potential predicted and foreseen by Dr. Sprague from the use of recurrent selection methods in Iowa Stiff Stalk Synthetic also was realized with the release of B73 (released in 1972) from the 5th cycle and B84 (released in 1978) from the 7th cycle of half-sib recurrent selection. Zuber in 1980, reported that 42.4% of the hybrid seed produced in 1979 for use in 1980 was produced on lines that included Iowa Stiff Stalk Synthetic germplasm.

Dr. G. F. Sprague's long and productive career in corn research has been recognized by his students, colleagues, and peers. His foresight, vision, and interests have had a direct impact on corn breeding methodology. He has been a recipient of many awards and honors in recognition of his dedicated service. He was elected Fellow, American Society Agronomy (1947); received the first Crop Science Award, American Society of Agronomy (1957); awarded Faculty Citation, Iowa State Alumni Association (1958) and Honorary Doctor of Science, University of Nebraska (1958); received Distinguished and Superior Service Awards, U.S. Department of Agriculture (1959 and 1965); elected president of the American Society of Agronomy (1960) and Corresponding Academician, Academia di Agricoltura di Bologna (1960); elected to membership in the National Academy of Sciences (1968); received the National Council Commercial Plant Breeders Award (1972) and DeKalb Distinguished Service Award (1980); and was awarded the Wolf prize in Agriculture, Wolf Foundation, Israel (1981). All these prestigious awards and honors were accorded because of his dedicated pursuit of knowledge in corn research. In addition to the awards and honors, his knowledge and counsel were sought as a member of many local, national, and international committees and organizations. His contributed time and efforts were freely given, and he made many significant suggestions and ideas that enhanced the future course of action.

Dr. G. F. Sprague will be remembered primarily as one of the great researchers in corn genetics and breeding. But he also is valued as a teacher, graduate student advisor, counselor, colleague, and friend. He is always available to discuss a broad range of topics related to corn improvement and contribute pertinent suggestions and ideas. He is the recognized leader of corn breeders in the United States and internationally. His accomplishments are many, and he has established standards in corn research that will be difficult to surpass. Personally, it has been a pleasure to have known him for nearly 28 years and to participate in the preparation of this volume of *MAYDICA* honoring Dr. Sprague and his 60 years of contributions to corn research.

25 Paul C. Mangelsdorf

A Lifetime in the Quest for the Origins of Corn*

By Surinder M. Sehgal[1]

This commemorative issue is a tribute to Dr. Paul C. Mangelsdorf, in recognition of his outstanding accomplishments as a researcher, educator, genetic conservationist, and humanitarian.

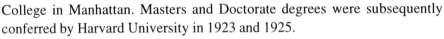

Paul Christoph Mangelsdorf was born July 20, 1899, in Atchison, Kansas. His father was a commercial seedsman and florist. His mother was a great plant lover. His brother, Albert, grew up to become a famous sugar cane breeder.

He received his Bachelor of Science degree in Agriculture in 1921 from what was then Kansas State College in Manhattan. Masters and Doctorate degrees were subsequently conferred by Harvard University in 1923 and 1925.

In college at Kansas State, he served as an assistant to small grain breeder John Parker. Mangelsdorf continued as Parker's assistant for the entire four years of his undergraduate education. It was this relationship which set the stage for his later apprenticeship with the preeminent maize breeders of the day. At Parker's urging, he read and evaluated the then new and influential volume by Edward M. East and Donald F. Jones, *Inbreeding and Outbreeding*. This volume described the authors' pioneering experimentation with hybrid maize varieties. Aided by a recommendation from Parker upon completion of his degree, Mangelsdorf won the opportunity to work with Dr. Jones who had contacted Parker for referrals for an assistantship at the Connecticut Agricultural Experiment Station in New Haven.

* *MAYDICA* 1985, Vol. 30, No. 2
[1] Vice President, Pioneer Hi-Bred International, Inc., President, Pioneer Overseas Corporation

Since Mangelsdorf was one of three Parker assistants and the only one really interested in maize, he was the logical candidate for the job. The title was assistant geneticist, and the position meant graduate work at Harvard University under Edward Murray East, as well as hands-on breeding work for six months at a time with various small grains under Dr. D. F. Jones.

East was a seminal figure in his own right. He had been mentor not only to Jones, but also to two other principal figures in corn breeding and genetics, H. K. Hayes and R. A. Emerson, who had spawned a whole generation of geneticists, including George W. Beadle and Barbara McClintock who won Nobel prizes in medicine. Mangelsdorf later described those five and a half years as a graduate student under East and Jones as some of the most fruitful years of his life. At Harvard, he learned applied genetics at the Connecticut Experiment Station, he put into practice what was then known about plant breeding and experimental hybridization. This became the framework from which Paul C. Mangelsdorf, the theoretical geneticist, applied plant breeder and evolutionary botanist would emerge.

At Harvard, he was for a time the roommate of another great student of plant science, Edgar Anderson. Anderson later became director of the Missouri Botanical Garden in St. Louis and a mentor of many prominent scientists. Mangelsdorf and Anderson engaged in a lifelong correspondence on various aspects of maize which was mutually stimulating and rewarding. This remarkable collection of letters was passed on to Dr. William L. Brown by Dr. Anderson prior to his death.

The environment at Harvard was certainly conducive to the study of plants and plant science. The relaxed atmosphere allowed Mangelsdorf to spend a lot of time investigating the available literature on hybridization between races of maize and between maize and its closest relatives. Through his own curiosity and initiative, he became acquainted with teosinte and *Tripsacum*—two plants that would later influence all of his theoretical thinking.

Within some months of his graduation from Harvard in the fall of 1926, he assumed a position at the Texas Agricultural Experiment Station in College Station. Mangelsdorf remained in Texas for thirteen years. His initial appointment was as an agronomist in charge of small grain breeding. Shortly, however, he began what became an obsessive study on maize, its origin and evolution, for the remainder of his career.

Within a year, Mangelsdorf was joined in College Station by maize cytologist Dr. Robert G. Reeves from Iowa State College in Ames. The arrival of Reeves set the stage for a cooperative research effort into the cytogenetics of maize and its relatives, teosinte and *Tripsacum*. The first crosses they made were between maize carrying marker genes on various chromosomes and teosinte. These were completely successful. However, a second set of crosses between maize and *Tripsacum* were not successful. Yet, another set made in the summer of 1929 was successful, resulting in the first-ever cross between these two genera. To achieve this cross, they employed an ingenious technique of shortening the silks of maize to about the same length as those of *Tripsacum*. Their observations seemed to confirm many of the things Mangelsdorf and Reeves had speculated. Their work would eventually be called the "Tripartite Theory" for the origin of maize and its closest relatives.

For two years, Mangelsdorf, in part on sabbatical at Harvard, and Reeves, in College Station, worked on the draft of a book: *The Origin of Indian Corn and Its Relatives*. Although some investigators later disagreed with the conclusions arrived at by Mangelsdorf and Reeves, all acknowledged it had truly been a novel enunciation of the possible origin of corn. It also proved to be a most effective stimulus and focal point for promoting intensive research on maize among breeders and theoretical plant geneticists for years to follow. The entire research and publication process took some thirteen years, from 1927 to 1940.

The Tripartite Theory suggested by Mangelsdorf and Reeves stated:
1. The ancestor of cultivated maize was a form of pod maize.
2. Teosinte, maize's closest relative, is not its ancestor, but instead a derivative of the hybridization of maize and *Tripsacum*.
3. Many modern varieties of maize have undergone genetic introgression from teosinte or *Tripsacum* or both.

More recently, after the discovery of diploid perennial teosinte in 1979, Mangelsdorf revised the Tripartite Theory by substituting for point two of the old theory a new one that designates annual teosinte as a derivative of the hybridization of maize and perennial teosinte, *Zea diploperennis*. The other two points of the Tripartite Theory remain intact.

The Mangelsdorf and Reeves' publication directly led to Mangelsdorf's appointment to Harvard as professor of botany and assistant director of the Botanical Museum in 1940. At Harvard, Mangelsdorf's attention turned toward

Mexico and the search for evidence to support his position on the origin of cultivated maize. In part, he enlisted help from Dr. A. V. Kidder, also at Harvard, who Mangelsdorf once described as the dean of American archaeology.

In his first year at Cambridge, Mangelsdorf described one collection of archaeological maize from Mexico containing prehistoric specimens of maize. Until that time, archaeological investigations had thrown little direct light on the problem of the origin of maize. Being convinced that archaeology could greatly add to the understanding on the origin and evolution of maize, he anxiously collaborated in such investigations in succeeding years.

In 1948, Herbert Dick, a graduate student in archaeology at the Peabody Museum at Harvard, and C. Earl Smith, a graduate student in Botany, excavated a cave in New Mexico called Bat Cave. There, they found cobs of maize at every level of the cave. At the bottom of the debris, they found tiny cobs of maize which, when dated by the radiocarbon determinations of associated charcoal method, were found to be 5,600 years old. This was a combination of both a popcorn and a form of pod corn as Mangelsdorf and Reeves had postulated earlier, providing the first direct evidence in support of the Mangelsdorf and Reeves theory on the origin of maize. Later, Dr. Richard S. MacNeish, while excavating caves in the Tehuacan Valley in Mexico, discovered cobs dated at ca. 5000 B.C. These appeared to be those of wild corn and a pod-pop type.

Convinced by studies of the earliest archaeological specimens that primitive corn was both a popcorn and a pod corn, Dr. Mangelsdorf undertook to produce a genetically reconstructed form by combining the principal characteristics of popcorn and pod corn. He succeeded in breeding a pod-popcorn which he would proudly show to visitors at his experimental plots. He described it as the world's most nonproductive corn which he had bred. In 1961/62, I, like several others, had the opportunity to see this corn planted by Mangelsdorf in a simulated wild habitat. It indeed behaved like a wild species.

Perhaps not as well known are the contributions Mangelsdorf made to maize improvement in Texas during his 13 years at the Texas Agricultural Experiment Station at College Station, Texas. According to John S. Rogers who was one of Dr. Mangelsdorf's graduate students, Dr. Mangelsdorf's two early accomplishments were to develop sweet corn varieties and yellow dent corn varieties adapted to Texas. The successful sweet corn varieties were named "Honey June" and "Surecropper Sugar."

Since a number of the better-adapted dent corn varieties at the time were white, he converted several of them to yellow. The more important yellow varieties he developed were identified as "Yellow Surecropper," "Yellow Tuxpan," "Golden Thomas" and "Texas Golden Prolific." These varieties were grown for a number of years prior to the advent of Texas hybrid corn in the 1940s.

Mangelsdorf also initiated the inbreeding of Texas corn varieties, which provided the basic inbreds for the Texas corn hybrids introduced from 1940 to 1950. While his original lines are probably no longer in use, they have provided germplasm for some of the current lines.

After joining the faculty at Harvard University in 1940, Mangelsdorf teamed with Karl Sax, director of the school's Arnold Arboretum, to breed a sweet corn hybrid that they named "Harvard Hybrid." Everyone who worked at the Bussey Institution of Harvard University until about 1962 not only helped to grow it but also enjoyed eating it.

Mangelsdorf in 1938 discovered the Texas cytoplasmic male sterility in maize while he was still in Texas. D. F. Jones in 1948 discovered fertility restoring genes at New Haven, Connecticut. Easily introduced into inbred lines through repeated backcrossing, this discovery had great application as an alternative to detasseling in the production of hybrid seed. Jones and Mangelsdorf applied for a patent on their inventions in 1948, but their application was turned down. In 1950, Jones applied for a second patent on the method of restoration of fertility through "restorer" genes. This was granted in 1956 despite strong opposition from academia, the American Society of Agronomy and the American Seed Trade Association. Thus Mangelsdorf and Jones helped pioneer the University Patent approach, seeking to return to research some of the financial rewards that the invention created. Although vigorously opposed then, many universities now encourage their faculty to seek plant patents and finance at least a part of their research through royalties.

Mangelsdorf's firsthand experiences in plant breeding were instrumental in his influencing the direction of breeding efforts by the Rockefeller Foundation, first in Mexico and later on in many other countries in Latin America and Asia.

In December, 1940, Henry A. Wallace, who had recently been elected Vice President, represented the United States at the inauguration of Manuel Avila Camacho as President of the Republic of Mexico. Traveling by car from Laredo, Texas, to Mexico City, he was appalled by the primitive agricultural methods which most of the Mexican farmers were practicing and the pitifully low yields they were getting.

Shortly after his return to Washington in January, 1941, Wallace, in a conference with Raymond Fosdick, president of the Rockefeller Foundation, suggested that perhaps the Foundation, a private philanthropic organization, might help Mexico in various ways to modernize her agriculture. He suggested expansion of the ongoing helpful programs and a fact-finding trip to formulate an accurate assessment of the level of Mexican agricultural sophistication. Fosdick and his staff then selected a team whose combined technical expertise represented the best of American agricultural thinking. Following its intent to select three individuals instrumental in the phenomenal agricultural revolution of the previous twenty years, the Foundation picked Dr. Richard Bradfield of Cornell, Mangelsdorf from Harvard, and Dr. E. C. Stackman from the University of Minnesota. Respectively, the team represented soils and agronomy, plant genetics and breeding, and plant protection.

Officially called the "Survey Commission of Mexican Agriculture," the group traveled through 16 of the 35 Mexican States, much of the time by horse, truck or on foot. In order to assess the situation accurately, they followed back roads to talk to the people themselves. The resulting program came to be called the "Mexican Corn Improvement Program." It was initiated in 1943. Chosen by Mangelsdorf to head this project was Dr. E. J. Wellhausen and one of Dr. Mangelsdorf's Texas students, Luis M. Roberts. In 1959, the Mexican program of the Rockefeller Foundation became the International Corn and Wheat Improvement Program, and in 1966, the International Maize and Wheat Improvement Center, or CIMMYT, was founded. The Mexican Program was a resounding success. It soon served as the prototype for others of its kind around the globe.

In 1952, the continuing success of the Mexican program prompted the Rockefeller Foundation to begin looking toward the Far East for its next agricultural improvement program. The committee for the initial survey included Warren Weaver, J. George Harrar and, once again, Paul Mangelsdorf. The group traveled extensively throughout Asia. The following year, Harrar and Weaver returned for final consultations with government officials from prospective countries. At home, the wheels were rolling to attempt to fund what was beginning to look like a rather extensive program. This led to the first cooperative project between the Ford and Rockefeller Foundations. The project, a rice research institute, was built on land donated by the Philippine government on the campus of the College of Agriculture of the University of the Philippines at Los Banos. The new facility, The International Rice Research Institute (IRRI) began operation in 1962. Due to the role Mangelsdorf played in the establishment of CIMMYT's predecessors and the IRRI, and subsequently in The Green Revolution, he is often associated with that agricultural movement.

In 1967, Stackman, Bradfield and Mangelsdorf published a book entitled *Campaigns against Hunger* which, in great detail, describes the efforts of the authors in improving agriculture in developing countries.

As a professor at Harvard University, Dr. Mangelsdorf, in collaboration with Dr. Evans E. Schultes and Dr. Albert F. Hill, taught the course "Plants and Human Affairs." It became the model for many courses now offered in United States universities which review the uses of plants in both developing and high-level cultures. Over the years, several thousand Harvard and Radcliffe students in plant sciences had the good fortune to take this course which was unorthodox in its approach to teaching economic botany. Instead of following the classical approach of teaching form and function of plants, the course went into great detail to explain the origin of food crops and the complete dependence of man on plants, not only for all the necessities of life (food, clothing, shelter and fuel) but also for many comforts and luxuries of life (perfumes, spices, caffeine beverages, medicines, and the like). It traced the impact of economic plants in shaping man's cultural patterns in the past, and it taught how man has now learned to shape them to his advantage. The approach to teaching the course was what Mangelsdorf used to call "anthropocentric."

Many students in the course prepared term papers which, over the years, have been published in the journal, *Economic Botany*.

In addition to the journal course, Mangelsdorf played a prominent role in graduate education at Harvard as chairman of the Institute of Experimental and Applied Botany. Probably every graduate student in genetics attending Harvard from 1940 to 1968 was influenced by his presence. His inquisitive mind and penetrating questions kept students on the alert. He often emphasized the power of deep thinking and engaged his students in the thinking process. His graduate students included James Cameron, John Rogers, John Edwardson,

William Hatheway, Alexander Grobman, Raju Chaganti, Gordon Johnston, H. Garrison Wilkes, Ramana Tantravahi and me. Drs. Walton C. Galinat, Y. C. Ting and Angelo Bianchi were associated with him as research associates.

To his students, colleagues and research associates, Mangelsdorf taught the necessity of discipline, integrity, perseverance and commitment in carrying out successful research. He was an outstanding mentor and, as could be expected, was demanding about the quality and integrity of research conducted under his supervision.

Dr. and Mrs. (Peg) Mangelsdorf took personal interest in the graduate students. It was common to be invited to their home during holidays, especially Thanksgiving. Mrs. Mangelsdorf was an outstanding hostess and an extremely caring lady. She was indeed genuinely interested in the welfare of the students, research associates, and their families, who were associated with Dr. Mangelsdorf.

As director of the Harvard Botanical Museum, Mangelsdorf was responsible for administration of unique educational resources, including the famous Ware collection of Glass Flowers, the Ames Orchid Herbarium and the Economic Botany Library. He became professor emeritus in 1967.

Among honors and degrees awarded to Dr. Mangelsdorf over the years are: Fellow, American Society of Agronomy; Fellow, American Academy of Arts and Sciences; Honorary Member, Genetics Society of Argentina; D. Sc. (Honorary), Park College, 1960; LL. D. (Honorary), Kansas State University, 1961; D. Sc. (Honorary), St. Benedict's College, 1965; and D. Sc. (Honorary), Harvard University, 1977.

After his retirement from Harvard in 1968, Mangelsdorf continued to be active as an educator, lecturing on useful plants at the University of North Carolina at Chapel Hill. Since his retirement from UNC in 1980 following his wife's death, he has continued to write, study and grow maize. His continued interest in everything about maize has never caused. At age 84, his most recent article, "The Mystery of Corn: New Perspectives," published in 1983 in the *Proceedings of the American Philosophical Society*, shows his clarity of thought, thorough understanding of the subject and a visionary foresight. A recent note in the *Maize Genetics Cooperation Newsletter* (1985) illustrates the continued thought process in generating ideas on the origin of maize.

Researcher, educator, humanitarian, philosopher, historian and genetic conservationist: this is Paul C. Mangelsdorf. Although he had opportunities to work elsewhere, he chose to spend his entire life in the academic field.

As a researcher, Dr. Paul C. Mangelsdorf is best known for his multidisciplinary approach in understanding the origin and evolutionary history of maize. For more than half a century, he has delved into the plant's phylogeny, and sought to unravel the genetic mechanisms governing its inherited characteristics. His life's work culminated in his personal synthesis of maize research in the book, *Corn: its Origin, Evolution, and Improvement*. Seldom have genetic researchers lead such a brilliant career with a single plant; even less frequently have they capped it off with an acclaimed personal accounting of their research and the hows and whys of their conclusions.

As an educator, Dr. Mangelsdorf was thorough, able and dedicated. During the majority of his 27 years tenure as professor at Harvard University, he was responsible for teaching Harvard's famous course, "Plants and Human Affairs." The course was unique in its approach to teaching economic botany and perhaps the only one of its kind offered by any university in the United States or elsewhere in the world at that time.

As a humanitarian, he was extremely concerned with issues pertaining to population growth and food supply. He argued actively for more research on the world's principal food plants of which corn is one. He often stated that there are only 13 species of plants which quite literally stand between mankind and starvation, and we should know as much about each of them as the medical profession, for example, knows about the world's principal diseases.

It has been the improvement in agriculture, both in quantity and quality, which has so far enabled the world to keep up, at least after a fashion, with its explosively expanding population. And, according to Dr. Mangelsdorf, it is further improvement in our agriculture which offers the greatest immediate hope of keeping the world fed until widely employed restrictions on fertility begin to come into greater play.

It is indeed satisfying to note that Dr. Mangelsdorf actively pursued both paths in his career. On the first point, he, as stated earlier, devoted his entire life to the study of the maize plant; and on the second, as a member of the Survey Commission to Mexico, he helped lay the foundations of what has become to be known as the "International Maize and Wheat Improvement Center" (CIMMYT) with a mandate for improving agricultural productivity of these crops in the developing countries.

As a genetic conservationist, Dr. Mangelsdorf is highly respected for his pioneering work in the collection, classification and preservation of races of maize found in Mexico. Not only by himself but by students and associates that he had trained, a series of publications on races of maize in various Latin American countries are indeed a lasting legacy.

Publication of his book, *Races of Maize in Mexico*, in 1951 (Spanish) and in 1952 (English), set the stage for subsequent collection, preservation, and description of races of maize from other countries in Latin America. More than 300 races are now recognized. These studies on races of maize preceded the wide recognition of the issue of genetic erosion. Dr. Mangelsdorf did as much or more than anyone else to put genetic conservation of maize on a firm scientific basis.

I feel fortunate to have been the beneficiary of his thoughtful insight during the years (1959–1963) that I have been associated with Paul C. Mangelsdorf, and it is with that thought in mind that I take great pride in telling about his life and some of his major contributions. My own interests in international agriculture in part trace back to my association with him.

26 Barbara McClintock

From the Beginnings of Maize Genetics to the Explosion of Molecular Activity*

An appreciation by M. M. Rhoades[1]

When Barbara McClintock began her graduate work at Cornell in the middle 1920s, the foundations of maize genetics had been firmly laid but comparatively little cytological work had been done. The carmine smear technique, which greatly facilitated cytological studies, had just been developed by Belling. McClintock quickly found that carmine smears of maize sporocytes at midprophase of meiosis yielded preparations of extraordinary beauty and clarity. Maize could now be used for detailed cytogenetic analyses of a kind heretofore impossible with any organism, and McClintock in  the succeeding years published a series of remarkable papers which clearly established her as the foremost investigator in cytogenetics.

Her first major contribution was the demonstration that the chromosomes were individually recognizable by their relative lengths and arm ratios, distinctive chromomere patterns, and deep staining knobs in characteristic positions. This was followed by such significant studies as the analysis of translocation heterozygotes, the correlation of cytological and genetical crossing over, the assignment of linkage groups to specific chromosomes, the physical location of gene loci by deficiencies, the formation of dicentric bridges and acentric fragments as a result of crossing over in inversion heterozygotes, the somatic and meiotic behavior of unstable ring chromosomes, the occurrence of non-homologous pairing, the structure and function of the nucleolus organizing region, the production of viable homozygous deficiencies

* *MAYDICA* 1986, Vo. 31 No. 1
[1] Department of Biology, Indiana University, Bloomington, Indiana 47405

that simulated gene mutation and formed a pseudoallelic series, and the genetic and cytological consequences of the breakage-fusion-bridge cycle. Her recent studies on the evolutionary history of races of maize as disclosed by the number and location of specific chromosome knobs have been conducted with typical precision and elegance.

Her consummate skill and versatility as a cytologist are perhaps best evidenced by the fact that in the few weeks she devoted to Neurospora resulted in what remains more than thirty years later as the definitive paper on the meiotic chromosomes of this fungus. So difficult cytologically is Neurospora, that not even the correct chromosome number was known prior to her studies, but McClintock showed that each of the seven chromosomes was cytologically distinguishable at meiotic prophase and she was able to demonstrate induced structural changes in the chromosomes.

Perhaps McClintock's most outstanding contribution is her analysis of the control of gene action in maize and the discovery of the two-unit interacting system. This concept was the precursor of the regulator-operon theory of gene regulation that won for its promulgators, Jacob and Monod, the Nobel Prize in 1965. It was during her studies of the breakage-fusion-bridge cycle that she detected an unexpected burst of unstable genes in her stocks. Genic instability had been observed repeatedly by others and it had been established that it was under genetic control but she discovered a new kind of gene element, regulatory or controlling, which had profound implications for our understanding of gene function and regulation in development. Her finding that controlling elements can move from place to place in the genome, that they can modify expression of a gene by insertion in or near that gene, and that gene expression can be restored when the controlling element is excised afforded a new and revolutionary insight into chromosome structure and gene expression. She pioneered a new era in genetic research. So unusual and novel were her findings that for a decade her conclusions were not accepted by many workers and it was not until comparable mobile genetic elements or transposons were found in a wide range of diverse organisms that the significance of her work was appreciated. She was ahead of her time and only in recent years has she received wide recognition for having developed a new field of genetic inquiry. Many honors belatedly came her way, including a score of honorary doctorates from famous universities. Capping the numerous accolades was the bestowal of the Nobel Prize in Medicine in 1983.

Some major conclusions about controlling elements reached by McClintock are as follows:

1. Controlling element systems may involve two units, a distant controller and a responding element adjacent to or in the affected structural gene. Mutable systems where a single element functions both as controller and responder are also possible.

2. Not only do the controlling elements regulate the expression of a structural gene but they are also capable of specifically inducing mutation in that structural gene.

3. The system of controlling elements acts in a spatially and temporally specific manner in the modulation of gene activity and the induction of mutations.

4. Sensitive genes can be preset by a controlling element leading to a change in function at a predetermined time in a later generation.

5. Controlling elements undergo changes in "state" which are revealed by modified regulatory and developmental properties.

All of her conclusions were based on convincing supporting data.

As a result of McClintock's studies it is evident that programmatic information can be encoded by DNA, that this information can be expressed in unexpected ways, including abrupt changes in levels of gene expression and that these effects are due to controlling elements capable of modifying the action of virtually any structural gene irrespective of that gene's specific biochemical function.

One of the remarkable aspects of Barbara McClintock's surprisingly beautiful investigations is that they came solely from her own labors. Without technical help of any kind she has by virtue of her boundless energy, her complete devotion to science, her originality and ingenuity, and her quick and high intelligence, made a series of significant discoveries unparalleled in the history of cytogenetics. A skilled experimentalist, a master at interpreting cytological detail, a brilliant theoretician, she has had an illuminating and pervasive role in the development of cytology and genetics. It is no exaggeration to say that hers has been one of the most influential minds in biology in the 20th century. Her work has been characterized by a sustained intellectual brilliance which made it possible for her to grasp the conceptual order underlying the origin of unstable genes in maize. These investigations led her to the concept of genetic regulation. In her discoveries and in her interpretation of their meaning, she was many years ahead of her time. Genetics would not occupy its present high estate were it not for her magnificent and pioneering contributions.

27 Charles O. Gardner

40 Years of Contributions to Quantitative Genetics*

An appreciation by L.M. Pollak[1] and M.A. Thomas-Compton[2]

This issue has been dedicated to Dr. Charles O. Gardner in honor of his accomplishments and in appreciation of his efforts on behalf of his students and associates. It has been planned to commemorate his 70th birthday, March 15, 1989.

Charles Gardner is a native of Tecumseh, Nebraska, where he grew up on a farm. After graduating from Tecumseh High School in 1936, he attended the University of Nebraska. It was at the University of Nebraska while working on the alfalfa breeding project of Dr. H. M. Tysdal that he gained an interest in plant breeding. He graduated with High Distinction in 1941 with a B.Sc. degree in Technical Science in Agriculture. After only one semester of graduate studies at the University of Nebraska under the Charles Stuart Fellowship, his education was interrupted by World War II. Service in the U.S. Army did, however, give him the opportunity to receive an M.B.A. degree in Business Administration from Harvard University in 1943.

At the end of the war, Dr. Gardner returned to Nebraska and the field of agronomy where he worked as an Assistant Extension Agronomist while completing his M.S. degree in 1948 in Agronomy and Botany at the University of Nebraska. For this degree he worked under Dr. T. A. Kiesselbach, who is well known for his description of the corn plant and its growth. His research

* *MAYDICA* 1989, Vol. 34, No. 1
[1] USDA-ARS, Iowa State University, Ames, Iowa 50011
[2] University of Nebraska, Lincoln, Nebraska, 68583

for the Ph.D. degree, which he received from North Carolina State University in 1951, was conducted under Dr. P. H. Harvey in collaboration with Drs. H. F. Robinson and R. E. Comstock. This research was important in refuting the theory that overdominance was an important cause of heterosis in corn. Instead, the bias in average degree of dominance for genes determining quantitative traits in maize mainly resulted from linkage rather than from overdominance or epistasis.

Dr. Gardner spent one year at North Carolina State University as an Assistant Statistician before returning to the University of Nebraska, where he worked as an Associate Professor of Agronomy until 1957, Professor of Agronomy until 1970, and at that time became a Regent's Professor to the present day. He also served as Chairman of the University of Nebraska Statistical Laboratory from 1957 to 1968, a role he reassumed on July 1, 1988, serving as Acting Head of the Biometrics Center of the University of Nebraska. He spent a sabbatical year as Visiting Professor of Genetics at the University of Wisconsin from 1962 to 1963, and in 1978, two months at the University of California in Davis using isozyme techniques to study corn populations.

The research accomplishments of Dr. Gardner at the University of Nebraska are many, and have led to our greater understanding of the quantitative genetics of crops, especially maize and sorghum. His work using mass selection in corn has shown that this population improvement system is simply done and is effective in maize by incorporating environmental control via the grid system. Mass selection has since been used to improve populations of corn as well as other crop plants in many areas of the world. He has been one of the scientists who designed and conducted long-term selection studies in maize which have clearly showed that recurrent selection increases the frequency of favorable alleles in maize populations. Inbred lines developed from these improved populations are superior to those developed from the original populations. He has developed genetic models to use in the analysis of quantitatively inherited traits that have been, and continue to be, widely used.

Along with theoretical advances in understanding quantitative genetics of population improvement systems, Dr. Gardner's research program has provided useful germplasm for quantitative traits important to corn breeders. When a genetic male sterile system in sorghum allowed the formation of random-mating sorghum populations, Dr. Gardner was instrumental in applying the quantitative genetic models and population improvement systems developed in maize to sorghum. He also realized the potential of using isozymes as marker genes to study mating systems and to identify associations with genes controlling quantitative traits by adding these techniques to his research program as additional tools to supplement field studies of maize populations.

Dr. Gardner has been an effective teacher of quantitative genetics at the University of Nebraska. His influence has extended to his graduate students, visiting scientists and post doctorates, as well as to the students in his classes and his associates. He has been an invaluable source of information on statistics, experimental design, plant breeding, and the influence of quantitative genetics in plant improvement. He has always been exceptionally helpful to those young scientists starting their careers in plant breeding, and is known to always be very fair. His influence also extends to the example he sets for hard work, attention to details, accuracy and honesty.

A few of the honors received by Dr. Gardner for his accomplishments are: the Gamma Sigma Delta International Award for Distinguished Service to Agriculture, 1977; Crop Science Award, Crop Science Society of America, 1978; Fellow, American Society of Agronomy, 1964; American Association for the Advancement of Science, 1978 and Crop Science Society of America, 1985; University of Nebraska Award for Outstanding Research and Creative Activity, 1981; the Crop Science DeKalb-Pfizer Distinguished Career Award, Crop Science Society of America, 1984; and the USDA's Distinguished Service Award for Education and Information, 1988. He has served as the President of the Crop Science Society of America in 1975, President of the American Society of Agronomy in 1982, and Chairman, Section O Agriculture, American Association for the Advancement of Science, 1987–1988.

Dr. Gardner and his wife, Wanda, have three sons and one daughter. Both Charles and Wanda have been kind and helpful in so many ways to students and associates. They have earned our gratitude and admiration.

28 John R. Laughnan

Over 40 years of Contributions to Genetic Concepts: Genetics from A to Zea in Three Score and Ten*

by Ed Coe[1]
with contributions from
S. Gabay-Laughnan, E.B. Patterson [2]

This issue of MAYDICA honors John R. Laughnan, a teacher who has brought the pleasure and stimulation of genetics to large numbers of students in the classroom, laboratory and field, and a scientist whose elegant research style has influenced lifetime students of the discipline of genetics, teaching them how to define and attack problems, and how to analyze and assess the resulting data.

When speaking of a scientist like John Laughnan, too serious a tone may be inappropriate. As the first syllable of our subject's last name is often mispronounced, we believe an audacious "laughdation" better suited to John Laughnan's wit, whimsy, and hyperbole than a lofty laudation enumerating his many scientific contributions and praising his undeniable virtues of dogged curiosity and creativity.

First, a bit of data would be desirable before any conclusions can be drawn. Our subject was born on a farm in a Wyoming township near Spring Green, Wisconsin, on Sept. 27, 1919. The family moved to Sauk City,

* *MAYDICA* 1989, Vo. 34, No. 3
[1] USDA-Agricultural Research Service and Department of Agronomy, University of Missouri, Columbia, MO 65211
[2] Departments of Plant Biology and Agronomy, University of Illinois, Urbana, IL 61801

Wisconsin, where schooling years were completed. His B.S. (1942) was in Plant Sciences at the University of Wisconsin, where he became acquainted with R. A. Brink and was advised to pursue a graduate career in maize genetics with L. J. Stadler at the University of Missouri. The research for his Ph.D. (1946) focused on the structure and function of the $A1$ locus of maize and involved chemical and histological studies of pigments associated with a series of alleles and dosages. A post-doctoral NRC fellowship at Iowa State with G. F. Sprague was followed by two years as Assistant Professor at Princeton and a Gosney Fellowship at California Institute of Technology with E. G. Anderson. In 1948, he was offered an Assistant Professorship by the University of Illinois Botany Department at the request of a new staff member, Professor M. M. Rhoades, who was seeking a congenial and able associate. Following the death of L. J. Stadler in 1954, John was called to Missouri as Stadler's replacement, but after one year was called irresistibly back to Illinois to chair the Department of Botany. From 1955-59, and again from 1963–65, he headed that Department. Teaching and research have been his delight throughout.

Genetics from A to Zea (recollected by E.H.C. and S.G.-L., with stimulation from other lettered colleagues): After exhaustive analysis of the $A1$ locus, John's research emphasis underwent an alphabetical transition, subtly and perniciously accompanied by an organismal diversion. In retrospect, it is clear that the transition to the next letter was signaled by detailed studies of the allele then called "A super B" (now known as $A\text{-}b$ or $a\text{-}b$). In that now-gone era, presentation of a model called for construction of contradicting arguments to be shot down, and his arguments were sometimes all too effective. Having argued effectively (1949) that certain A alleles in maize were not simple but were part of compound tandem repeats that underwent unequal recombination between duplicated components, he faced a new problem. A number of events persistently occurred in the absence of recombination, which appeared to be due to gene conversion. This too-simple answer called forth from him an alternative model ("intrachromosomal exchange"). Evidence by which one might rule out the exchange model could be sought from the other mecca of genetic and cytological analysis, *Drosophila melanogaster*, where a compound tandem-repeat locus, Bar (B), was ready for exploitation. To the surprise of all, this effort happily failed to contradict the model. Despite his commendable attempts to eliminate the model, this study serves as a precedent for a new kind of chromosomal behavior (curiously, occasional reports can be seen still, even in maize or *Drosophila* literature, that attribute non-recombinant events to gene conversion or mutation). In future studies with loci affecting disease resistance and the many functions involved in quantitative traits, these analyses of the genetic behavior of duplicated segments may occasionally be remembered. Incidentally, B of maize was graciously passed over and left for a former student to examine; so was C (that is, the nuclear gene, not the cytoplasm, cms-C, to which he did in fact make his own way). So much for an undeviating path.

Duplication in alphabets: Not confined to one path, our subject at one and the same time has advanced in the Greek alphabet as well as the Roman alphabet, some of which is known only to intimate insiders. Pressing forward with alpha and beta but foregoing gamma, he soon employed delta as a marker in chromosome 3 (delta: an arcane symbol for translocations learned from E. G. Anderson while waiting for smog to clear for Caltech pollinations). Another segment, on chromosome 10, was marked with E, but that gene and those experiments are known only to a discreetly limited audience. Yet another controversial marker, nonetheless, is etched in the memory of a large audience. Study of F, H, or I is unrecorded, but g and j took extended roles with E. Because others used K as a marker on chromosome 10, little reason was left to hob-knob with K except to argue its validity. Then there was *lg2*, which results in ligulelessness. But enough guilelessness; the reader at this point in the alphabet must be tolerant, as the trail only rises to general consciousness again at S (about which more later) and at *sh2* (next).

A gustatory digression: According to historical sources [E.H.C.], serendipity played a part in a practical discovery (1953) from which many sweet corn worshipers now benefit. Soaking and chewing upon a corn seed to aid in concentration is a pervasive but minor indiscretion in the profession, generally conducted surreptitiously and especially embraced when seeking rare mutations or recombinants. Muttering, so it is said, "that's shrunken, too," then "super, it's sweet!" our subject came upon the now popular and widely grown, high-sugar Super Sweet type. When next the reader has a table ear with butter (or better, corn oil margarine) and salt, it might be gratefully remembered that the *sh2* factor is so close to *A1* that it was originally attractive as a marker in intensive genetic analyses—else it might yet be only a phenotypic curiosity. The gene of importance is now reversed, but a is still present in Super Sweet strains despite crosses and crosses. If this obscure recessive has any influence on flavor, our subject has never defined this by taste tests on recombinants, though he did propose that the A gene did not do what it is now known it does. But this was in the era when speculations on gene functions were permitted by reviewers; so much for consistent serendipity.

On clairvoyance (or, being on the bandwagon before there is one): By the time of the 1970 epidemic of *Helminthosporium maydis* race T, our maize geneticist already had set out to identify and to characterize mutations in cytoplasmic male sterility traits, and was doing so. To the extent that organelle-inheritance research in higher plants was popular at the time, it consisted of conceptual black boxes whose dissection, in the absence of molecular tools, was conducted only in the realm of "operational thinking."

If black boxes were too simple and one wished to undertake study of a system whose biology interferes with analysis, then as challenging as any would be mutation studies involving a maternally transmitted character expressed only in the male inflorescence (so much for a quick and easy route to fame). The analysis entailed finding fertile exceptions in cms-S strains that had occurred at a time such that the male and female inflorescences would both carry the genetic change. The cases were followed up by reciprocal progeny analysis sufficient to establish whether the mutation was nuclear or cytoplasmic. Teachers of genetic reasoning and design take note: a student who can follow such reasoning should be warned that a career may be looming on the horizon, and that many, but by no means all, colleagues in the profession will follow the reasoning. A student in whom this tendency is recognized might well be given this further problem: construct experiments that will establish whether nuclear constitution (1) influences the frequency of fertile exceptions, and (2) influences the relative frequencies of cytoplasmic vs. nuclear events. After the student has figured out how to do this, the publications and data of our subject should be casually made available. Given the mutation data, if the student dares to suggest that an episome might be involved, and offers experiments to challenge the idea, release the student upon the genetic world. Today efforts are proceeding to catch up on all the molecular basis of what has been found; tomorrow there will be more genetic designing required, and our subject is each summer and winter up to his ears doing that.

A teacher's cunning: Effective teachers often set out mazes for the student to explore, and then stay available to help when they miss a corner. "I [Ellen Dempsey] remember Jerry Kermicle's story of how he became fascinated with genetics after John Laughnan, who taught the class for undergraduates, took the time to help him untangle his neglected *Drosophila* cultures. John set a good example for worthwhile teaching...."

About true grits: "Some twenty years ago, there were a number of successive years in which I [E.B.P.] accompanied John Laughnan to pollinate winter plantings of maize in south Florida. One trip stands out in my memory because it involved breakfast menus and the fact that John was on a diet. He usually ordered eggs for breakfast and invariably was given the accompanying choice of hash browns or grits. Actually, he wanted to avoid both. Declining the potatoes was no problem, but he was concerned that if he declined the grits he might hurt the feelings of the waitress. So he conceived the idea of ordering grits enthusiastically, but leaving them on his plate. This spotlighting of grits, however, led to a lengthy discussion of their popularity and the immense logistical problem of delivering them to all the vast number of restaurants and lunch counters throughout the South. We could only shake our heads in wonder. Then one day John whooped in exultation as he hit upon an explanation that cleared everything up for him. As soon as he shared his speculation, I had to admit that it appeared to offer a satisfactory solution. Proceeding down the highway were two cement mixer trucks, their barrels turning busily in hominy."

More alphabet: In 1970, John was on leave from U of I, serving a sojourn with AEC (now DOE). This period may aptly be referred to as a rare instance in which he took time to go fission.

About being organized: "I [E.B.P.] have never known anyone who would become immersed so completely and so tirelessly in painstaking field research during the long days of summer for a period that regularly extends to some fifty consecutive days. During these times of unrelenting daily work schedules, any interruptions or distractions easily evoke feelings of frustration. And yet these interruptions are inevitable: short-deadline reports, visitors to be met, committee meetings, phone calls to be returned, messages to be relayed, information to be provided, essential errands, unpredictable minor crises.

"Just to keep track of all these items requires a bewildering array of reminders detailing time, place, subject, information to be assembled, etc. During the period when the field nursery is the constant base of operation, these reminders frequently take the form of notes or messages hastily scrawled on whatever writing surfaces are immediately at hand, and then the notes may be thrust into any one of several pockets, tucked into pollinating aprons or stashed in momentarily plausible locations in field vehicles.

"John Laughnan has wryly and unerringly recognized a central truth. Before we can address any task, we must first be reminded of it. Recently he was asked how things were going. He glanced around cautiously, then sidled up with an air of great confidentiality and affirmed reassuringly, 'I have it all written down.'"

Organizational activities (source a "ditto" letter found while screening colleagues' files and minds for suitable material): "An informal discussion... led to the suggestion that we consider an annual, informal get-together of maize geneticists... The purpose of this letter is to indicate that we would be happy to sponsor such a meeting at Illinois and would suggest that January 8 and 9, 1959, might be satisfactory dates. We think there should be maximum opportunity for participation by graduate students... May we have your comments at your earliest opportunity?" John R. Laughnan, November 26, 1958). The Annual Maize Genetics Conference began that January and has grown from a cozy group of two dozen to a cozy group of over 300, including students, post-docs, technicians, long-timers and worldwide participants. There is still no formal process for the get-together, no fixed mailing list, no "leader" but plenty of "helpers." Our subject may be seen standing out of the crowd in the accompanying 1960 photograph (Part A, Fig. 23). So much for grand contributions to organizational structures.

Lest we may have strayed too far, we return to this central point: This issue huskily honors John R. Laughnan, teacher and scientist, cooperator in maize genetics, colleague and friend, whose enthusiasm for science is an inspiration.

29 Hugh H. Iltis

An Avid Investigator and Searcher for the Origin of Corn*

An appreciation by B.F. Benz[1]

Hugh H. Iltis was born in Brno, Czechoslovakia in 1925. He moved to the United States with his family in 1939. Stimulated by his father's interest in botany and genetics, he very early fashioned his own interest in taxonomy and evolution. His career as a botanist began in his formative years and flowered when he began his own herbarium at the budding age of 14. As an undergraduate at the University of Tennessee under A. Sharp, Iltis' interest in biogeography and the North American Tertiary Floras blossomed, stimulated no doubt by nearly ten years of plant collecting in his boyhood haunts in Virginia and brief tours of France and the Pacific Northwest. While maintaining his interest in North American floristics, his graduate work at Washington University (St. Louis) and the Missouri Botanical Garden (1949–1952) introduced him to the diverse beauty of the tropics where he set himself to work on the taxonomy of the Caper family (*Capparidaceae*). Since the time that he was a graduate student at Washington University studying with Edgar Anderson and collecting *Tripsacum* mutants with Paul Weatherwax, Prof. Iltis has doggedly pursued an interest in Indian Corn (Maize, *Zea mays* L. subsp. *mays*) and its relatives.

After three years of teaching at the University of Arkansas, Iltis took up residence in 1955 at the University of Wisconsin-Madison as the herbarium curator and assistant professor. During his 35 year tenure there, he has been the driving force in expanding and improving the herbarium, teaching courses

* *MAYDICA* 1990, Vol. 35, No. 2
[1] Laboratorio Natural Las Joyas, Universidad de Guadalajara Apto. Postal 1-3933, Guadalajara, Jalisco, Mexico C.P. 44110

in evolution, biogeography and taxonomy, especially of grasses, and advising 21 Master's and 20 Ph.D. students. Having spent six years doing graduate work with Prof. Iltis, I recognize his skills as an educator, though sometimes had misgivings about his methods. His often recited motto of "the harder you squeeze an apple pip the further it will fly," provides a glimpse of how many of his graduate students have felt. Apart from his insuperable generosity and support, the end seems to justify his means; many of his former students are now directors, curators or collectors in many of the most prestigious herbaria and arboreta in the U.S.

Aside from his taxonomic work, Dr. Iltis has also dedicated much of his professional life to conservation goals. His is recognized as one of the leading conservationists in the U.S. He has made strong impressions on students and politicians, as well as the public in general, of the need to conserve nature and thereby maintain or improve the human environment. With this volume, we laud Prof. Iltis' achievements in the pursuit of understanding the origin of maize and the taxonomy of the genus *Zea*, and for his indefatigable promotion, to often inconvertible audiences, of a sound environmental ethic.

This volume contains selected papers presented at the International Symposium on *Zea diploperennis* and the Conservation of Genetic Resources, sponsored by the University of Guadalajara and the World Wildlife Fund in Guadalajara, Jalisco in December of 1988. This symposium commemorated the eleventh anniversary of the discovery *of Z. diploperennis* Iltis, Doebley and Guzman at La Ventana in the Sierra de Manatlan, Jalisco, Mexico. Scientists and students were brought together to celebrate this anniversary and to discuss the biology, ecology and conservation of *Z. diploperennis* and its relatives, their importance as a genetic resource as well as the controversies surrounding genetic resource conservation. The results presented in this volume are a reflection of Iltis' dedication to a cause and to the resolution of a mystery. Much of the research presented in this volume can be related to Prof. Iltis' involvement, is the result of research planned and/or directed by him, or was stimulated by his often ultra-enthusiastic demeanor.

Prof. Iltis' contributions can be highlighted by two basic themes. First, his recognition and emphasis of the relationship between maize and the other members of the genus *Zea* (the "teosintes") led he and his student J. Doebley to finally establish a common-sense taxonomic order to the genus. Moreover, Iltis' Catastrophic Sexual Transmutation Theory on the origin of the maize ear has spurred research in such diverse disciplines as molecular genetics, developmental morphology and archaeology, and most of the research to date (including some of the papers published herein) has yet to falsify his thesis that the maize ear evolved from the central raceme of a teosinte tassel that terminates a lateral branch. Second, he recognized very early the economic potential and therefore the great importance of *Z. diploperennis* as a perennial diploid relative of the cultigen maize, using this very successfully to stimulate research and call attention to the importance of the regional biotic diversity, thereby providing the impetus for the creation of a local program (the Laboratorio Natural Las Joyas of the University of Guadalajara) dedicated to the conservation and study of this biota. His heartfelt and felicitous pleas were instrumental in the purchase of land to conserve in perpetuity one of only four populations of *Z. diploperennis* known to exist in the wild. This ultimately lead to the creation of the Sierra de Manatlan Biosphere Reserve in 1987, and to its inclusion in the U.N.E.S.C.O.—Man and the Biosphere system in 1988. The impetus he provided is beyond value and future generations will be forever grateful.

Prof. Iltis' promulgation of a strict conservation ethic is perhaps best gauged by his indefatigable pursuit of converting the inconvertible. His acuity and dedication is demonstrated amply by his response to the perennial question posed in juxtaposition to his plea for nature preservation, "how can I use it? What is it worth to me?" His reply, "what good are you!" presents a very clear message, viz., that human and especially personal well-being is too often placed over the well-being of our support system—the biosphere—and the time is upon us to rid ourselves of our homocentricity. What he has written in regard to human's need for nature needs to be taken to heart.

"Man's love for natural colors, patterns, and harmonies, his preference for forest—grassland ecotones which he recreates wherever he settles, even in drastically different landscapes, must be the result... of ...natural selection through eons of mammalian and anthropoid evolutionary time... Would it not then be incredible indeed, if savannas and forest groves, flowers and animals, the multiplicity of environmental components to which our bodies were originally shaped, were not, at the very least, still important to us? Would not such a concept of 'nature' be a major part of what might be called a basic optimum human environment?

"Here, finally, [is] an argument for nature preservation free of purely utilitarian considerations: not just clean air because polluted air gives cancer; not just pure water because polluted water kills the fish we might like to catch; not just saving plants or ducks because they could be 'useful' or 'edible,' but preservation of the natural ecosystem to give body and soul a change to function in the way they were selected to function in their original phylogenetic home. The ultimate argument for nature preservation, as well as for landscape architecture or urban planning, would then rest squarely on evolutionary principles".

The now legendary account of the discovery of *Z. diploperennis* provides additional insights into Iltis' character. His own, now habitual, "New Year's" greeting card (it usually arrives in June) addressed to the "Flowers, Butterflies, Birds and Whales" was the stimulus underlying the rediscovery of *Z. perennis* and the discovery of *Z. diploperennis* and characterizes Prof. Iltis' unrelenting curiosity and the real sense of urgency that he associates with extinction. This particular greeting card, proclaiming *Zea perennis* extinct in the wild, was sent to Ma. Maria Luz de Puga, Professor of Botany at the University of Guadalajara thereby initiated a very long chain of events including the rediscovery of *Z. perennis*, the discovery of *Z. diploperennis*, and eventually the creation of the Sierra de Manatlan Biosphere Reserve. His interest then, and continued involvement now, demonstrates his curiosity about and preoccupation with the status of a single species, both the result of vocational experience as a teacher, taxonomist, evolutionary biologist and witness to the ever-increasing rate of species extirpation, especially in the tropics. Whether due to interest and persistence or serendipity, Prof. Iltis' involvement was and is that of a "keystone mutualist."

This essay would not be complete without lauding Iltis as a teacher and a speaker. As those of us who have had the opportunity to listen to his lectures will attest, he possesses all the skill, knowledge, understanding and commitment to bring together an informative and entertaining discourse. His rapport with audiences is somewhat mystical and definitely enviable. He can provoke bitter anger and protest (all the while saying he isn't there to please anyone), coax you into outbursts of laughter, or bring a tear to the eye and win you over to support his environmental ethic. At once he is actor, orator and intellectual, promoting a cause that at the very least will halt human population growth, all the while subscribing to whatever means necessary to establish parks, reserves and the like to preserve what remains of nature for the enjoyment and enrichment of future generations. All of his students have appreciated these qualities.

While numerous species of plants bear his name to honor him as the first time collector of said species or genus, and numerous species of *Capparidaceae* denote him as the authority who conferred on them their scientific name, here also we laud his contribution to science and humanity.

30 Oliver E. Nelson, Jr.

A Pioneer in Physiological Maize Genetics*

An appreciation by M. Tomes[1]
and L.C. Hannah[2]

It is with great delight and in sincere appreciation that we, his former students and present and past colleagues, dedicate this issue of MAYDICA to Oliver E. Nelson on the occasion of his seventieth birthday. His seminal studies of maize, dealing primarily with the seed, have had profound impact on our understanding of the genetics and the biochemistry of seed constituents as well as the nature of the plant gene and its associated variability. His studies serve as guides for us in our present and future endeavors.

Oliver Evans Nelson, Jr., the first child of Oliver Evans Nelson and Mary Isabella Grant Nelson was born August 16, 1920, in Seattle, Washington. Referred to as "Doon" (a child's attempt at junior) by his younger brothers Grant and Frank, and subsequently by all family members, Oliver spent his formative years in the vicinity of New Haven, Connecticut where he attended Hopkins Grammar School and New Haven School. Clearly recognized as a precocious child, Oliver's exploits were legendary within the Nelson family. An early reader, Oliver was once promised a financial reward by his grandfather if he could read a rather lengthy book. After two days, Oliver returned to collect his reward. It was at that time that he faced his first unexpected oral examination. After being quizzed in great detail by a disbelieving grandfather, Oliver passed with flying colors and collected his reward!

Oliver was first introduced to genetics when he served as a summer assistant in the Department of Genetics, Connecticut Agricultural Experiment

* *MAYDICA* 1990, Vol. 35, No. 4
[1] Purdue University, Lafayette, Indiana, U.S.A.
[2] University of Florida, Gainesville, Florida, U.S.A.

Station. The year was 1937—the same year he graduated from preparatory school. At the Connecticut Station, Oliver worked under D. F. Jones, a noted corn breeder and geneticist. This association continued at various times until he completed his formal genetic training. Oliver spent the summer of 1940 with the Pioneer Hi-Bred Corn Company in Johnston, Iowa. He was awarded an A.B. from Colgate (his father's alma mater) in 1941, where he graduated magna cum laude with honors in Botany. From 1941–46, Oliver served as a Fellow for the Eastern States Farmer's Exchange at the Connecticut Agricultural Experiment Station. He completed his formal genetic training under E. W. Sinnott, who was most influential in developing Oliver's early interest in genetics. Oliver did his doctoral research under D. F. Jones. Prior to completing the doctorate, Oliver spent six months as a plant breeder with the Robson Seed Company in New York. He earned the M.S. in 1943 and the Ph.D. in 1947, both from Yale University.

In 1947, Oliver accepted a position as Assistant Professor of Genetics in the Department of Botany and Plant Pathology, Purdue University. He was promoted to Associate Professor in 1949 and to Professor in 1954. He moved to the University of Wisconsin in 1969.

At Purdue University, Oliver's first research responsibility dealt with corn breeding with emphasis on popcorn. Oliver's charge also included organizing and teaching a graduate course in physiological genetics. Basic research was done on an "as-time-permits" basis.

Oliver's popcorn breeding program was quite successful. Some of the lines he developed in this early effort are still in commercial use. The course in physiological genetics also proved to be quite successful; the content of which changed rapidly to keep pace with developments concerning the nature of the gene at the chemical level.

During this early period, Oliver's basic research was confined mainly to phenogenetic studies of morphological mutants and to the study of gametophyte factors in maize. The genetics of, and the mechanisms underlying non-Mendelian ratios, as exemplified by the *Ga* locus, were actively pursued. He also outlined how *Ga* factors could be utilized in a practical sense to maintain isolation among special types of corn. It was also during this time that he grasped and endorsed the use of mutants to study pathways leading to normal development.

In an effort to move his studies to the biochemical level, Oliver spent a sabbatical year (1954–55) as a visiting investigator at the Department of Genetics, Swedish Forest Experiment Station and the Biochemical Institute, University of Stockholm. The year was spent trying to determine the biochemical basis for such extreme morphological mutants as *Knotted*, *Ragged* and *Corn Grass*. Biochemical differences were found, but the causal lesions were elusive. The experience, however, was excellent training for later studies on lignin, starch, protein and anthocyanin biosynthesis.

On his return to Purdue, Oliver was relieved of most of the applied portion of his position. Purdue hired an additional breeder and Oliver was free to expand his work in basic genetics. Experiments at Purdue dealing with the nature of the rII region of phage T4 prompted Oliver to ask whether similar types of answers would be found with a gene in a higher organism. Oliver exploited the finding of R. A. Brink and others that the *waxy* gene of maize is expressed in the pollen. Using a single pollen grain as the unit of observation, it was possible to evaluate populations large enough to monitor recombination within a gene. Oliver's studies with *waxy* plus those done with the *rosy* locus of Drosophila, were seminal to modern genetics in demonstrating that recombination, indeed, can occur within a eukaryotic gene. While the *waxy* system clearly had many advantages, the lack of outside markers also expressed in the pollen, as well as the seemingly non-additive genetic distances within the gene, proved troublesome for detailed studies. It is interesting to note that some of the alleles studied in the early experiments turned out to contain large insertions.

A question tractable with the *waxy* system concerned the placement of transposable elements within a gene. Because non-autonomous elements of two-element systems are stable in the absence of the trans-acting element, Oliver realized that such elements could be mapped within the gene. A series of elements within *waxy* were then mapped and Oliver came to the conclusion, somewhat surprising at the time, that the elements mapped throughout the gene. An idea then in vogue was that transposable elements functioned like ordinary control elements of genes and thus might be expected to map at one or the other end of the gene. While the recombinational analysis of *waxy* and its interesting alleles provided unexpected observations, Oliver was quick to realize that a deeper understanding would come only when the gene or its immediate product could be analyzed directly.

To develop competence in enzyme and protein chemistry, Oliver took his second and final sabbatical leave at the California Institute of Technology from October 1961 to September 1962. There followed a series of studies on starch synthesis and on lignin synthesis in maize. In the latter, the abnormal lignins produced by the *brown midrib-1* mutants were defined.

Sabbatical leaves had more than one dramatic effect on Oliver's life. During his stay at the University of Stockholm, he became friends with a particular young lady. As fate would have it, their paths crossed again in California when Oliver was doing his second sabbatical. Believing that this was more than random chance, Oliver married his young acquaintance, Gerda, in 1963. It is interesting to note that Oliver has not been allowed a third sabbatical!

Studies of starch synthesis at Purdue dealt with a comparison of genotypes such as *waxy, shrunken-2,* and *brittle-2,* with wild type. It was during this time that the lesion associated with the *waxy* locus, a starch-bound ADP-glucose glucosyl transferase was determined. This was one of the first instances associating the enzymic lesion with a classically defined gene in a higher plant. The data also provided definitive evidence that the two major polymers of starch—amylose and amylopectin—were synthesized via separate pathways. Work from the Nelson laboratory showed that the bound enzyme increased in activity as a function of the number of functional alleles at *waxy*. However, one dose was all that was necessary to get wild type levels of amylose. This ruled out the possibility that the lack of the bound enzyme was due to a lack of amylose. This hypothesis had been suggested as an alternative explanation for the function of the *waxy* gene.

Perhaps the greatest notoriety and acclaim given to Oliver came for the work done during the period 1962 to 1969. These studies, done primarily in collaboration with Ed Mertz of the Biochemistry Department, Purdue University, showed that levels of essential amino acids could be enhanced by mutation. Studies on *opaque-2, floury-2,* and other mutants in maize demonstrated that protein quality could be altered genetically. The discovery that certain amino acids, such as lysine, could be enhanced by breeding was of tremendous importance. These amino acids are required by man and animals, and they are universally sub-optional in plant proteins. These studies stimulated widespread research in genetics, breeding, and nutrition and they had worldwide implications for the improvement of food for both humans and animals.

Oliver left Purdue in 1969 for the University of Wisconsin as a result of a phone call from a colleague and close friend, R. A. Brink. Upon Brink's retirement, he called Oliver and asked if he knew of a person interested in coming to Wisconsin. Oliver surprised and delighted Brink when he expressed an interest in the Wisconsin position. At Wisconsin, Oliver and Brink were housed in adjacent offices and many an hour were spent observing and discussing ears and kernels. Fortunately, because the Brink seed room also housed Oliver's graduate students and post-doctorates, the budding young geneticists could observe and absorb many of the interesting discussions between Oliver and R. A. Brink. It should be noted that Brink developed an interest in starch defective mutants towards the end of his life. No doubt, interactions with Oliver were instrumental in Brink's last professional activities.

Upon Oliver's move to the University of Wisconsin in fall of 1969, major emphasis was placed on the development of a system whereby the influence of a transposable element on the function of the gene could be assayed at the protein level. (To the young reader, we hasten to point out that this was before the days of recombinant DNA). Studies with *shrunken-2* and then with *bronze* led to the conclusion that a structurally altered protein was being produced when the transposable element dissociation was incorporated into the locus. Although somewhat surprising to some followers of transposable elements, this finding was in accord with the earlier recombination data with *waxy* in which it was shown that the transposable elements mapped through the gene.

With the advent of recombinant DNA technology, studies of transposable elements quickly moved to the DNA level. Because of the previous studies at the protein level and because of several interesting insertions and subsequent revertants, the bronze locus was targeted for cloning in the Nelson laboratory. The cloning of *bronze*, done in collaboration with Nina Fedoroff, was of significance for a couple of reasons: (1) it represented the first case of so-called "transposon tagging" in maize, and (2) it opened up the interesting transposable element mutants of bronze for dissection at the DNA level. Subsequent studies have focused on many of the *Spm(En)* and *Ac/Ds*-derived bronze mutations and their interesting revertants.

On a second front, the Nelson laboratory continued the investigations of starch synthesis in many of the known mutants as well as the determination of the enzymic lesion associated with these mutants. Since moving to Wisconsin, the Nelson laboratory has identified the lesion associated with *shrunken-4*, *shrunken-1*, *sugary*, *brittle-1*, *dull* and at least one other, less publicized, mutant.

Oliver's interest in the development of the maize seed and the many alterations resulting from mutation is well known. On a personal level, Oliver is quite modest in spite of his many accomplishments. Rarely does he volunteer information from his laboratory, not because he is fearful of someone "stealing his thunder," but because he does not believe that the world revolves around his every finding. Oliver is a gentleman. He clearly cherishes his independence and is extremely respectful of the freedom of others. The standard practice in his laboratory is to provide as much freedom as possible to the graduate students and post-doctoral associates as they pursue their research interests. After initial visits concerning the nature of the project, people are free to pursue their work on their own basis. Pressures of publication, grantsmanship, etc., rarely, if ever, find their way to the people in his laboratory.

Oliver has received a number of prestigious awards during his career. These include the Herbert Newby McCoy Award (1967), the John Scott Medal (1967), the Hoblitzele National Award in the Agricultural Sciences (1968), and the Commemorative Medal of the Federal Land Bank System (1968). He was elected to the National Academy of Sciences (1972) and was awarded an honorary Doctor of Agriculture by Purdue University in 1973.

31 Drew Schwartz

An Era of Molecular Biology*

An appreciation by M. Freeling
and A. Pryor

Professor Drew Schwartz nurtured 18 graduate
students to Ph.D.s in Genetics at Indiana University.
If we students could merge our memories and
insights, there would be a monumental image of our
professor that Drew himself would probably
disavow. The genetics community sees Drew as, in
the words of Marcus Rhoades, "a theoretician par
excellente." Drew's many important research
publications, usually made in collaboration with one
of his graduate students, assure him a place in the
history of genetic research. We students saw another
aspect of Drew. Each of us was privileged to enter
Drew's scientific family, to experience his enthusiasm
for discovery, his urgency when causes for aberrant results were not apparent,
his intensity when others did not immediately agree with his interpretations,
and his criticism whenever faulty logic was evident. Drew carefully avoided
public criticism of patently absurd models for *Adh* gene organization, published
by a competing laboratory, for fear of giving legitimacy to the ridiculous.
However, as post-doc Jacobs recalls, these same absurdities were often the
object of Drew's lessons in illogical deduction.

Thanksgiving with Pearl and Drew, a game of dictionary, an atmosphere
of caring and philosophical discussion; one could then imagine an integrated
life where family matters and religious dedication to science might happen as
naturally as Drew carving turkey.

Before any student could get to lab in the morning, Drew had already
found the data books, idealized the data on the blackboard in his office, and sat

* *MAYDICA* 1991, Vol. 36, No. 2

waiting. He may have left the carving knife home, but it was there nevertheless. It was hard to defend our data before we had even seen it. Our experiments, however naive or mundane, seemed larger than life and well worth fighting about. These debates led, at least once, to a month of silence between Drew and a student. Our talents may be bred in the bone, but not so with our enthusiasms. Drew Schwartz gave a bit of his own fire to all of his students. We can feel it when we meet.

"Drew's talk is why I come to this meeting every year". "Absolute BS." "There is no way a position effect can travel 60 map units down a chromosome." "His data seem good to me, but..." "I'd like it better if he weren't measuring methylation patterns." "Brilliant experiments; I think Drew has discovered an entirely new mode of cis-acting regulation." So went the comments at the lunch following Drew's research presentation at the 1990 Maize Genetics Meetings held in Delaven, Wisconsin. That Drew's research should be intensely discussed, appreciated and derided, and remembered vividly is nothing new. Although Drew was 71, the subject of age never came up.

After receiving his Ph.D. at 31 with Marcus Rhoades at Columbia University, 1950, and after a year at the University of Illinois, Drew spent 11 years as a Senior Biologist at Oak Ridge National Laboratory. These were glorious years when genetics acquired a molecular basis. Drew's interests in chromosome structure and gene regulation, and his talents as a thoughtful experimentalist, fit in well with colleagues like Atwood, W. K. Baker, Doerman, Lindsley, Novitski, Sandler, Vokin, the Grells and the Russells. Bill Baker recalls:

"Back in the early 1950s the door of the *Drosophila* laboratory at Oak Ridge would open in the middle of almost every morning and in would walk Drew. You could tell by the expression on his face what was up: If there was a grin you knew that he had rolled up a big score when bowling the night before. If his countenance was serious and he had draped around his neck a couple of long strands of different colored plastic tubing, you knew you were about to learn his latest theory as to how DNA replicated within the chromosome.

"Drew is an inveterate model builder and what made our association so rich was his insistence on thinking about important cytogenetic problems and his skill in devising experiments to get evidence for or against his ideas—an imaginative mind but one bounded by experimental verification."

Even so, most of Drew's students had more than enough model building during their time at Indiana University. Freeling won't make models with moving parts, and Birchler is committed to pursuing science with no models at all. Drew's friends and colleagues at Indiana—especially Marcus Rhoades, Ellen Dempsey and Carlos Miller—have spent a significant part of their lives considering Drew's models. It is rumored that Drew is working with coach Bobby Knight on a model to fully understand and reproduce IU basketball victories. Perhaps Drew's most surprising model-building experience was when Pryor fudged some data for April Fool's Day, 1970. Of course, Drew had the data digested and modeled by the time Pryor arrived in lab. Pryor had "maliciously" added isozymes at various times during an electrophoretic run to mimic a truly mind-blowing result. Drew assembled his lab and began a logically sound explanation. A model evolved. Even though "April Fools" was announced clearly, silence prevailed until Drew finished explaining his model and retired to his office. To this day, none of us knows if Drew thought the experience funny, and we revelers felt somewhat sacrilegious. Science was important in Drew's lab.

All of us students had difficulties penetrating Drew's formal, serious style. Laughner was overheard asking "Dr. Schwartz, why do I have to call you Dr. Schwartz?" "Why don't you call me Drew, Bill"? was the reply. "Oh, thank you, Dr. Schwartz!" Drew occasionally stooped to a practical joke, like the time he suggested that Birchler's invaluable Ph.D. Qualifying Examination, which had been carefully pushed under Drew's office door, had been removed with the garbage.

Drew is fond of invoking "serendipity" as the ultimate cause of scientific success and failure, and is certainly a devotee of the analytical pursuit of experimental truth generally referred to as "the scientific method." The idea is, perhaps, that scientific success is simply the combination of good luck and good logic. Were we students to accuse Drew of actually being a mystic, with a "second sense" about how to maximize good luck, and magical feelings about which questions were really important, he would most certainly take issue.

Drew founded the field of plant molecular biology with his work on the esterases, the shrunken-1 protein (later identified as sucrose synthase) and the alcohol dehydrogenases. Most of this work was done since 1964, the year he joined the faculty of Indiana University as a Professor. There he served as mentor to 18 students who got their Ph.D.s and 3 post-doctoral students. These students have conferred the Ph.D. onto at least 20 students, and some of these students now have students of their own. For all of the differences among Drew's students, and grandstudents and great-grandstudents, the "way" that Drew conducted science is a legacy perhaps even more important than his considerable research accomplishments. The current trend among molecular biologists to collect data first and decide their meaning later had no place in the Schwartz lab. Every experiment was to test a hypothesis, no matter how elaborate.

Graduate students make unreliable friends, here today and gone tomorrow. Few of us knew very much about Drew's family. According to Marcus Rhoades, Drew was raised in a traditional Jewish family in Pennsylvania. He married Pearl, who was from New York City, and after a two-year hitch in the U.S. Army with action in the Burma-India theater, had two children, Alan and Rena. During our Indiana University days, Rena was studying in Israel, where she now lives. Alan died of a heart ailment at the age of 17. We students discussed this tragedy and wondered where Drew and Pearl found the strength to give others so much. Pearl died in 1985. Pearl's love of beauty, gracious kindness, lively intellect, liberal bent, skill at mediation, and calm control of movements—somehow portrayed in her painting—are memories we share.

We students are gone on our ways. We meet Drew from time to time, at meetings or when passing through Indiana University. Drew's recent research on *Ac* transposition, patterns of methylation, and on using *Ac* as a marker for chromosomal events affecting methylation pattern remains exciting and of much interest to the genetics community. There were times when there were so many students in Drew's lab that we worked shifts. Now there are fewer distractions, thoughts must have more time to pile up with one another. This volume of research papers—contributed by several of Drew's students, a post-doc, and colleagues in research—symbolizes our appreciation and our best thoughts for the future.

32 Peter A. Peterson

A Feeling for Maize Transposable Elements*

An appreciation by H. Saedler

As told to the writer, Peter A. Peterson was born March 17, 1925 in Bristol, Connecticut, U.S.A. Bristol was a strongly, multiethnic-oriented community. It had many parochial schools, and, in fact, there was a sizable segment of ethnically based parochial education (1920s–1930s) since this was largely a first generation immigrant community. It was a small industrial town closely surrounded by a number of larger towns in middle Connecticut.

His family moved to Portland, Connecticut, following the severe economic depression of the early 1930s. Portland was a small rural community that was situated high above the Connecticut River Valley. This move brought new opportunities and opened new aspects to his life. Here in this rural community along the Connecticut River, he worked in an ice cream dairy during his early years. This included before school, after school, and during the summers. There was time only for a small amount of vegetable and flower gardening in the backyard. His strong area in Portland High School was history, especially ancient Greek history. There he played varsity soccer and basketball.

In mid-year 1943, he was admitted, as were many others because of the war urgency, into a mid-year class into Tufts College in Medford, Massachusetts then, a small liberal arts college in the Boston Area. He undertook a biology-chemistry area of study. As for extra curricular activities, he played both soccer and lacrosse and earned varsity letters in both. This was not to last long since the war in the early 1940s made a demand on recruitment, and for Peter Peterson, along with many thousands, there was a call into the U.S. Navy. With much free time anticipated though not necessarily

* *MAYDICA* 1991, Vol. 36, No. 4

realized in Navy "after-hours," he brought along his Sinnott and Dunn's genetics text recruited from his genetics class (Professor Paul Warren) since this Columbia University originated text was the standard text of that period. This was a chance to fill in "Navy free time."

In the Navy he was assigned to the Sampson naval training base in the northern New York area at the Lake Geneva region during the mid-winter, a very cold and bone-chilling winter, especially in naval training rowing exercises on the lake. These were the days before down clothing was in general use. With the assignment of a specialty, he was sent to USN pharmacist's school. First, he went to Bainbridge, Maryland, and then on to Bethesda, Maryland, at the U.S. Naval Hospital for further training. It was at this time while in the Washington, D.C. area that he took night courses at the American University in order to fill out some requirements that would be necessary when he returned to Tufts, if World War II was to ever end. Yet, finally while at Bethesda during August 1945 the war ended and Washington D.C. was in a sea of frenzied celebration following the Japanese surrender.

Following his service in the Navy, he returned to Tufts in the fall of 1946 to finish out his degree. This was a time to find a career direction. He was helped in this by a sympathetic and kind professor, Dr. Paul Warren, a geneticist who originally did his Ph.D. at the University of Michigan on the genetics of peppers and tomatoes. It was here where he undertook an honors project and Dr. Warren guided him in looking at the chromosomes of the "Lily of the Valley" (*Convallaria majalis*). Tufts did not at the time have an extensive research capability, so the only resource was a venture into Dr. Warren's backyard along the garage wall to find flowers of the emerging plants containing the meiosis of the flower. This was the beginning of his interest in botany and also things genetic and cytological. Yet, the "D" (departure) from the cloisters of a secure college life was imminent. And, no vocational instinct arose. The period in the naval service was still an overwhelming feature of one's life and future direction was difficult to conceive. Dr. Warren, recognizing this, suggested that he go for a year to the Carnegie Laboratories at Cold Spring Harbor in Long Island, NY. This was a most fortunate choice since it opened up such new vistas that they have not found their end yet.

Arriving at Cold Spring Harbor in September 1947, he was assigned to a carcinogen project with Dr. M. Demerec in the *Drosophila* genetics lab testing the effects of various carcinogens on the induction of mutation in fruit flies. (The war had uncovered many chemicals beginning with the infamous, nitrogen mustard). This was a very busy lab and one that attracted many visitors that came to see the Carnegie labs and, especially the carcinogen lab, possibly because it was the only lab at Carnegie with a machine that ran: namely, an aerosol generating machine to spray the flies with the carcinogen. Such visitors included Vannevar Bush who was then President of Carnegie Institution, and who before this period during the war had been a scientific advisor to President Franklin D. Roosevelt. Dr. Bush was most interested in anything scientific and this was well illustrated at the end of the war in his 1945 report to the President (Science—the Endless Frontier) in recommending the founding of a new agency for research that eventually became the National Science Foundation. Every major geneticist from Europe or Japan who landed in New York and found themselves on the East Coast, came to Cold Spring Harbor and, of course, to visit along with Dr. Demerec, others at the Carnegie Lab, including Barbara McClintock, Evelyn Witkin, B. Kaufmann, McDowell, and E. Caspari. Other visitors included Dr. H. Muller, Dr. T. Dobzhansky, and others of the Columbia group including Frances Ryan, Franz Schrader, A. Pollister and many others. Professor R. A. Fisher came to visit while he was a representative to the United Nations for England at the time. Always they came to the *Drosophila* laboratory to see among others the aerosol generating machine on fruit flies.

The seminar series in Cold Spring Harbor was filled with every preeminent geneticist of the time. This was a pre-DNA period (1947–1948) but in large part preempted with the early DNA reports of A. Mirsky and Hans Riis, the future course of events.

A continuing feature in Cold Spring Harbor at the Carnegie Institution was the exchange of seminars with the Columbia University group which was then the strongest genetics group in the United States. They all came, faculty and students, to exchange scientific ideas by advanced seminars on new ventures in genetics and then enjoy an inexpensive spaghetti dinner afterwards. Some of the students included Ruth Sager, Max Levitan, Colin Pittendrigh, Monty Moses, David Perkins, Ellen Dempsey, Drew Schwartz, and Peggy Lieb. Bruce Wallace was also part of the Columbia group but he was also in charge of the carcinogen lab in Cold Spring Harbor and being such he was a frequent host to T. Dobzhansky who was his mentor, and who always came to visit him. It was also here at Carnegie that Peter met his future wife, Sally Rohrer, who had just graduated from Vassar College in Biology-Chemistry, and joined the fly laboratory. They eventually went together to Illinois to pursue their graduate studies in the Botany Department.

But, this is getting ahead of the story. Now is the time (1948) to go to graduate school, and Dr. Demerec was anxious that his assistant go into the new area of "molecular" genetics which included bacteria and bacteriophage. At this time Dr. Demerec had shifted to *Drosophila* (from maize) and was well on his way with bacteria (*Salmonella*). The problem was that Peterson was attracted to the endeavors of Barbara McClintock who was out in the maize genetics nursery among her corn plants. (This is the same field where 44 years previously, George Harrison Shull was growing his corn that led to his theories on inbreds and hybrid vigor). Not only that, she enjoyed talking about it and had a very enthusiastic attitude about all that she was doing. Since her laboratory was upstairs from Peterson's dormitory room, there were frequent talks during the early evenings with Sally and Peter about things in general and about something that then was very exciting in the corn field (note: this is spring, 1948). Unfortunately, both Sally and Peter couldn't quite grasp the extent of it though it was so engrossing as McClintock tried to explain. But, later on during that year they knew that she did write to Marcus Rhoades who was then in Paris about something in the variegated corn that was "jumping." The rest is history.

In any case, despite Dr. Demerec's entreaties on where to go and the enticement of the new approaches in gene studies with bacteria and phages, Peterson chose to go into a plant field. Here, Barbara McClintock had some suggestions. With a visit to Professor Marcus M. Rhoades at Hastings-on-the-Hudson cornfields which was then the Columbia University research plot, he had a chance to see the group working in cytogenetics. During the summer of 1948, Marcus Rhoades was invited by Professor O. Tippo who was the Botany Department's Chair to come to the University of Illinois at Urbana and develop a cytogenetics group. Both Sally and Peter followed him there along with his Columbia group that included Drew Schwartz, Ellen Dempsey, and E. Dollinger. This was the core corn cytogenetics group until John Laughnan, who was then at Princeton, joined Rhoades' group in the latter part of the fall to mark the beginning of maize cytogenetics at Illinois and the development of many students in maize cytogenetics. In fact, it was Dr. Tippo (Head of the Botany

Department) who felt it essential that the Midwestern states should have a strong corn program and in this manner brought in Marcus M. Rhoades and his group to Urbana. And because MMR was there, the Corn Corporative was moved from Cornell to Urbana.

What then, to do for a research problem in 1948? Following the Bikini Atoll atom bomb test, Dr. E. G. Anderson, who was then at the California Institute of Technology, received from the U.S. Navy in 1946, large amounts of seed that were exposed to the A-bomb radiation at the Bikini Atoll. This seed was planted in benches at Pasadena, and Dr. Anderson, along with other interests in various mutants, selected out ones that had variegated phenotypes and sent them to M. M. Rhoades. (Remember, M. M. Rhoades had recently described unstable dotted system as well as some pale green alleles inherited as duplicate loci). One of them was a pale green mutable and this was the originating source of the *En* system which was eventually cloned more than 35 years later (1985). This turned out to be Peterson's research problem and in this research he described and reported on the *En* system in 1953.

At this time in the late 1940s, the McClintock ideas on unstable systems had been very much a topic of conversation among Professor Rhoades' group. One may recall that they (Rhoades and McClintock) were close friends since Cornell days and being in the New York city area they were close neighbors. Professor Rhoades was a confidant and most knowledgeable listener of McClintock's, and she felt he was a good interpreter of her ideas. Further, they had long associations since their Cornell days in the late 1920s, and as such, maintained a close professional and social relationship. One readily sees this in Evelyn Fox Keller's McClintock biography (*Feeling for the Organism*) where this relationship is discussed.

With this in mind, the research problem began (1948). This, at a time before there was available any meaningful interpretation of genes (Hershey, Watson, and Crick were still years away). Yet Barbara McClintock was busy at this time unraveling the mysteries of elements that moved in the genome. There were striking patterns of timing and frequency changes. She attempted to interpret her developmental patterns and her heritability events and these became the basis of her presentation in her 1951 Cold Spring Harbor Symposium talk. Professor Rhoades' was one of two or three (Evelyn Witkin, for example, was another) that Barbara McClintock entrusted with her ideas or felt that her ideas could be adequately understood.

For Peterson and his fellow graduate students, the University of Illinois scene in the early 1950s was a very active one for young graduate students in genetics. The biology and biochemistry group brought in a number of outstanding geneticists to join various departments including Salvator Luria and Sol Spiegelman and their graduate students and postdocs. In addition, Dr. Gunsalus came into biochemistry and this made a very exciting genetically oriented research group, especially, with all the geneticists that the genetics group attracted to the Illinois campus for seminars and visits.

Following the granting of his Ph.D. degree (Botany-Genetics) in 1953, he ventured to California (accompanied by Sally, and two daughters, Sara and Susan) and took his first position at the Citrus Experiment Station in Riverside, California. Here he was assigned the genetics and breeding of the avocado (*Persea americana*) as well as the genetics of the pepper (*Capsicum*). These were highly botanically oriented projects, and from this, he published several papers. It was the description of the diurnal periodicity of the avocado flower that most attracted him, and he found mutants that affected both male sterility and the differences of the avocado flowering (diurnal periodicity; three papers). Also, during this period in his work with *Capsicum*, he discovered a male sterile mutant as well as the first linkage group in *Capsicum*. One mutant was a single gene control of size, which today would be considered a QTL. In the meantime he could use weekends to carry on his corn work and continue the study of the *En* System.

In the fall of 1955, Dr. Gowen at the Iowa State University Genetics Department wrote him and asked if he would be interested in coming to Ames to join the Department of Genetics. Leaving the Citrus Experiment Station and California wasn't an easy decision but he accepted. The attraction was the opportunity to return to full time maize genetics and genetics teaching. At first he was reluctant to leave California, but the opportunity to carry on a maize genetics program in a Midwestern setting (Iowa) seemed very attractive. So, the family moved to Ames in January 1956 and he began in the Department of Genetics.

The study of *En* continued, and a fortunate discovery back in 1952 was also available; namely, a variegated tassel was observed in the field, and this turned out to be the *A1* allele where *En* was found. It was shown to be controlled by the *En* system, which later in 1985, was shown to contain a full-sized *En*, and was the basis of the tagging of the *A1* locus from which was eventually isolated as a cDNA and put into the Brick Red Petunia—a controversial feature in 1990 in the Germany political system. So, the initial tassel sector found in a small group of plants in 1952 found its way from that variegation to the new petunia variety, namely, Brick-Red Petunia.

In 1960 Dr. W. Pierre in the Agronomy Department wanted a cytogeneticist. He induced Peterson to join the Agronomy Department and to continue some of his ongoing work and look into aspects of practicality of cytogenetics in the plant-breeding group, yet, still with the opportunity to maintain ties to genetics. Here in the Agronomy Department in the 1960s and 1970s the subject of mobile elements was not especially appreciated nor welcomed (note: this was the 1960s). It was unclear to his peers why one would continue such a study when so many more practical subjects were available. Thus, against these odds, Peterson continued to study mobile elements and overcame obstacles that were prevalent with regard to the continuation of such a study. Eventually and much later, along with Professor Francesco Salamini (Bergamo, Italy), they were able to demonstrate the presence of transposons in corn populations and this led to further studies which in 1990 were continuing as to the role of transposons in driving the variation seen in maize populations.

About the same time (late 1950s) the corn conference where maize geneticists convened for two-day meetings at Allerton (Illinois) was started. This was a very exciting period for maize genetics because the early meetings were small and they generated a great deal of debate. One can envisage 12–15 geneticists around a table in a long library room at Allerton House in Monticello, Illinois. Dr. Brink had unraveled mutable pericarp genetics and was in the midst (in 1956) of his paramutation study. Out of the numerous debates (since there was very close eye contact around this table) the question was raised about the control of patterning in variegation that R. A. Brink had just described at the *P-vv* locus—whether it was "position" that determined it or the qualitative difference in the elements themselves. (Of course at that time elements were not well conceived and were largely whatever one wanted to consider them). However, a consequence of this was that Peterson, in 1960, began targeting the *C1* and the *A2* loci to see if a pattern controlled by *En* that was known and identified at the originating site could then be transferred to the target site (*A2* or *C1*) in a changed (position) or unchanged (qualitative). Of course, the consequences of this was that many *C1* and *A2* mutants were found with inserts and these later became of interest in tagging the *C1* locus as well as describing the *C1* protein. A consequence of the availability of *En* in *C1* was that Javier Paz-Ares could readily clone the *C1* locus in 1986 because of the availability of the *En* tag. Later, with the available *A2* alleles, the *A2* allele was tagged with an *rcy* element because of an *a2m* mutant that arose in 1962.

In 1964, Peterson was invited to teach a course in Cytogenetics at Stanford. This was an interesting class. In this were some of Yanofsky's and Perkins' students, including Tom Creighton and Noreen Murray (a post-doc in the Perkins' lab). This period at Stanford was a pivotal move since for the first time his corn material, presented in seminars, was challenged by the molecular genetics group at Stanford. Actually, the whole concept of current studies in maize variegation was novel to this group. It was such that the molecular answers that were so readily available to Stanford geneticists (C. Yanofsky, D. Kaiser, for example) were not readily forthcoming in this maize material (note: this was 1964).

As a consequence, Dale Kaiser suggested to Peterson in 1964 that it was advisable to go to some place with bacteriophage studies where there would be a dialogue that would generate some meaningful experiments on these mobile elements in maize (neither *Drosophila P* elements nor bacterial *IS* were known then). He chose to go to the Karolinska Institute in Stockholm to work with the P2 phage in 1968 with Dr. Joe Bertani. With Joe Bertani he explored the ideas coming out of bacteriophage studies where phage inserted into genes and was especially evident with *Mu* phage.

Was there a commonality between the inserts in bacteriophage as those in lysogenic bacteria and those by the elements in corn? It was Peterson's idea that these inserts were foreign to the gene itself but not necessarily foreign to the genome, and they were not part of the regulatory apparatus of the gene but artifactual inserts. Thus, they were more like bacteriophage inserting into genes in bacteria, such as in *E. coli*, for instance. From this dialogue, he published a paper relating maize elements to lysogenic bacteria.

Though his experiments to move the phage P_2 into some targeted genes were not successful, this stay in Stockholm was a very meaningful sojourn since the Karolinska Institute was a "must" visit for many visiting scientists, and many advancing new ideas could permeate through the community. During this stay in Sweden, he was invited to a number of laboratories in Germany, including Cologne, to present seminars. Especially meaningful were the IS2 studies of the Cologne Genetics Department (Starlinger, Saedler and Jordan).

Following the catastrophic infection of field corn of the Texas cytoplasm containing phenotypes by the spread of *Helminthosporium maydis* in early 1970, it was imperative at the Iowa State Experiment Station to look into the problem. Research teams were quickly developed. So, for a brief period, Peterson embarked on the study of mitochondria and their effect by the toxin at ISU. To learn more about mitochondria, he went to Vienna at the Department of Biochemistry with Professor Hans Tuppy who was well known for his studies with mitochondria. There he learned some aspects of mitochondrial handling and manipulation. From Vienna, he went on to join Dr. Richard Flavell at the Plant Breeding Institute in Cambridge (now at Norwich) where an assay was developed to detect toxin-induced effects on mitochondria. In all these endeavors he was enthused to meet so many people in so many different endeavors and to become acquainted with new technologies.

But, this was taking him away from the cornfield and he returned to the study of transposons. This was accentuated by the discovery of the IS elements in bacteria in 1967–1968 which in itself for the first time authenticated the maize mobile elements and gave them some credibility to the biological world. Following a meeting in Leeds (England) where many different geneticists were assembled by Professor Fincham in the fall of 1973, a commonality of this mobility among a whole series of plant types was explored. There he met and discussed with Professors Starlinger and myself the nature of *IS* elements and maize mobile elements. Furthermore, a number of prokaryotic geneticists including Starlinger and myself were encouraged to turn to maize genetics as a further opportunity to explore the mobile units.

Only three years later, an opportunity with a German DAAD Fellowship made it possible to join my group at Freiburg for a brief study of *IS* elements. There Peterson undertook a study of *IS* elements with the Freiburg group. This turned out to be a longer stay and continued for two years with some brief interruptions back to the maize genetics nursery summers in Ames.

The question was whether he should embark on molecular studies with bacteria or stay in corn. This was resolved by my invitation to move to Köln where a genetic component was needed to supplement an ever-growing plant molecular capability that was developing at the Max-Planck Institüt. In any case, the technology (molecular) had grown quite rapidly and such a union was not only desirable, but absolutely necessary.

With a goal to get into plants, our laboratory undertook a study of maize transposons and Peterson joined forces with us in a collaboration that continues today.

The first and most urgent goal was the molecular isolation of *En*. *En* was known genetically and it had many attractive attributes with the capability to "turn on and turn off" genes. An unlimited number of alleles were present with inserts of *En*. But how to get into these alleles with no known product such as the *A1* locus or the *C1* locus, or the *A2* locus and isolate *En*. Available was the chalcone synthase probe from parsley that was isolated by Professor Klaus Hahlbrock which could be used as a heterologous probe.

Could *En* be isolated from such a probe? The only options in 1980 were the chalcone synthase cDNA which was isolated from parsley. The connection was made that the chalcone synthase was related to the then understood flavonoid synthase which was controlled by the *C2* locus. And at the *C2* locus there was an insert in the form of the *c2-ml* allele. Following two years of frustrated attempts, no success was possible using this as a heterologous probe. The break came with the availability of the *wx* probe by Zsusanna Schwarz-Sommer. The *wx* probe made the *Wx* locus a necessary target for *En*. Finally the insertion of *En* was identified in the *wx884* allele. It was then possible to isolate *En* by Peterson's student, Andy Pereira, and with the availability of *En* as a tag, it quickly became apparent that *C1*, *C2*, and *A1* could readily be isolated in a few months. Thus, several genes now were available for study and occupied the laboratory for the next several years.

Peterson continues to do research on transposons, and is currently continuing the pursuit of various aspects of transposons. He is also involved in the tagging of important genes in controlling disease resistance in corn as well as developmental mutants. He continues his teaching duties with his course *"Cytogenetics in Relationship to Plant Breeding"* and has had approximately 10 graduate students during each of the last 10 years. And to sustain his momentum in the field, he is known to maintain a year-round fitness regimen.

In the now more than ten years of our collaboration and more than thirteen years of association, Dr. Peterson has neither become tired of maize nor of us. My warmest feelings for him personally are intermingled with the admiration I have for his research effort. If I had to summarize this life full of scientific achievements the following contributions come to mind:

The discovery of the *En* transposable element system in 1953 was followed by its thorough analysis for over almost four decades. Here two items seemed to be prominent: states of activity of the element and position effects. In any case, these studies revealed that a receptor element (now we know that these are mostly deletion derivatives of autonomous regulatory elements) reacts to functions emitted by the regulatory element. Dr. Peterson recognized and used this operational test and thus discovered other very diverse transposable element systems, like Cuna (*Cu*), Cycler (*Cy*) and Ubiquitous (*Uq*), one of which (*Cy*) has been shown molecularly to be related to the mutator family of elements.

Dr. Peterson's notion of a correlation between performance of maize populations and the activity of transposable elements might reveal the true biological functions of these elements, i.e. the contribution of the plants' flexible response to changing endogenous and exogenous environments. This observation on the significance of mobile elements in natural maize populations hence might also open a new avenue of investigation for molecular biologists.

33 Arnel R. Hallauer

Quantitative Genetics Integrated into Corn Breeding*

An appreciation by G.F. Sprague[1] and K.R. Lamkey[2]

This volume is dedicated to Arnel R. Hallauer. For the last three decades, he has made significant contributions to quantitative genetics with particular reference to corn breeding. His research effort has focused on the evaluation of quantitative genetic theory, and the evaluation and utilization of recurrent selection for population improvement and the efficient use in corn hybrids

Arnel R. Hallauer is a Midwesterner. He was born in Kansas on May 4, 1932 at Netawaka. His interest in corn research began when he was 14, in the fall of 1946, when he had a part-time job with Dr. Lloyd A. Tatum harvesting experimental corn trials in Northeast Kansas. He continued working part-time with Dr. Tatum while attending high school (1946–1950), planting experimental corn trials in the spring and harvesting them in the fall. During the summer months of his high school years, he detasseled corn for the production of double-cross hybrid seed for Mr. Carl Overly of the Kansas Crop Improvement Association. These associations with Dr. Tatum and Mr. Overly led him to Kansas State University where he received his B.S. degree with honors in 1954 with a major in Plant Science. During his undergraduate studies, he was continuously employed part-time on an hourly basis with the cooperative Federal-State corn breeding project at Kansas State University directed by Dr. Tatum. During this time in his undergraduate studies, he experienced all aspects of the breeding project including preparing seed, planting,

* *MAYDICA* 1992, Vol. 37, No. 1
[1] University of Illinois, Department of Agronomy, Urbana, IL 61801, U.S.A.
[2] Field Crops Research Unit, USDA, Agriculture Research Service, Department of Agronomy, Iowa State University, Ames, IA 50011, U.S.A.

data collection, harvesting, pollinating in the breeding nurseries, and analyzing and summarizing data of experimental trials.

After finishing his undergraduate degree, followed by two years of military service (1954–1956), Dr. Tatum, impressed with his young assistant's dedication and work ethic, urged Arnel to attend graduate school for an advanced degree in plant breeding. After considering offers from several universities, Iowa State University became his choice. This offer was chosen after consultation with Dr. Tatum because the offer included working with Dr. George F. Sprague, a plant breeder who was highly regarded for his research program. Dr. Hallauer completed the requirements for the M.S. degree in plant breeding in 1958. He accepted a position as a research agronomist with the U.S. Department of Agriculture, Agriculture Research Service (USDA, ARS) in the summer of 1958. The major responsibilities of this position were conducting host-plant resistance studies with Mr. Ferd Dicke at the USDA, ARS corn insects laboratory in Ankeny, Iowa. While working for the USDA, ARS, he continued on in graduate school and received his Ph.D. in 1960. Dr. W. A. Russell served as his major advisor for the Ph.D. degree after Dr. Sprague was transferred to Beltsville, Maryland in 1958 to become USDA, ARS Investigations Leader for corn and sorghum research.

With the completion of his Ph.D. degree he was transferred to North Carolina State College as a USDA, ARS Post-Doctoral Research Geneticist to spend a year with Professor C. Clark Cockerham. In 1962, he was transferred back to Ames as a USDA, ARS Research Geneticist, where he continued in this capacity for the next 27 years. In December of 1989, Dr. Hallauer retired from the USDA, ARS and accepted a position with Iowa State University as Professor of Plant Breeding. In 1991, he was promoted to C. F. Curtiss Distinguished Professor in Agriculture. The research interests throughout his career have been in the extension and evaluation of quantitative genetic theory as it relates to corn breeding. Before detailing his accomplishments, it is necessary to put the development of corn breeding and quantitative genetic theory in historical perspective.

Quantitative Genetics: A Historical Perspective

With the beginning of the modern era of corn breeding (1920–1950), hybrids rapidly supplanted the previously grown open-pollinated varieties. In the major corn producing states, this transition was completed in about 10 years with an accompanying substantial increase in per acre yields. Following this initial success, corn breeders continued their sampling of open-pollinated varieties but to an increasing extent devoted their efforts to "second-cycle" breeding; the development of new lines from the best parents of the then current hybrids. When such lines began to reach the evaluation stage, the results were generally disappointing. The newer lines tended to be higher yielding than their original parents, but the hybrids among them were not markedly superior to those first released (Hallauer, 1990).

During this period (1920–1950), some corn breeders interpreted this experience as indicating that the big gains had been achieved and that further progress would be limited and difficult to achieve. Others interpreted the results as indicating a conceptual limitation of the methods in use and directed their efforts toward the development and evaluation of new concepts and procedures. Such was the beginning of quantitative genetics as applied to corn breeding.

Era: Quantitative Genetics Integrated into Corn Breeding

The quantitative genetics approach was not universally accepted. Reservations as to the utility of the method were due, in large part, to the perception that the mathematical restrictions of the model might limit its biological utility. The assumptions of the model causing most concern were: no mutation, no linkage, and no epistasis. These restrictions were imposed to simplify algebraic manipulations rather than the belief that such phenomena either did not exist or were of limited importance. Classical genetics had devoted considerable effort in detailed studies of precisely these topics.

New mutations would be expected to have only minor effects in the development of stable lines but could be of importance in long term selection experiments or synthetic populations developed by the breeders. The effects of linkage bias would be minimized if quantitative traits were conditioned by large numbers of genes with random genomic distribution. Where linkage equilibrium can be assumed, linkage bias would be minimized, but unfortunately for many populations used, F_2s, backcrosses, etc., linkage bias would be at a maximum. The failure to accommodate epistasis also could introduce a bias. Epistasis had been well-established in classical genetic studies. Presumably it might be equally important for quantitative traits though this had yet to be established.

With the limitations arising from these and other sources the question remained: did the mathematical models encompass sufficient detail to provide a useful picture of the biological complexities involved? More specifically, would the models provide estimates of population parameters which would be useful for predicting response to selection or for comparing relative efficiency of alternative breeding systems? Answers to such questions could come only from detailed experiments. This was the situation when Dr. Hallauer entered the field, and his career has been devoted to seeking answers to these and related questions.

Obviously the relative importance of different types of gene action (additive, dominance, epistatic) involved in the inheritance of quantitative traits was of prime importance. Dr. Hallauer has contributed greatly to understanding gene action in corn through a variety of experimental approaches, some of which will be discussed briefly in the sections which follow.

Hallauer's Contributions

He conducted an extensive study relating mean performance for a series of quantitative traits to decrease in heterozygosity achieved under three systems of inbreeding: self-fertilization, half- and full-sib matings. The differences in rates of approach to homozygosity were found to be of limited importance. For most of the traits studied, the linear regression relating mean performance and level of inbreeding accounted for 98–99% of the variation. This would be the expectation with independent loci and any degree of dominance. Departures from linearity would be evidence for epistasis. The quadratic component for the trait, yield, was highly significant but accounted for only 0.5% of the total variation for yield (Hallauer and Sears, 1973; Good and Hallauer, 1977).

In quantitative genetics, epistasis is measured by the nonadditivity of effects among loci. This nonlinearity could arise from additive x additive, additive x dominance or dominant x dominant interactions. Several methods have been used for the estimation of epistasis.

Using a series of random lines derived from the corn population Iowa Stiff Stalk Synthetic (BSSS), Dr. Hallauer conducted both North Carolina Design I and Design II studies to obtain variances for the additive (A) and dominance (D) components and the three epistatic components (AA), (AD), and (DD). The additive component accounted for 93% of the total genetic variance for yield. When dominance was included, the two accounted for 99%. Thus epistasis, as measured, was a minor contributor to the genetic variance for yield (Silva and Hallauer, 1975).

On the possibility that epistasis might vary with the level of performance of the parental lines, four groups of lines were chosen: first cycle, second cycle, good, and poor. Within each set, all possible F_1 hybrids were obtained and through selfing and backcrossing heterozygosity levels of 0, 25, 50, and 100% were established within each set. Evidence for epistasis was examined by using departures from linearity in the regression of performance on heterozygosity. Evidence for epistasis was greatest among the "poor" set but was relatively unimportant for most traits measured within each group. Thus evidence for epistasis exists but this type of gene action appears to be relatively unimportant with the methods of estimation currently available (Martin and Hallauer, 1976).

The quantitative genetic approach has proven very useful in providing theoretical expectations for improvement from selection under alternative breeding systems. Corn breeders throughout the world have benefited from Dr. Hallauer's book. Here, he has performed a great service in preparing a detailed survey of his own studies and the extensive work of others as it relates to those expectations in his excellent book that he co-authored with J. B. Miranda, Filho entitled *Quantitative Genetics and Maize Breeding*. The breeding systems covered range from mass selection, selection within and among inbred families through the various modifications of recurrent selection. All of these systems are cyclical with the primary distinctions being the type of progenies evaluated and the time required to complete a cycle of selection.

Recurrent selection studies, with selection pressure primarily applied for yield, have been conducted with populations of differing genetic backgrounds; synthetics, composites and open-pollinated varieties. Response to selection is a function of allelic frequency and genetic variability as well as the selection pressure applied. Because there is no easy method for the estimation of allelic frequencies conditioning a quantitative trait, evaluation of response to selection has relied on changes in means and other population parameters over cycles of selection.

Estimates of additive (s^2_A) and dominance (s^2_D) variance provided a measure of the contribution of these types of gene action to genetic variability; these types of gene action being those most responsive to selection. Results from a number of studies, for all traits measured, indicated that a s^2_A was consistently greater than s^2D, which was often zero or negative.

With these general findings as a background, genetic expectations were developed for a series of alternative breeding systems. These differ in cycle time and the precision of estimates may be influenced by replication over locations and years. Dr. Hallauer emphasized that each method might have relevance in a breeding program depending on the objective to be achieved.

Dr. Hallauer is well known for his ability to investigate, interpret, and relate quantitative genetics to efficient development of germplasm populations, inbred lines, and hybrids of corn. He developed the reciprocal full-sib selection (FR) method to combine recurrent selection procedures for population improvement with inbred and hybrid development. Experiments have been designed to evaluate the actual genetic improvement that could be realized by this procedure. He initiated FR selection in 1963 in two prolific corn populations, BS10 and BS11. Eight cycles of selection have been completed and evaluated for direct [BS10(FR)Cn x BS11(FR)Cn] and indirect [BS10(FR)Cn and BS11(FR)Cn per se] response. The eight cycles of selection were effective for increasing grain yield of the population cross (6.5% per cycle), BS10 (2.9% per cycle) and BS11 (1.6% per cycle). These experiments have demonstrated that FR selection can be used effectively to improve populations and develop improved inbred lines simultaneously. The results from these experiments, along with data from other experiments that he has conducted, have also shown that inbred lines can serve effectively as testers in the development of corn hybrids. Most corn breeders in both the public and private sector have adopted the use of inbred testers in their breeding programs.

The Iowa corn breeding program has been characterized by one unique feature, quantitative genetic theory has been evaluated at two levels; by changes in population parameters as they are modified by the recurrent selection system imposed and by an independent evaluation of their applied utility. The lines chosen to form each new cycle are routinely evaluated for their possible commercial utility. The well-known and highly utilized lines B14, B37, B73, and B84 were identified under this system. In a survey of the parentage of seed available for 1980 plantings, these four lines were involved in 19% of the total US seed supply. Publicly released recovered strains of these lines accounted for another 15%. The total use, 34%, may be an underestimate of the total usage of these germplasm sources because proprietary modified versions of these lines would not have been included in the survey. Impressive as these survey results are, they represent usage for only a single year, whereas, these lines developed out of the cooperative Federal-State Iowa corn breeding program have been in use for a number of years. Thus, the cumulative contribution of these lines to U.S. and world maize production has been tremendous.

Dr. Hallauer's primary contributions to agriculture have occurred, directly and indirectly, because of his research in corn breeding. He has had a leading or supportive role in the development and evaluation of more than 30 corn synthetics and 18 inbred lines that have been released to the seed industry by the Iowa Agricultural Experiment Station during his years as leader of the Federal-State corn breeding research project. A survey sponsored by the American Seed Trade Association in 1980 showed that inbred B73 was used more extensively than any other inbred line in the production of hybrids in the U.S. B73 is also used widely as a parental line for hybrids in several European countries and Asia. The corn synthetics and inbred lines are also used by private company breeders in the development of new inbred lines. Corn breeders in the private sector frequently seek his counsel relative to breeding methods and the application of quantitative genetic theories to breeding procedures. Several of his graduate students are now employees in the seed industry, where they are engaged in the development of improved corn hybrids for agriculture in Iowa and the U.S.

Recurrent selection has been used effectively by Dr. Hallauer as a procedure for integrating exotic germplasm into existing breeding programs. He showed that recurrent phenotypic selection for maturity and plant type was effective for adapting foreign germplasm to U.S. Corn Belt environments. He has successfully adapted Latin American cultivars Eto Composite, Tuxpeno, and Antigua, and is currently working with Suwan 1 from southeast Asia, Mexican Dent, and Cateto x Caribbean Flint. These populations are used extensively in the tropics. The improved germplasm has good yield potential and a greater frequency of individuals with disease and insect resistance compared with populations used for the development of current commercial hybrids. His concern with genetic vulnerability and interest in the use of exotic germplasm has led to the development of a germplasm research position in the USDA, ARS corn research project at Iowa State University.

In addition to his exemplary research career, Dr. Hallauer has also been an outstanding educator. He has devoted much attention to the training of graduate students who have come from the U.S. and 16 foreign countries. He has been major advisor to 47 students (16 M.S. and 31 Ph.D.) and 12 postdoctoral or visiting scientists. Currently he serves as major advisor for 13 students (2 M.S. and 11 Ph.D.). Additionally, he has served on the graduate study committee for over 95 students in plant breeding and related sciences. Many foreign scientists have spent varying periods of time with him to observe and discuss his research studies in maize breeding. He currently teaches one graduate level course in plant breeding.

Dr. Hallauer has received numerous honors and awards in recognition of his research contributions. Among these are: Fellow (American Society of Agronomy, 1979); Fellow (Crop Science of America Society, 1985); Crop Science Award (Crop Science of America Society, 1981); Agronomic Achievement Award–Crops (Agronomy Society of America, 1989); Applied Research and Extension Award (Iowa State University, 1981); Distinguished Fellow Award (Iowa Academy of Sciences, 1985); Northrup King Recognition Award for Research in Corn Breeding (1985); Breeding and Genetics Award (National Council of Commercial Plant Breeders, 1984); Scientist of the Year Award (U.S. Department of Agriculture, 1985); Iowa Governor's Science Application Medal (1990); member of the National Academy of Sciences (1989); and DeKalb Crop Science Distinguished Career Award (Crop Science Society of America, 1990).

Due, in large measure, to Dr. Hallauer's research activities, plant breeding is now firmly based on quantitative genetic principles, and given a knowledge of population parameters prediction can be made as to the relative effectiveness of alternative breeding systems. The impact of his research on corn breeding and production has been enormous. We find it both a pleasure and an honor to participate in honoring Dr. Arnel R. Hallauer for his brilliant and innovative contributions to corn breeding. The assemblage of papers in this volume is a testimony to this.

34 Earl B. Patterson

Friend, Colleague, Scholar—A Tribute: Four Decades of Dedicated Service to the Maize Community*

By J. Laughnan[1]
with contributions from J. Day,[2]
S. Gabay-Laughnan,[3] M. Sachs,[4] D.B. Walden[5]

In dedicating this issue of MAYDICA to Dr. Earl Patterson we honor him for his singular contributions to genetics that have enriched the entire maize genetics community. His name is synonymous with the Maize Genetics Cooperation Stock Center whose current thriving status is attributable, in large measure, to his unstinting effort in its behalf. I have been a close and admiring colleague of Earl Patterson for over 40 years, and a friend indeed. The following tribute will certainly err by abbreviation, understatement, and even exclusion of pertinent information, but we hope not in accuracy.

Earl Patterson was born on a farm in southeastern Nebraska near the town of Reynolds, on July 21, 1923, the youngest of nine unusually gifted children in a closely knit family of four girls and five boys. This was a resourceful, industrious, education-oriented family with carefully assigned and observed responsibilities for all. There were occasional disputes

* *MAYDICA* 1993, Vol. 38, No. 2
[1] Department of Plant Biology and Agronomy, Emeritus, University of Illinois, Urbana, IL 61801
[2] Departments of Plant Biology and Agronomy, University of Illinois, Urbana, IL 61801
[3] Department of Plant Biology, University of Illinois, Urbana, IL 61801
[4] USDA-Agricultural Research Service and Department of Agronomy, University of Illinois, Urbana, IL 61801
[5] Department of Plant Sciences, University of Western Ontario, London N6A-5B7 Canada

to be settled; Earl's father, from whom Earl must have drawn heavily for his highly developed sense of humor, once remarked that "he never had to strike any of his children except in self defense." Earl has the distinction of having had his brother Kenneth and his brother Fred, both only a few years his senior, as his seventh and eighth grade teachers, respectively, in country school. He attended the University of Nebraska where, after serving three years in the U.S. armed services during WWII, he received his B.S. degree in 1947. Dr. Frank Keim, long-time head of the Department of Agronomy at the University of Nebraska, and a genetics teacher who was familiar with Earl's interest in the subject, and with his excellent qualifications, encouraged him to pursue advanced studies with Dr. E. G. Anderson, himself of Nebraska origin, at the California Institute of Technology in Pasadena. Upon Dr. Keim's recommendation, Earl's application was accepted and his graduate years were spent in the Biology Division at Cal Tech with Dr. Anderson as his mentor. He received his Ph.D. degree in genetics at that institution in 1952, and stayed at Cal Tech for another year as a post-doctorate fellow. It is worth digressing to note that among E. G. Anderson's colleagues in the Biology Division at this time were George Beadle, head of the division, Max Delbruck, Sterling Emerson, Norman Horowitz, E. B. Lewis, Ray Owen, and A. H. Sturtevant. More of that later.

In 1953 Earl accepted a position in the Departments of Botany and Agronomy at the University of Illinois in Urbana, where he was responsible for the Maize Genetics Cooperation Stock Center which had just been moved from Cornell University to Urbana. Two years later, in 1955, he became project leader of that program in the Department of Agronomy. In 1966 Bob Lambert assumed responsibility for Stock Center activities and served for 16 years in that capacity until 1983 when Gil Fletcher was chosen to succeed him. When the position again became vacant in 1986, Larry Schrader, then Head of the Agronomy Department at Illinois, persuaded Earl Patterson to resume management of the Stock Center. It was to the great benefit of all maize researchers that Earl returned to that position at a time when future support and direction of the center were uncertain.

This year is the 40th anniversary of the Maize Genetics Cooperation Stock Center at the University of Illinois, and the 61st since its founding at Cornell University by Dr. R. A. Emerson and his colleagues in 1932. It also marks 20 years of outstanding, dedicated service to that organization by Earl Patterson, who has described the chief purpose of the Stock Center as "the maintenance and distribution of genetic stocks of maize." One might conclude from this concise definition that the activities of the center are routine since it gives no indication of the special insight, careful planning and many, many hours of work that have been involved. The word "maintenance" is especially misleading since during the earlier years of the center's existence a great deal of work had to be devoted to "finding and locating genes, chromosome mapping and developing new genetic tester combinations. The free interchange of stocks and information which resulted contributed in great measure to the rapid progress in maize genetics which was achieved." Moreover, earlier maintenance of the maize genetic stocks at Cornell led to selection of strains that were adapted to the short growing season at Ithaca but only poorly suited to culture in the Corn Belt and most other corn growing regions in this country and elsewhere. As a result, Earl Patterson's first task in his new position at Illinois was to commence the conversion of these many genetic stocks to inbred and hybrid backgrounds that were better adapted to most corn growing regions.

As for "distribution" of seed stocks, Earl has always given that procedure very special attention. On each request for seed he brings to bear his encyclopedic knowledge of maize genetics lore. A request for seeds often results in the shipment of more packets than requested because of Earl's uncanny ability to anticipate needs and problems associated with growing and handling the items requested. A particular stock needed as a tester might not have been requested, or, why not try this particular combination of markers because they have this or that advantage? I can imagine Earl writing "If you grow this tassel-seed stock you will improve the chances of a satisfactory seed set if you remove the tassel before or after pollination..." or, "You probably know that you don't have to wait to see the tassel on this vestigial-glume strain because these plants are also liguleless and you can score for that in seedlings." All manner of useful suggestions will likely be found in the letters that accompany the packets of seeds requested. There is no doubt that a collection of letters that Earl has sent in response to seed requests over the years would itself be a valuable resource for maize geneticists.

While the Maize Genetics Cooperation Stock Center is today well supported and a thriving organization, it was not always so. In its earlier years at Illinois funds for its operation were uncertain and often meager. Moreover, the Stock Center, over a period of years, did not enjoy a very high priority among programs in the Department of Agronomy. Its current high-priority status in the department is due to Larry Schrader, who recognized its importance not only for maize geneticists and breeders, but for agriculture in general, and to Gary Heichel, the present Head of the Agronomy Department who, recognizing the significance of this program within and beyond Illinois borders, has made a strong commitment to its support. Financial support of the Stock Center is also now secure. With an improved internal status for the Stock Center in recent years has come increased support from the Agricultural Research Service of the United States Department of Agriculture, and recently this agency assumed responsibility for operations and funding of the program. Marty Sachs has been appointed as the new director of the center and a curator is soon to be appointed. To Earl Patterson, whose labors, and sometime frustrations, have been so closely associated with the development of the Stock Center, the strong position that has recently been achieved for it should be a source of great satisfaction and pride.

I first met Earl Patterson in the summer of 1948 when, at the kind invitation of Dr. E. G. Anderson (Andy, as I came to know him later), I had the unusual opportunity to join my summer corn nursery with his and his students' plantings at Arcadia, California. It was a most exciting summer for me; I marveled later at the wonderful opportunity afforded Earl Patterson and his then student colleague Don Robertson, to pursue (enjoy) graduate studies in such an environment. Not only were the faculty in the Division of Biology at Cal Tech accessible through a seminar program that summer but there were others who were invited to have their nurseries at Arcadia. Edgar Anderson, with whom I roomed that summer on Huntington Boulevard at the foot of Mt. Wilson, was there too, growing and tending a variety of Asiatic and other exotic strains of maize. Herschel Roman, then located at the University of Washington in Seattle, was also a part of the Arcadia group that summer; by far the most spectacular part of his nursery, or any one else's for that matter, were the plants growing upside down just outside the laboratory. Most maize geneticists will remember that Herschel did the pioneer studies on B-A translocations, and these plants, along with their right-side-up controls, were part of his study to determine the effects of gravity, if any, on the phenomenon of preferential fertilization by the differing sperm cell products of nondisjunction in the microspore. The outcome of these experiments will not be related here on grounds that it might detract from the romance of the situation. Alexander Brink visited the Arcadia corn nursery that summer; he was just commencing his revisitation of the research on the *P* locus by R. A. Emerson and his colleagues, including E. G. Anderson himself, and we all know how that turned out. Dr. A. E. Longley, the other partner in the maize translocation research enterprise at Cal Tech was at Arcadia that summer too. One could climb the steps to the second floor of a nearby frame house and find him at the microscope, pipe lying idle nearby, working his magic on the seemingly endless samples of sporocytes of maize strains that carried chromosomal aberrations induced by radiation at the Bikini and Eniwetok bomb test sites. Dr. Longley was responsible for the assignment of chromosomal breakpoints in over a thousand such strains. Many maize geneticists have used, and continue to use, these well-defined aberrational strains as both tools and objects of their research.

For a maize geneticist the corn nursery at Arcadia was unique. There were no irrigation pipes to carry, no overhead sprinklers to contend with. The slope of the San Fernando Valley is so slight and so constant that water delivered at one end of the field would flow gradually in trenches alongside each row of corn plants. In some respects it was like a Garden of Eden; there were fruit trees—orange, peach, apricot, even a fig—growing in this nursery, but it was not forbidden to partake of this fruit between pollinating missions. The one unforgivable sin was to whistle in the garden; Earl had forewarned me about Andy's sensitivity to high-pitched sounds. I must confess that, on one occasion, I fell out of grace in this respect, realizing my transgression only after I had encountered the silent rebuke of Andy's stern gaze upon me.

The smog that plagued inhabitants of the valley that summer was a boon for the maize group at the Arcadia farm. It was so heavy, the light so meager, that anthers exerted on plants in the morning did not dehisce until 10:30 or 11:00 a.m., which afforded the opportunity for seminar sessions in the morning before field work. Earl's major professor was the chief presenter at these sessions, but we all took our turns. Not all was science at these gatherings, as we also heard stories from Andy about the period he spent in the *Drosophila* (fly) lab at Columbia University where T. H. Morgan presided, H. J. Muller visited, and two promising young graduate students named Bridges and Sturtevant trained. Most of these stories have never been recorded in print, and I doubt that they ever will, or should be.

Earl and I have on many occasions reminisced about those days, and I know that the nostalgia that summer remains with Earl as it does with me. I remember thinking at a later time that the move from his graduate student environment to the land of corn and beans in Illinois must have been quite a shock.

In 1958 Ed Coe, Jerry Neuffer, Earl Patterson and I talked about the possibility that we might have an annual informal get-together of maize geneticists and their graduate students. The detail of that discussion I hardly remember but it was surely a prophetic one for maize researchers. The first meeting was in January, 1959, and took place at Allerton Park, a part of a farm facility owned by the University of Illinois and located just outside of Monticello, Illinois. There were about twelve of us at that first meeting, so few that it could be held in the quite small Oak Room in Allerton Park House. These maize meetings as they came to be called were delightfully informal and grew in numbers of participants over the years. They were presided over by Earl Patterson. He made all the arrangements for use of the facility and dates of the meetings each year. He sent out notices of meetings to potential participants and arranged for ground transportation from Willard airport in Champaign, and from the university campus in Urbana, to Allerton House. There was no prearranged program of speakers; participants would arrive on Friday evening and at that time or early the next morning Earl would talk with people interested in sharing their research experiences and in that way developed a program for the get-together. There was some attempt to assign speakers to categories such as biochemical genetics, cytogenetics, evolution or origin of maize or just plain genetics. It really didn't matter that much. Presentations dealing with likely maize relatives were allowed, and even an occasional talk about fruit fly genetics slipped by. At first there was no need for a microphone, even for the most soft-spoken among us, but as the meetings grew in size we moved to amplification. Earl introduced the speakers, adjusted the microphone, operated the overhead, arranged for the right kind of soft chalk and erased the blackboard, all with a special finesse that earned for him the position of permanent chair of all sessions. In addition to all these things Earl presided over the gene mapping sessions usually held on Saturday evenings after the "tea" party and dinner. As the meetings grew in size it was recognized that some modest level of organization was needed and Earl's suggestion that a steering committee for the annual meetings, and an advisory committee for the Stock Center, be established was approved by the maize group. Today these committees continue to serve an important function in the Maize Genetics Cooperation. I should mention here that the Allerton maize meetings served as a model for the later establishment of similar meetings by Drosophila, Neurospora, yeast and other researchers, and for the Illinois Corn Breeders School that is sponsored each year by the Department of Agronomy at the University of Illinois.

After 25 years the maize meetings grew to such a size that Allerton House could no longer accommodate them and so, regretfully we were obliged to move the meetings from this treasured site.

In March of 1993, the 35th annual meeting of maize geneticists, now called the Maize Genetics Conference, was held at the Pheasant Run convention center in St. Charles, Illinois, with over 300 teachers and researchers in attendance. Younger members of the maize genetics group are probably not acquainted with Earl Patterson nor aware of the reverence in which the Allerton meetings are still held by their predecessors. But they should know that it is Earl Patterson who established the original format for these meetings and successfully propagated the informal atmosphere that is still recognizable in our present-day meetings, in spite of their size.

Earl Patterson's matriculation to emeritus status [the summer of 1993] signals retirement in name only. We know he will continue his active role in the maize stock center and in some of the research he has been obliged to put off because of that commitment. In noting Earl's indispensability to the maize stock center I am put in mind of the milk company in Urbana that a long time ago made the transition from door-to-door home delivery by horse and wagon, to a motorized vehicle for this purpose. The company found it necessary to purchase a truck large enough to carry the horse, as well as the milk, so that the driver would know where to make the stops. If the reader doesn't believe this story, how about the following comments, in the same vein, made by Marty Sachs, the newly appointed director of the maize stock center?

"It appears to have been 'predetermined' that Earl Patterson and I collaborate on running the Maize Genetics Cooperation Stock Center. Among the signs are that Earl and I share a birthday (July 21) and I was born in the year that Earl first became involved with running the Stock Center (1953). I really appreciate Earl's help, advice and patience since I moved to Illinois and assumed my position as Director of the Maize Genetics Cooperation Stock

Center. I am enjoying our collaboration and look forward to working with Earl for many years to come. I am confident that, with Earl's help, the changes that we are planning for the Stock Center will be very successful and its importance as a resource to maize scientists world-wide will continue well into the 21st Century."

35 Arthur Lee Hooker

Distinguished in Combining Plant Pathology, Corn Breeding and Genetics*

by D. White[1]
with contributions from D. Smith, S. Lim, R. E. Ford and S. Pataky

This issue is dedicated to Arthur Lee Hooker former Professor of Plant Pathology and Agronomy at the University of Illinois. Dr. Hooker died at his home in St. Charles, IL on July 4, 1991. His had a distinguished career as a plant pathologist, corn breeder and geneticist. He devoted more than 40 years of his life to research, teaching, and service to the corn seed industry. He is greatly missed by his students and associates.

Art Hooker was born October 12, 1924, on a farm at Lodi, Wisconsin. Following graduation from high school, his undergraduate studies were interrupted by service in the U.S. Army from 1942–46. He earned B.S., M.S. and Ph.D. degrees from the University of Wisconsin in 1948, 1949 and 1952, respectively. He married Hellen Margaret Zimmerman in 1950. His first position was as Assistant Professor of Botany and Plant Pathology at Iowa State University from 1952–54. In 1954 he became a Plant Pathologist with USDA ARS at the University of Wisconsin. In 1958 he joined the faculty of the newly formed Department of Plant Pathology at the University of Illinois.

Dr. Hooker was a Fellow of the American Phytopathological Society, the American Association for the Advancement of Science, the American Society of Agronomy and the Crop Science Society of America. He served on the Council of the American Phytopathological Society. He was a member

* *MAYDICA* 1993, Vol. 38, No. 3
[1] Pathology Dept., University of Illinois, Nebraska, IL

of the panel that studied genetic vulnerability of major crops for the Agriculture Board of the National Research Council, National Academy of Sciences. He received the University of Illinois College of Agriculture Funk Award, which is the highest recognition at the University of Illinois for outstanding service to agriculture through research and teaching. He was commended in Senate Resolution No. 176 by the 77th General Assembly of the Illinois Senate for his outstanding work during the southern corn leaf blight epidemic. He was a Guggenheim Memorial Foundation Fellow, and a member of Gamma Alpha, Phi Kappa Phi, Gamma Sigma Delta, Sigma Xi and Alpha Zeta. He was an invited speaker and consultant in 19 countries, and authored or coauthored over 160 publications, symposia articles, and book chapters.

Dr. Hooker is best known for his research and teaching at the University of Illinois which focused on preventing plant diseases through the use of genetic resistance. His approach, as he would explain it, was rather simple and straightforward, but required organization and diligence. The first step was to identify important diseases and monitor potentially troublesome pathogens. The second step was the development of inoculation and rating systems that were both practical and easily used by breeders with limited access to laboratories. His numerous inoculation techniques and rating systems are still widely used by pathologists and corn breeders. His third step was the evaluation and identification of resistant germplasm in greenhouse and field studies. Dr. Hooker accumulated more than 1,000 different inbreds for evaluation of resistance. During his career, he identified and used genes for resistance to numerous diseases including northern corn leaf blight, southern corn leaf blight, common corn rust; southern corn rust, and anthracnose leaf blight. He also identified sources of resistance to Stewart's wilt, northern corn leaf spot, downy mildew, anthracnose stalk rot and Diplodia and Gibberella stalk rot. The fourth step was the determination of inheritance of resistance with emphasis on sources of resistance that were simply inherited, and thus readily utilized in practical breeding programs. His research identified and studied in detail more genes for resistance in corn than any other researcher, which supplied plant breeders and pathologists throughout the world with information and seed stocks. The fifth step was the incorporation of resistance into currently used germplasm. Dr. Hooker's view of his work at the University of Illinois is best expressed in "A plant pathologist's view of germplasm evaluation and utilization" (Crop Sci. 17: 689–694). As he wrote "I believe that it should be the primary mission of some pathology-genetics programs to transfer known genes and cytoplasms for resistance into elite lines and populations... This program would not be one of varietal or inbred line development, but one of building the parents of variances or lines." Dr. Hooker, his students, and colleagues conducted extensive studies in the 1960s, on the inheritance of resistance to *Puccinia sorghi*, common corn rust. Five different gene loci, occurring on three chromosomes, were identified as possessing alleles for specific resistance. The most important locus was at the end of the short arm of chromosome 10. In a classical study of this locus, Rp_1, Dr. Hooker and associates showed that it has at least fourteen different dominant alleles conditioned resistance and one recessive allele conditioned susceptibility. Recombinant at very low frequencies in segregating progenies demonstrated the complexity of the locus. Currently, studies of resistance to *P. sorghi* are being continued at the molecular level by the younger generation of corn geneticists. These studies are possible because of the initial studies by Dr. Hooker and his associates.

Dr. Hooker is probably best known for his research dealing with northern and southern corn leaf blights. In the early 1960s, he described a single dominant gene for resistance to *Exserohilum turcicum* which causes northern corn leaf blight. The gene for resistance was called the Ht_1 gene but was known to many Illinois farmers as the Hooker gene. This gene reduced lesion size and sporulation of the fungus and provided good control of the disease. By the late 1960s, the Ht_1 gene had been backcrossed into numerous commercial inbreds and was estimated to be present in 70% or more of the corn hybrids grown in the Midwestern United States by 1970. He knew that biotypes of *E. turcicum* would eventually overcome Ht_1 resistance, and was continually evaluating fungal isolates collected throughout the United States. Race 1 of *E. turcicum* (virulent on corn with the Ht_1 gene) was identified on corn in Hawaii in 1974 and in Indiana in 1980. Although the Ht_1 gene provided control of northern corn leaf blight for nearly 15 years its widespread use has been discontinued. From the late 1960s until his death he continued to search for and find additional sources of resistance *to E. turcicum* including two single dominant genes designated Ht_2 and Ht_3.

Dr. Hooker is equally well known for his research which aided in solving the southern corn leaf blight epidemics of 1970 and 1971. A year before the "year of the blight," Hooker and associates identified a new race of *Bipolaris maydis* in the Midwest. His co-authored publication describing race T of *B. maydis* is a Citation Classic of Current Contents as one of the most frequently cited scientific papers in agricultural journals. This research dramatically revealed the importance of plant cytoplasm in susceptibility to disease and proved that the diligence of routine evaluations (which Hooker often championed) could reduce the risk that plant diseases pose to crops.

Dr. Hooker's collection and evaluation of fungal isolates lead to the discovery of race T of *B. maydis*. In the fall of 1969 on a drive to Macomb, Illinois, to teach an off-campus Diseases of Field Crops course, he noticed a field that had been badly damaged by leaf blight. Collected samples revealed the pathogen as *B. maydis*. This was particularly interesting to him because southern corn leaf blight usually did not occur with such severity that far north. Furthermore, he noted that some plants in the field were severely damaged and others were not. In retrospect, it is likely that the field contained a blend of normal and cytoplasm male-sterile source from Texas (cms-T). The leaf specimens were kept until January of 1970 when they were sporulated and resulting spores used to inoculate plants. Dr. Hooker gave instruction to his student, Dave Smith, to inoculate both normal and cms-T plants. Dr. Hooker must have suspected the involvement of cms-T since this was not done routinely. When reactions were evaluated it was apparent that the newly collected isolate was much more pathogenic on plants with cms-T than on normal plants. As southern corn leaf blight devastated the corn crop in 1970, Dr. Hooker had already alerted the seed corn producers of the potentially damaging effect of the new race on hybrids using cms-T. By 1972, cms-T was no longer widely used in the United States. If not for Dr. Hooker's vigilance, commercial seed companies may not have started increased production of normal cytoplasm hybrids and the blight epidemic may have lasted another year.

During his tenure at the University of Illinois he directed graduate research of more than 30 M.S. and Ph.D. students. Many of his students are well-known commercial plant breeders and pathologists, while others are at universities and research institutes worldwide. While at the University of Illinois, he also developed and taught two courses in plant pathology/agronomy. A course on Diseases of Field Crops was oriented to provide students with a practical understanding of field crop diseases with particular emphasis on control strategies. Along with a thorough introduction into methods of inoculation and disease evaluation, students gained practical insight into breeding for disease resistance. Many of the students from his classes are currently utilizing techniques learned from Dr. Hooker in various plant breeding programs. He also developed a course on Genetics of Plant-Pathogen Interactions which used host genetics to study the genetics of pathogens. Dr. Hooker was fascinated with the interactions of hosts and their pathogens and his contagious enthusiasm was conveyed to students. Both courses are currently taught by others using the base developed by Dr. Hooker.

In 1980, Art resigned from the University of Illinois to join Pfizer Genetics in order to fulfill his desire to do more corn breeding. He moved to De Kalb, Illinois when De Kalb Seed Company and Pfizer Genetics merged in 1983. With De Kalb-Pfizer Genetics he was the Bioscience Director and Principal Scientist and directed a large research staff. During that time he developed inbred lines through conventional breeding that had superior performance in company trials. As Bioscience Director, he developed an understanding and appreciation of biotechnology and was able to direct research to achieve practical results. In 1986, he retired from De Kalb-Pfizer Genetics and was instrumental in establishing the Plant Molecular Biology Center at Northern Illinois University. During the time, he also worked as a corn breeder with Hughes Hybrids, a family owned seed company. At the Hughes research farm, he developed several superior inbreds which ranked first in company yield trials. He also identified sources of genes for endosperm hardness which will be extremely useful in developing opaque-2 corn hybrids. Always interested in plant pathology, Dr. Hooker also was investigating northern corn leaf spot which had become a problem in seed production fields.

Dr. Hooker was an intensely motivated man who expected maximum performance from himself, his students and colleagues. He was generous with his time in helping the development of young scientists with interests in breeding for disease resistance. His advice was actively sought throughout his career by students, colleagues, seed corn companies, extension agents and administrators. He was instrumental in the establishment of the Crop Advisory Committees to the National Plant Germplasm System and served on the Corn Advisory Committee until his death. He also served on scientific advisory committees of the Illinois Crop Improvement Association and Illinois Foundation Seeds, Inc. His presentation of five invited papers at the Annual Corn-Sorghum Research Conference, of the Americans Seed Trade

Association, is further evidence of the recognition by the seed industry of his research with corn diseases.

The most remarkable aspect of Dr. Hooker's career was the intense focus of his work. From the early years of his career at Iowa State University when he developed inoculation procedures for *E. turcicum* and *P. sorghi* and screened several acres of corn accessions for sources of resistance, through the identification of genes for resistance to *E. turcicum* and *P. sorghi* at Illinois, to the exciting discovery of Race T of *B. maydis*, to his uncompleted work with northern corn leaf spot and opaque-2 at Hughes Hybrids, Dr. Hooker was determined to improve maize performance by breeding for improved disease resistance and other traits. His special contributions will be greatly missed.

36 Myron G. Neuffer

His Contributions to Maize Genetics*

An appreciation by William F. Sheridan[1]
and Ming T. Chang[2]

This volume is dedicated to Myron G. Neuffer. It is an acknowledgment of over 40 years of scholarly contributions to maize genetics. Since entering graduate school in 1947 as a student of L. J. Stadler, Professor Neuffer's productive career has spanned the period from prior to the widespread recognition of DNA as the hereditary material up to the present day. Throughout these nearly 50 years he has pioneered basic investigations in several areas of maize research.

Gerry Neuffer was born in Preston, Idaho, on March 4, 1922. He grew up on a farm where he helped his father in hoeing and irrigating their sugar beets. He lived an active life in the country and graduated from Preston high school. He loved horses as did his future wife, Margaret, and he courted her by taking her out horseback riding. Early during World War II he joined the Navy but, following a disabling accident suffered by his father, Gerry was discharged and sent home to work on the farm for the war effort. He graduated from the University of Idaho in Moscow, Idaho, in 1947 with a B.S. degree in Agronomy and then, with his wife Margaret, moved to the University of Missouri in Columbia, Missouri. The Neuffers have resided in Columbia since their initial move there. They have four sons and three daughters, and at the time of this writing they have twenty-six grandchildren.

M. G. Neuffer was appointed Assistant Professor of the Field Crops Department at the University of Missouri in 1951. When Professor Stadler died in 1954, John Laughnan left Illinois and came to Missouri as Stadler's

* *MAYDICA* 1994, Vol. 39, No. 1
[1] Biology Dept., Iowa State University, Ames, IA
[2] ICI-Garst, Slater, IA

replacement. But the following year Dr. Laughnan left Missouri and returned to chair the department of Botany at the University of Illinois. Dr. Neuffer then took the position originally held by Stadler and has remained at Missouri throughout his career. He was promoted to Associate Professor in 1956 and to Professor in 1966. From 1967 to 1969 he chaired the newly formed Department of Genetics. From 1959 until 1970 he had both teaching and research responsibilities but has had full-time research duties since 1970. Upon his retirement in 1993, Dr. Neuffer was granted the title of Professor Emeritus in the Department of Agronomy.

Gerry Neuffer's research in maize genetics can be divided into four time periods. Prior to 1954 he followed in Dr. Stadler's footsteps and mainly worked on radiation genetics. During this period he reported with Stadler on the compound nature of the *R1* locus (1953), and the discovery of the aleurone and plant color factor, *bz2*, which he cytologically localized to chromosome arm 1L by use of a radiation-induced interstitial deletion of the normal allele (1954). From 1954 onward he maintained his interest in radiation genetics but expanded his studies to include the study of highly mutable alleles of the *al* locus controlled by the *Dt* transposable element (1955, 1961, 1966). During this second period he demonstrated by the use of appropriately genetically marked stocks that tetrasporic embryo sac development could occur in maize (1964). About 1966 Neuffer moved into the third period of pioneering research when he began working with chemical mutagenesis. In 1955 Dr. Ed Coe joined the University of Missouri as a Research Geneticist with the USDA ARS. He discovered that maize pollen would remain viable and could be used to pollinate silks after suspension in paraffin oil (1966). Together Neuffer and Coe developed the paraffin oil method for treating corn pollen with the chemical mutagens, ethyl methanesulfonate (EMS) and nitrosoguanidine (NG). This method was later published in this journal in 1978 but Neuffer began using it in 1966. In the meantime, he prepared with L. Jones and M. Zuber the first edition of the reference work titled "Mutants of Maize" which was published in 1968. During the early years of using EMS in paraffin oil to treat maize pollen, Neuffer successfully induced and isolated hundreds of new mutations of corn. These were a bountiful reward of a search for new mutant alleles that were being sought for the *al* locus. The search for the new alleles at this particular locus resulted in Neuffer moving into a fourth period of research which has been especially rewarding, not only for him but for numerous maize researchers in this country and abroad.

This period began in 1973 when Gerry Neuffer visited Bill Sheridan's laboratory on an October day in 1973. He suggested that Sheridan's maize tissue and embryo culture experience might be used to screen some of the defective kernel (dek) mutations recently induced by EMS treatment of pollen and isolated by Neuffer. The suggestion was well received and led to a collaborative effort to search for auxotrophic mutations that spanned more than ten years. A total of 110 lethal mutations were tested by culturing immature mutant embryos on basal and enriched media, and five proline auxotrophs were identified and reported in 1980. These were all proven to be allelic to the proline requiring mutant first reported by Gavazzi and co-workers in 1975. In 1974 Neuffer described the protocol and initial results, demonstrating that the EMS treatment yielded a mutation at any particular locus with a frequency of 1 out of 900 pollen grains treated. Neuffer realized early on that he was in a position to isolate and recover mutations at loci throughout the maize genome. This fourth period, which began with the search for new *al* alleles and then gained a much broader focus in the search for auxotrophs, has now spanned twenty years and continues today. During this period Neuffer has isolated thousands of mutations affecting a great variety of processes and features of all stages of the maize life cycle. In addition to isolating these mutations he has sought to determine their chromosome arm location. So far he has identified and named 122 dominant mutations and, using the waxy translocation series, located 54 of them to 18 chromosome arms. He has identified and named 2524 recessive mutations and located 706 of them to 19 chromosome arms.

Neuffer has always been known as a generous and considerate individual both in his personal life and as a scientist. He has always shared his materials and genetic expertise with students and other investigators. This has been very beneficial to many researchers but especially those individuals who did not receive their graduate training in maize genetics. Such was the case with Bill Sheridan. Neuffer and Sheridan analyzed 207 defective kernel mutants and identified 14 putative developmental mutants. The subsequent analysis of these dek mutants was carried forward by Sheridan and co-workers. This work eventually led to the successful search for embryo specific mutations, amongst materials made available by Dr. Don Robertson as well as by Neuffer. Another fruitful collaboration whereby workers new to maize benefited from their association with Neuffer involved the study of disease lesion mimic mutants. Together they have identified 23 dominant and 24 recessive lesion mutants, with Dave Hoisington and Virginia Walbot.

In addition to these collaborative projects involving the EMS induced mutations, numerous other researchers have utilized them for their basic investigations. These include the study of high chlorophyll fluorescence (*hcf*) mutants by Dr. Donald Miles, dwarf mutants studied by Dr. Bernard O. Phinney, leaf and ligule mutants studied by Dr. Michael Freeling, physic acid mutants studied by Dr. Victor Raboy, and the orange pericarp mutants studied by Dr. Allen Wright. The large collection of EMS-induced mutations developed by Neuffer has been utilized by many researchers besides those named above. Dr. Neuffer is well known for his attitude about the results of his research. His policy has always been to cooperate, to help, to advise, and to share. These characteristics have led him to make his mutant collection available to any researcher who requested material, or to any researcher who cared to visit his laboratory to select materials of interest.

One additional feature of Dr. Neuffer's long professional career at Missouri warrants further consideration. Throughout the nearly forty years since Ed Coe came to Missouri a year after L. J. Stadler's death, Gerry Neuffer and Ed Coe have enjoyed and mutually benefited from a continuous professional interaction. This scientific relationship has always been marked by their mutual congeniality and cooperative natures, and in a number of instances, fruitful collaborations. In addition to the paraffin oil-EMS mutagenic technique, the latter include their joint research on determining cell destinies in the developing corn plant and their joint authorship of the comprehensive review of maize genetics published in 1977 and in revised form in 1988. It also includes the considerable assistance provided by Coe to the preparation of the 1968 edition of "Mutants of Maize" and the co-authorship (along with Dr. Susan R. Wessler) of the new, greatly expanded edition of "Mutants of Maize" to be published by Cold Spring Harbor Press in 1997. Together, Neuffer and Coe have provided the leadership and worked together to build the maize genetics research group at Missouri into the largest and strongest such group in America.

An additional aspect of Neuffer's research interests also merits comment. The results of the long collaborative research of Neuffer and Sheridan and coworkers were summarized in a 1986 article by Neuffer, Chang, Clark and

Sheridan. In this paper the authors proposed an explanation for the paucity of auxotrophs (other than proline-requiring mutants) recovered from their extensive screening effort. They suggested that duplicate factors might control the biosynthetic pathways at the various steps leading to the synthesis of the amino acids, nucleotides and vitamins. Consequently, the isolation of most auxotrophic mutants would require the isolation of double mutants. This notion was subsequently proved correct by Neuffer's genetic analysis of the orange pericarp phenotype and the identification of the *orp1* and *orp2* loci. With Allen Wright he showed that the double mutant is a tryptophan auxotroph and this led to a collaboration of Neuffer and Wright with Dr. Karen Cone that resulted in the cloning of the tryptophan synthase genes in maize.

Maintaining genetic records on this collection of mutants became more and more difficult as the collection increased. Following Dr. Stadler's system, Dr. Neuffer had from the beginning maintained a card file on each genetic stock he maintained. On each card was recorded the source, pedigree, planting date, number of seeds planted, pollination direction, plant characters, harvest records, and other notes regarding the stock. In addition, each card listed where, in future years, that stock had been planted. Thus it was possible to trace a complete lineage of every stock. Dr. Neuffer's ability to remember the details of each stock in his collection was phenomenal; however, as time went on it became apparent that the process would have to be computerized. In this he was fortunate to have, as a post-doctoral student, Dr. David Hoisington, who understood, as few people did at that time, the possibilities of computerizing data; in addition, he possessed the programming skills to develop a complete set of programs for recording and tracking the data, in a system used for many years by everyone in Curtiss Hall.

Willingness to try new ideas is characteristic of Dr. Neuffer. He also treats everyone—staff, colleagues, students—equally and fairly. Dr. Neuffer is also extremely supportive of the people who work for him—evidenced by the fact that so many of his employees have worked for him for many years. The project foreman, Henry Lee, worked for Dr. Neuffer for over 30 years until illness forced him to retire. Dr. Neuffer's Research Specialist has worked for him over 12 years.

Dr. Neuffer's advice to his graduate students was always useful and to the point. He told them that a research project had to be simple and the results had to be clear; if the student could not organize his or her thoughts in a simple fashion it would not be possible to design a good experiment to address the basic question. Clear-cut results were necessary to make your idea convincing. Dr. Neuffer suggested that the student write down their every idea, research the literature to make sure the idea was sound, then design a genetic test to prove it.

Dr. Neuffer had six Ph.D. students: K. S. Hsu (1960); Gyula Ficsor (1965); Lino S. Cortes (1970); William S. Rafaill (1978); Rodney R. Higgins (1983); and Ming-Tang Chang (1983). He had seven post-doctoral fellows: G. Y. Kikudome (1960–62), O. P. Sehgal (1963–1965); David Hoisington, who also accepted a position as Research Assistant Professor until leaving for his current position at CIMMYT (1980–1986); Robert McK. Bird (1982–1985); Craig Echt (1984–1988); Ming-Tang Chang (1984–1986); and Allen Wright (1986–1991). He also sponsored three IAEA Fellows: S. E. Pawar (1979–1980); T. Lavapaurya (1987); and S. Lamseejan (1988).

In recent years, Dr. Neuffer has been studying genetic instabilities. He has been attempting to obtain a set of small ring chromosomes, each covering a specific chromosome segment, and a set of chromosome-breaking Ds stocks covering all twenty of the chromosome arms. The small ring chromosome did not recombine, but frequently got lost during mitosis. Therefore, the homozygous lethal defective mutant seeds with a ring chromosome (*dek/dek/*+-ring) should be viable. These plants would yield chimeric sector tissues due to loss of the ring chromosome. This would provide the opportunity to observe the mutant tissue phenotype, and also to observe feeding reactions between normal and mutant tissues. The chromosome-breaking *Ds* worked as well for the same purpose, using the chromosome-breaking *Ds* that was inserted at the position very close to the centromere. When the *Ds* element is activated, the whole chromosome arm breaks and is deleted, resulting in the loss of the normal allele, forming many chimeric sector tissues.

A significant acknowledgment of the value of Dr. Neuffer's work has been the very high degree of support he has received from granting agencies: of 36 applications, 28 received an award, 1 was withdrawn, and only 7 were not funded.

Dr. Neuffer has always been an "early bird" and was frequently in his office by 7:00 a.m., usually after jogging to work. Since his retirement he has begun to take it easy, often not coming into the office until 8:00.

Dr. Neuffer's warmth, his willingness to cooperate, his good, pragmatic advice, and his friendly manner have been much appreciated by his colleagues and associates. His wife of over 50 years, Margaret, is very much like him in manner and values. When family needs did not prevent it, Margaret has been of frequent help to Dr. Neuffer. One great asset is that Margaret, unlike Dr. Neuffer, is not color blind!

We wish Dr. and Mrs. Neuffer the very best, and hope that retirement will prove rewarding and fulfilling.

37 Ercole Ottaviano

A Feeling for Quantitative Genetics*

An appreciation by Mirella Sari-Gorla[1]

This issue is dedicated to Ercole Ottaviano, Professor of Genetics at the University of Milan, who died suddenly on June 7, 1991.

Ercole Ottaviano was born at Vasto (Italy) in 1937 and began his studies in Agricultural Sciences at the Bologna University. His subsequent life and career was decisively influenced, at the end of the 1950s, by the start of an important project, supported by Felice Ippolito, Secretary of CNEN (National Committee for Nuclear Energy, currently named ENEA), which provided funds to the Italian Institutes of Genetics for the study of the effects of ionizing radiations. This led to the arrival at the Institute of Genetics of Milan University, directed by Professor Claudio Barigozzi, of Angelo Bianchi, at that time the leading maize geneticist in Italy, coming from Pavia University. In the context of this research project, some fellowships for undergraduate students were created, one of which was awarded to Ercole Ottaviano, who came to Milan to carry out his thesis work on the genetic effects of X-ray in maize under the guidance of Bianchi, and graduated in Agronomy in 1960.

In 1964, during a period of post-graduate study under Kenneth Mather at the University of Birmingham, his interest extended to the field of biometrics and quantitative genetics and, on his return to Milan, he was appointed professor of Statistics in the Faculty of Science, a teaching activity that, after 1975, when he became Full Professor of Genetics, was continued by members of his group: Mirella Sari-Gorla and, later, Alessandro Camussi. This was the starting point for the development of a succession of researches of

* *MAYDICA* 1994, Vol. 39, No. 2
[1] Dept. Genetics, University of Milan, Milan, Italy

methodological character in the field of quantitative genetics for crop improvement: analysis of genotype-environment interaction, hybrid performance prediction, use of multivariate methods for the genetic analysis of complex traits, estimation of genetic distances. His expertise as a quantitative plant geneticist has been decisive in the planning and realization of the first project specifically devoted to maize improvement in Italy, which was sponsored by Ministero Agricultura e Foreste in the person of Giovanni Marcora, Minister of Agriculture at the end of the seventies, and coordinated by Francesco Salamini, Director of the Experimental Institute of Cerealiculture at Bergamo.

A fruitful scientific collaboration in the field of quantitative genetics developed between Prof. Ottaviano's group and a group of Polish biometricists. In 1972, Tadeusz Calinski, Professor of Mathematical and Statistical Methods at the Academy of Agriculture of Poznan, was invited by Ettore Marubini, Director of the Institute of Biometry and Medical Statistics of Milan University, to give a course on multivariate analysis. During this first meeting Calinski and Ottaviano established the basis of their collaboration, with the cooperation also of Zygmunt Kaczmarek, of the Institute of Plant Genetics of Polish Academy of Sciences. This collaboration is still in course in the framework of a cooperative program of the CNR and the Polish Academy of Science, and has led to the production of numerous important joint papers, as the one published in Genetics (1985).

Experiences in the field of biometrical genetics were influential in determining the character of Prof. Ottaviano's research group. In those years the use of the statistical methodology for the study of biological problems was uncommon in Italy, in spite of the pioneering work in this field made by Cavalli-Sforza, and, above all, extraneous to the mentality of most of the biologists. The credit for being the first to convey the importance of a rigorous treatment of experimental data and a correct planning of experiments is undoubtedly due to Giulio Maccacaro, who dedicated his attention to the medical fields, while Ottaviano, not without a great personal commitment, achieved a perhaps even greater success in getting these principles accepted and applied in the area of biological and agronomical sciences. His group thus came to play an important didactic role in this regard, with the organization of courses for researchers on behalf of the Biometrical Society (Italian Region), of which Ottaviano was a member of the committee and president for the Italian Region, and other scientific societies.

The 1960s saw the institution of the School of Applied Genetics at Milan University, in which Ottaviano played a major part; many Italian researchers now active in plant genetics and breeding in Italy came from that school, as well as young researchers from abroad, Latin America and Albania in particular, who came to the Milan School to complete their training in this field.

In the 1970s Ottaviano began to be interested in the genetics of pollen: expression and molecular analysis of pollen-specific genes, detection of genes expressed in both pollen and plant by isozyme analysis, methods of gametophytic selection and their application for genetic improvement of crops. Since until a few years earlier the pollen of higher plants had been considered genetically silent, these studies led to major changes in this field, suggesting new formulations for population genetic models of higher plants and offering unexpected possibilities for the use of pollen as a particularly powerful tool for genetic manipulation. In this research area Ottaviano produced many important papers, confirmed by the quality of the journals in which they were published, such as *Science* (1980) and *Advances in Genetics* (1989). The two works cited above were the result of a scientific collaboration with David Mulcahy, of the University of Massachusetts (USA), which began in 1975, when Mulcahy came to the Institute of Genetics of Milan for his Sabbatical year, and soon developed into a lasting friendship. Together with Mulcahy, Ottaviano established the practice of organizing periodical international meetings dedicated to these research topics, which were attended by increasing numbers of researchers.

Since the first International Symposium organized in 1975 and dedicated to "Gamete competition in Plants and Animals," this plant research area and the number of scientists involved has greatly expanded, as is testified by the frequency of the international meetings on this subject, by the intensive courses organized on the various topics, such as the EEC courses organized in the context of the EEC BRIDGE program by Mauro Cresti at Siena University, and by the publication of international journals specifically dedicated to this field, such as Sexual Plant Reproduction, founded by Linskens in 1989. Its editor-in-chief is currently Joseph Mascarenhas, perhaps the most competent of pollen molecular biologist, to whom Ottaviano sent young researchers of his group to complete their preparation in this field.

More recently, another area of research that received Ottaviano's attention concerned the adaptability of plants to environmental stresses and the use of molecular markers for the analysis of the structure of the genome and for the localization of useful genes. Indeed, crop improvement for yield stability under stress conditions, particularly with regard to the possibility of colonizing marginal lands, has become in recent years a priority breeding target. In addition to the research on tolerance to high temperature, chemicals and pathogens in maize, the starting point for the studies on sorghum species was the establishment of a collaborative program, still operating, with ICRISAT (International Crop Research Institute for Semi-Arid Tropics) at Patancheru, India, which Ottaviano visited many times.

To these purposes, a powerful tool for the identification and the chromosome localization of single genes, but in particular QTLs (quantitative trait loci), emerged in the eighties. With the idea of utilizing this methodology, based on molecular markers, and of promoting its diffusion among Italian plant geneticists, Ottaviano invited Ben Burr, who had published a fundamental paper on this subject in Genetics (1988), to the XXXII annual Meeting of the Italian Society of Agricultural Genetics (SIGA), of which he was President at that time, to give the opening lecture on gene mapping with recombinant inbreds in maize. After this, Ben Burr generously gave Ottaviano a particularly useful material: a population of recombinant inbred lines, constructed and typed for about 200 molecular markers (RFLP) by the Burr's group. Unfortunately, he had time to carry out only one work on this material, published in Theoretical and Applied Genetics in 1991, but in the following years many good papers based on this material have been published by the members of his group.

The last topic with which Ottaviano concerned himself only a short time before his death was the study of the genetic structure of, and environmental stress effects on, forest tree populations, by means of molecular markers. The first phases of the work consisted in solving the technical problems of RFLP typing in this material, and setting up a new class of molecular markers, RAPD, proposed by the Scott Tingey group of Dupont in 1990, with whom later collaboration was developed. Unfortunately, Ottaviano did not live to see the paper on the first linkage map of molecular markers of spruce *(Picea abies)*, to be published in Theoretical and Applied Genetics.

Ottaviano's scientific activity gave rise to over 100 papers, some of which have become classics in their particular sector, and to a number of books. He was also a member of the Italian Genetic Association (AGI) and of EUCARPIA, where he was councilor of the "Statistics and Plant Breeding" section.

Last but not least, mention should be made of his passionate dedication to teaching: in all his intense didactic activity (Statistics, General Genetics, Plant Genetics, Breeding Methods). His aim was always to make a proper cultural contribution, providing a wide and up-to-date view of the subject in all its aspects. And he was with his students, in the lecture hall, when he died, during the final lesson of his course of Genetics, leaving a void in the lives of all who knew him.

38 Donald Sage Robertson

A Long-Time Contributor to Maize Genetics; A Benefactor to Many*

An appreciation
by Peter A. Peterson[1]

Geneticists through the first decades of the 20th Century and well into the latter part of it had a significant impact in enhancing the progression of successes in molecular biology. These geneticists were able to uncover most of the mutants (in fact, almost all mutants) currently being published in molecular biology journals by laboratories practicing molecular biology. These include a long list of genes being investigated and reworked by a succession of molecular-oriented postdocs and students in defining the ultimate description of these genes. Don Robertson stands tall among these early geneticists. He has worked diligently and tirelessly perusing corn ears, seedling rows and field rows from the mid 1950s and 1980s in describing and defining gene interactions and gene loci. These findings have aided geneticists in general and have led to some of the current successes in molecular biology. His discovery late in the 1970s of the Mutator transposon has been a focal point of his studies, and this has benefited many laboratories, and this will be discussed later.

Early Years and Genetics Training

Don Robertson came to Iowa State University (then Iowa State College) in 1957. He arrived here to find a rural America, much different from the sprawling urban Los Angeles area that he left, and became suitably at home in the original

* *MAYDICA* 1994, Vol. 39, No. 4
[1] Dept. of Agronomy, Iowa State University, Ames, Iowa, U.S.A.

prairie that is now overflowing with endless rows of corn and soybeans in the surrounding county. Ames would not be very different from his early upbringing in Campbell, California, which is in the heart of the Santa Clara County, then a vast fruit growing valley. This is where Don grew up on a farm where the family settled when he was 9 years old. (He was born June 27, 1921). The farm included almond trees and numerous other fruits and vegetables, and, in addition, the family also raised ducks. Throughout America, and especially in California, this period in the 1930s was in the middle of an economic depression, and, as for many other families, this depression wiped out most of the Robertson holdings.

His father had been schooled in agriculture and had attended the University of California at Davis about 1914 when it was a men's agricultural college. Don's father married during the First World War and raised his family on this farm. During Don's early teenage years, he accompanied his father on camping trips in the nearby Sierra Mountains. He was very close to his father. With the depression, however, his father took what options were available; for instance, selling rides on an airplane and picking apricots with his sons. These were Don's teenage years. The depression was relentless, but his father during the 1930s was quite energetic and inventive and, being such, developed the rotary brush for aerial crop dusting and finally succeeded in that business during a period of expanding California agriculture, especially in the Central Valley. Indeed, he was later recognized as the Father of Aerial Spraying and was given an award at the National Agricultural Aviation Association and installed into their Hall of Fame. The depression brought out the entrepreneurial spirit in many families throughout the United States, and certainly, Don Robertson's family was no exception.

There probably were many opportunities for farm labor in the Santa Clara County, and Don participated in these while he was attending Campbell High. After his graduation from Campbell High, he attended San Jose State College in north-central California. His studies were interrupted, however, because World War II began for the United States with the bombing of Pearl Harbor in December 1941. As did many of his peers across the country, Don joined the Army Air Force in 1942 and served for 3 years. During this period in the Armed Forces, he married his high school sweet heart, Roxana. After the war, he received certain veterans' benefits (the GI Bill) and was able to proceed with his undergraduate education at Stanford University and was graduated in 1947.

Stanford was then in a vigorous period of genetic expansion because George Beadle had become head of the biology group there. This was only a few years after Beadle and Tatum had presented their one-gene, one-enzyme theory, following their investigations of the Neurospora mutants, which was later to be the basis of their being awarded the Nobel Prize in genetics. Certainly the atmosphere in biology at Stanford must have been electric. Don was able to take his genetics units under George Beadle and, with a farm background, was quickly imbued with the subject. With this inspired introduction to genetics, Don decided to pursue further studies in genetics. It was at this time that George Beadle was invited to head the Biological Sciences group at the California Institute of Technology, and with some encouragement from George Beadle to continue his studies, Don was advised to study under Dr. E. G. Anderson at California Institute of Technology.

This was a propitious time for plant genetics at Cal Tech, and E. G. Anderson was to embark on an extensive program. This was only one year after the atomic bomb test at Bikini Atoll in the South Pacific, where a large amount of seed, in addition to various other genetic materials, was exposed to the nuclear radiation. The United States Navy and the Atomic Energy Commission wished to have this exposed corn seed material analyzed for genetic changes and decided that the Cal Tech laboratory under Dr. E. G. Anderson would be the central focus of study of changes induced in the corn seed during these tests. Dr. E. G. Anderson collected a number of people at the "farm" to aid in this study and enlisted Dr. A. E. Longley from the United States Department of Agriculture to do the cytology of the numerous breaks and to correlate that with the genetic findings. He could not have found a more experienced cytogeneticist for this purpose. (See Laughnan's discussion of his days at Cal Tech in Chapter 34.) In addition, there were many visitors to the "farm" at Cal Tech: Edgar Anderson, John Laughnan, H. Kramer, Howard Teas, Norton Nickerson, and B. Phinney to name a few. And it was an extremely exciting time for all there.

These visitors included biochemists, genetics, botanists and agronomists, and because of the diversity of their background, E. G. Anderson would gather them around at lunch time and have someone lead a discussion on a botanical or genetic subject. This group would often hold noon-hour sessions with a portable chalkboard under a lattice shade for protection from the California sun to discuss corn in general, especially with the enthusiastic Edgar Anderson of the Missouri Botanical Garden. Because of the discovery of the "blue fluorescent" mutant and analysis by Howard Teas, (the standard procedure for cut tassels, to look at pollen for translocations, was to pass them under the UV light for the detection of additional mutants). Many were analyzed, linkage and cytological maps were correlated, and numerous seedling tests were made. Thus, with Don's arrival to study with Dr. Anderson (E.G.) there were available a limitless number of mutants to be genetically analyzed. Earl Patterson was there simultaneously with Don in these studies. Earl and Don relaxed by playing badminton in the evening outdoors under the lights. Earl recalls fond memories of this light exercise and light-hearted conversations with Don and Roxy on those pleasant cool evenings in Pasadena.

Don's work on vivipary in maize was the main focus of his doctoral study. This was a seminal piece of work that became the basis for numerous investigations in later decades, providing the genetic fidelity for the later molecular aspects of vivipary, which the molecular studies showed to be a main regulatory gene in controlling the anthocyanin pathway (D. McCarty).

After graduation from the California Institute of Technology, Don sought a new opportunity and began the development of the Science Department at Biola College, a Christian College now identified as Biola University. This Science Department was seeking accreditation. While teaching there, Don continued his corn genetics research with E. G. Anderson at Cal Tech during the summers.

The Move to Ames

In 1957, the Genetics Department at Iowa State College was increasing its plant section and asked Don to join their small department. The Genetics Department was seeking to increase its plant initiatives and sought a plant geneticist to join Professor Joseph O'Mara, Peter A. Peterson, James D. Smith, and Kiyoshi Sadanaga, who were then in that department. The head of the department was Professor John Gowen who commanded a massive research program that ranged from *Drosophila* genetics to a very extensive mouse colony. Gowen had a long and illustrious career in genetics and was successful in garnering research funds from the Atomic Energy Commission and the Rockefeller Foundation. He added building space to the small Genetics

Department (now housing an animal ecology group) that Dr. E. W. Lindstrom had founded in the early 1930s. Thus, the total department staff included J. Gowen, J. O'Mara, P. A. Peterson, K. Sadanaga, J. H. D. Bryan, Willard Hollander, Jim Smith and D. Robertson. They were very involved in teaching genetics to the large number of undergraduate students in the agricultural college, as well as in the whole university. The genetics seminar was well attended, including the Agronomy Department plant breeders and the animal breeders from Animal Science (Jay Lush had a larger number of students and visitors) in addition to the Genetics staff. In the meantime, Don along with the others, carried on their research program.

His early work at Iowa State University, following the initial studies with vivipary, continued with the many mutants that were uncovered from Atomic-bomb test. These included many carotinoid mutants. With these, he enlisted a physiologist and thus collaborated with Dr. I. C. Anderson in genetic-physiological studies at Iowa State University. He continued these studies for a number of years, uncovering temperature-sensitive alleles, and a number of pigment deficiencies of endosperm and seedling. He incorporated the *wx*-linked translocations cytogenetic technique and B-chromosome translocations to determine linkage relationships in some studies and attempted to transfer intact segments of maize chromosomes as a method of plant improvement. In these studies of chloroplasts mutants, there were numerous uses of seedling benches. Further, a number of investigators, knowing Don's interest in seedling mutants, sent him newly uncovered mutants of seedlings that a geneticist uncovers in a seedling bench. This was the period of the 1960s and 1970s when molecular biology had not yet invaded the geneticists' arena; thus, the chloroplasts mutants were effectively used in looking at chloroplast development, and there were numerous mutants that could be applied to these studies.

Discovery of Mutator

In this atmosphere of free exchange of materials to seek the most expeditious location for the use and analysis of mutants, a unique and fortunate event occurred. Knowing that Don would make allelic tests with newly derived and uncovered mutants, Dr. Jerry Kermicle at the University of Wisconsin sent him a line containing a yellow mutant later identified as *Y9*. This was a key delivery because in attempting to outcross and make various allelic studies, Don noticed a very high explosion of mutant phenotypes unrelated to the *Y9* mutant. In further analysis of this series of mutant, he identified and defined those mutating maize lines. In pursuing this discovery, he found that these lines exhibited a very high mutation rate and that this mutation rate generated new mutations at other loci at rates nearly 40–50 times the standard of maize lines. This was a fortunate observation and an astute one because the original line was not actively studied in the Wisconsin genetics program. Thus, in the spirit of cooperative investigations and effort that has been the main feature of maize geneticists beginning with the Emerson Laboratory at Cornell, this mutation line was distributed and was to occupy Don's research efforts in a very active study for the next 15 years.

A note can be added about his observation of high mutability. His discovery of this high mutability did not come from having assistants do the counting and observation in the greenhouse bench. How often one reads in grant reviewers' comments that undergraduates could do the counting. Don comes from a long line of active doers and participants in the research process. This is consistent with his mentor, Dr. E. G. Anderson at Cal Tech. This was nearly 20 years after he began observing mutants in a seedling bench, and from this long experience, he was quite able to quickly recognize that there was something most unusual in this observation. In further studies, he found that this line yielded many frequent mutations in subsequent generations. Further, there was a heritable feature to this mutability. Not only did the mutants arise, but a large number of them expressed the instability, as evident from the green striping on the yellow backgrounds or white stripes on green backgrounds. Because of the high frequency of mutability, he identified this as Mutator. Here were Mutator maize lines exhibiting a heritable capacity to generate new mutations manifold greater than the standard stocks. It did not qualify as a system at that time because the principal components of a system could not be identified, nor could a single locus be assigned as the cause of mutability induction. This was to come more than a decade later.

As this time, there was an extensive research effort in examining mutability systems, and some 20 years after, the McClintock *Ac/Ds*, the Brink *Mp* and the *En/Spm* of Peterson were firmly established. Thus, Don embarked on a program to study and understand the hereditary basis of this Mutator activity. Could there be a regulatory component as well as a receptor component? There were vexing problems of male versus female transmission and Mendelian inheritance of the Mutator activity. This occupied his research effort for many years. But what was unmistakable was that the mutation activity of this Mutator was many-fold greater (20–30 times) than that of other systems. Even viruses were suspected, and Don enlisted electron microscopists to examine corn tissues where virus-like rods were found.

With the same spirit of cooperation that began with the Emerson school, this Mutator line was distributed. Several laboratories began studies of this, and it was some 12 or 13 years later that the Mutator element was finally cloned by three laboratories. The mutability was so high that it was readily available to isolate the first mutant to be isolated with this line; namely, the *Adh1* allele. With the distribution of this Mutator line to the Freeling laboratory, a large number of studies were initiated, and many mutants isolated. The Stanford laboratory under Virginia Walbot was also a recipient of this. The Bennetzen laboratory beginning at California and then of Purdue was also very active in this area. Many laboratories and postdocs and graduate students were fully occupied for some years and will continue to be occupied with the use of this Mutator in future studies in isolating genes, all of which is evident from the publications that have emanated from this activity. It was finally used in "reverse genetics." This use of Mutator has resulted in three reviews that describe several aspects of Mutator in terms of its use in tagging and other experimental procedures. Finally, the autonomous Mutator transposable element was described in three different laboratories. It was named after Don (*MuDR*), though it is said that Don in all modesty would have preferred that his name not be used (*Mu* might be confused with the prokaryotic Mu element).

His research activities were not totally precluded by his administrative duties as head of the Genetics Department in the middle 1970s and proceeding until 1980. Mary E. Johnson, the long-time Genetics secretary found him an enjoyable "boss" to work with—trying to help him extricate himself from the frustrating bureaucracy and paper work that befalls an administrator. He was very thoughtful toward Mary and her colleague Eileen Muff, making a special personal effort to select an appropriate Christmas gift. He has served on a number of editorial boards and has contributed to a number of scientific advisory boards.

In 1984 he was the recipient of the Iowa Governors' Science Medal for scientific achievement for his work on Mutator. His colleagues honored him in 1988 by holding a Genetics and Molecular Biology Workshop series at Iowa State University. At this workshop, 50 or so scientists came together to review their work with Mutator and bring their accomplishments up to date. He was honored at this workshop for his work that helped to make these research opportunities possible.

His collaborators and associates speak highly of his work ethic and selfless collaboration. Brent Buckner, a contributor to this volume, joined him in the quest of the *Y1* gene. Brent found Don a thorough, hard-working scientist. Virginia Walbot found him a model collaborator, being extremely generous with his time and materials. His *Mu* story was an exciting entry for Walbot into mobile elements. And it is typical of the array of molecularly oriented researchers into maize genetics. This follows the same story of Nina Fedoroff, utilizing McClintock's vast array of material, and Starlinger, who also exploited the McClintock largesse and then the Max-Planck laboratory of Saedler who used the Peterson materials into the decade of the 1980s. All this occurred and brought maize materials into the hands of the molecular laboratories and provided the many students, post-docs and other colleagues with abundant material for examination. Of course, it should not be forgotten that this was uncovered by judicious analysis and painstaking observation by these plant geneticists.

Don, now Professor Emeritus of Zoology and Genetics, retired in 1991. Though diagnosed with Hodgkin's disease, he recovered from a serious illness after chemotherapy. He maintains his office and genetics activity in the department.

Don enjoys his family and, now, grandchildren. During the Biola years, Don and Roxy's dream of becoming parents became a reality with the birth of Mark in 1951, Leanne in 1955, and Bill in 1957. Don also enjoys the ISU sport teams and is a true loyal ISU rooter. In 1992, he and Roxy celebrated their 50th wedding anniversary. They enjoy their three grandchildren, Jessica, Laura and John, who in Don's retirement years are a joy to both of them.

Don committed himself to the Christian faith soon after he was graduated from Stanford. This has continued to this date, and he is an active member of the Evangelical Free Church. As a student of the Bible since age 25, on leaving Stanford as a teacher by profession, he now enjoys teaching the Bible to those that are so interested. When asked how he compromises his deep faith in Christianity and the study of science, particularly genetics, his response was that God had given him a brain to understand the biology and genetics of maize. According to Brent Buckner, Don feels "that maize is a truly wondrous creation of God and that he can glorify God in attempting to understand and explain his creation."

39 Edward H. Coe, Jr.

An Advocate for Green Power*

An appreciation by Sheila McCormick[1]

In 1992, Dr. E. H. Coe, Jr. was awarded the Thomas Hunt Morgan medal from the Genetics Society of America for his "lifetime of contributions to the field of genetics." Ed thus joined the illustrious ranks of the other geneticists who have been awarded this medal, including Barbara McClintock and Marcus Rhoades. I know that Ed is extremely proud of this award. I and the other authors of this volume (his students, post-docs, and colleagues he helped convert to corn) hope that Ed will be similarly pleased with this token of our esteem for his tireless efforts in maize genetics and to the maize genetics community.

This issue not only commemorates Ed's career work in maize genetics, but marks the 20th anniversary of his editorship of the Maize Genetics Newsletter. I was a beginning graduate student with Ed at that time, and I remember well how excited he was to take over the challenge of editing, but that he was also nervous about being able to maintain the standard set by Marcus Rhoades, the previous editor. However, there is no question that the maize newsletter improves in quality each year and, speaking from the perspective of someone who no longer works on maize, I wish that all plant newsletters could be as thorough and useful a resource as the MNL.

Ed was born in San Antonio, Texas, on December 7, 1926. Although he left Texas at the age of 6 months, there was apparently some environmental influence. Actually, Ed's usual hat in the field was one of those "gimme" hats from some seed company. Perhaps more evidence of his Texas origins is that for some reason, to get each other's attention in the field, the field crew would repeat an imitation of a roadrunner—"beep-beep"—until we managed to meet at the end of a row. Upon meeting up, the crew member (me, for example)

* *MAYDICA* 1995, Vol. 40, No. 1
[1] Plant Gene Expression Center, USDA-ARS UC Berkeley, 800 Buchanan St., Albany, CA 94710

might advocate a break to the air-conditioned shed, while Ed would seemingly prefer to stand in the field and discuss some intricate crossing scheme that he had drawn out on a pollinating bag. It is amazing to me that all those who worked in the Missouri cornfield during that time don't have high blood pressure, considering the number of salt pills we used to eat in attempts to survive the heat and humidity of the field.

But I digress—back to his origins. Ed's father was an army engineer, so Ed was an army brat, and moved around a lot as a child and teenager. His father was stationed in Washington, D.C., during and after WWII, so Ed lived in Arlington for awhile and attended high school where he met his future wife, Mary Longley. Mary's father was A. E. Longley, so perhaps Ed had some exposure to maize genetics even in high school.

I'm told that Ed was interested in plants as a teenager, helping his mother maintain their garden, as well as holding summer jobs planting in other people's yards. This gardening avocation persists—when I was a graduate student, I'd often come into the lab to see Ed peering into the dissecting microscope with rapt attention. Since he was at that time working on the embryo cells and their destinies project, I usually thought he'd found some neat new sectored kernel—however, what he was really doing was picking blackberry bramble thorns out of his fingers!

Ed started college at George Washington University, where he was a self-described "not serious" student. He then went to Korea for two years with the Army occupation forces. Since he had learned to type in high school, his position was as company clerk. He returned to George Washington University for one more semester and then finished his bachelor's degree at University of Minnesota, selected partially because his father had obtained his degree there. After graduation in 1949, he and Mary were married. He stayed on at Minnesota to do a M.S. with Charles Burnham, finishing in 1951. He had heard that John Laughnan had recently started at the University of Illinois, and so he went there and completed his Ph.D. with John in 1954. After a short postdoc at Cal Tech with E. G. Anderson, he joined the USDA-ARS plant genetics group in Columbia, Missouri, in 1955, where he has remained ever since (although there was a sabbatical stay at the University of Minnesota in 1983—84).

Over the years, Ed's research has run the gamut from A (anthocyanins) to Z (Zea-land, of course). Beginning with his M.S. (*Pr*) and Ph.D. (*A1, A2* and

Bz) work Ed made major contributions to the understanding of anthocyanin biosynthesis. His development of the pathway and careful documentation of tissue-specific expression provided the framework necessary for the subsequent isolation of the genes in the pathway, and for many studies with controlling elements. He made several contributions to gametephyte function, including the characterization of a maize line that exhibits a high frequency of haploids, the discovery that white pollen (*c2*, *whp*) plants are male sterile because of the absence of flavonols, and the discovery that gametophytically expressed genes can disrupt precise developmental events during microsporogenesis. He made contributions to both non-Mendelian inheritance (paramutation at the B locus) and to extrachromosomal inheritance (*ij* and *ncs*). He defined the cell lineage patterns of the maize plant, facilitating many developmental genetic analyses. In recent years, he has devoted a lot of time to the maize genome effort (becoming a computer nerd in the process). At all times he has made a consistent effort to make this crop plant a (the most?) useful model organism by spearheading the effort to compile the genetic information and gene interactions available within the research community.

Ed is in love with his organism: a true maize chauvinist. He tells the story that Edgar Anderson told him that one couldn't appreciate maize until "you would lie on your back in the field, look up and think like a corn plant"— another version of "A feeling for the organism," and I'm quite sure that Ed took Anderson's advice.

Ed is a notorious punster. His lab is filled with such paraphernalia as a small box labeled "anthers" sitting next to a large box labeled "quethions." More evidence of these horrible puns was once written on his blackboard: on it he had written "Claude Toad, the antinucleolar xenophobic leftist"—his version of *Xenopus laevis*, and his exhortation of Occam's razor "William of Occam—for close shaves."

Ed is exceedingly generous with his time, and was always willing to discuss new ideas; he also made a point to include students in discussions with illustrious visitors. From 1959 to 1981, Ed taught a graduate level class called Genetic Techniques—the philosophical perspective and subject matter I learned in that class continues to be incorporated into my work and in the courses I teach. My own students and post-docs think I am something of an anachronism when I tell them that I really liked my advisor and that we never had a disagreement—this is surely due more to Ed's even-temperedness than to mine!

Ed was an early and regular attendee at the maize meetings at Allerton Park. A photograph of those attendees who had been at the first meeting was taken at the 25th anniversary. He especially enjoyed the Fu dog statues on the grounds of Allerton, but I've heard that some of his roommates in the dormitory-style accommodations there were not fond of his snoring! Ed has always been an inveterate Maize Booster—several will recall the cheerleading session Ed led at a maize meeting in 1992, exhorting everyone to shout "GREEN POWER." Although the unwilling participants looked somewhat non-plussed, Ed loved it.

Ed is not daunted by adversity. Ginny Walbot tells the story of a drive back from Allerton Park to Columbia in a government issue car that had increasing electrical problems as the journey progressed through an increasingly severe rainstorm. But Ed could not be swayed from the idea that the government car would be able to make it all the way home. Ginny says that for the last 20 miles the trip was accomplished by having Jack Beckett lean out the door with a flashlight pointed on the white line of the highway, while Mary Polacco and Ginny operated the windshield wipers from the back seat, by pulling on shoestrings that had been hastily attached to the wiper blades!

We dedicate this volume to you, Ed, with affection, for your devotion to corn and to those who have worked with you.

40 Marcus S. Zuber

A Feeling for Maize Breeding*

An appreciation by L.L. Darrah,[1] B. Dean Barry,[2] and E.H. Coe[3]

Marcus S. Zuber's stimulation of ideas, and his motivating of others towards collaborative research work, have contributed to the effectiveness of the maize (*Zea mays* L.) breeding community for decades. He always had an idea for another little experiment if there was land left unplanted. His publications are in diverse areas of research, ranging from experimental design to plant standability to resistance to insects and fungal mycotoxins. Maize inbred lines released from his USDA-ARS project at the University of Missouri, especially Mo17, contributed to the rapid adoption of single-cross

maize hybrids having wide adaptability. His development of techniques for measurement of root and stalk strength provided more effective selection tools for hybrid and population improvement and now applied in public and private breeding programs.

Marcus S. Zuber was born in Gettysburg, South Dakota, on January 10, 1912, and raised on a family-operated grain farm. He received the B.S. degree from South Dakota State College in 1937, and the M.S. in 1940 and Ph.D. in 1950 from Iowa State College (now Iowa State University) with G. F. Sprague. From 1937 to 1942, he served as an Agent of the USDA in Ames, Iowa. In

* *MAYDICA* 1996, Vol. 41, No. 1
[1] Plant Genetics Research Unit, Agricultural Research Service, USDA and Department of Agronomy, University of Missouri, Columbia, MO 65211, U.S.A.
[2] Plant Genetics Research Unit, Agricultural Research Service, USDA and Department of Entomology, University of Missouri, Columbia, MO 65211, U.S.A.
[3] Plant Genetics Research Unit, Agricultural Research Service, USDA and Department of Agronomy, University of Missouri, Columbia, MO 65211, U.S.A.

1942, he joined the U.S. Army, serving on the Special Staff and earning the Army Commendation Medal. He left the service in 1946 with the rank of Major. Rejoining the USDA in 1946, he was employed as an Associate Agronomist, Agronomist, Research Agronomist, and Supervisory Research Agronomist and Research Leader located at the University of Missouri in Columbia. Following retirement from the USDA in 1976, he remained on a formal one-half-time appointment by the University of Missouri as Professor of Agronomy until August 1982, when he again retired as Professor Emeritus. Even after his second retirement, he continued to interact with maize researchers for nearly five more years.

Zuber's Contributions—A Sampling

Marcus S. Zuber began his maize-breeding career with USDA at the University of Missouri in Columbia in 1946. Zuber's research encompassed root and stalk lodging resistance, selection for corn earworm (*Helicoverpa zea Boddie*) and European corn borer (*Ostrinia nubilalis Hubner*) resistance, pipe corn hybrids, food-maize hybrid testing, high amylose starch and modified amino acid mutants, screening for resistance to stalk rots, and screening for aflatoxin resistance. Along with development of tools and techniques, basic research questions were answered. Often, genetically improved populations or inbreds resulting from the particular research were released to hybrid maize breeders for utilization and improvement of hybrids grown by farmers. Zuber authored or co-authored more than 250 publications. Of these, he was sole author of only 17 because of his unique ability to engender cooperative research. His co-authors included graduate students, plant pathologists, plant physiologists, entomologists, geneticists, chemists, animal nutritionists, statisticians, and soil chemists from universities, the USDA, and industry.

Experimental design

Zuber was one of the earliest maize researchers to investigate the efficiency of incomplete block designs The average of all designs without recovery of the inter-block information showed a gain of 25% in favor of incomplete block designs over the randomized complete block. With recovery of inter-block information, the gain realized was 36% as compared to the randomized complete block. This was convincing evidence of the need to use these designs for increased precision of maize experiment

Stalk strength

Work on improvement of stalk strength received major impetus with development of a technique for vertically crushing dried, 5.1-cm stalk sections by using a hydraulic press. Later improvements included an electric motor to power the hydraulic press in place of an automotive jack. A study of total ash and potassium content of stalks gave physiological insight into the association of high ash content with stalk lodging susceptible lines. Following four cycles of recurrent S_0 plant selection for high and low stalk crushing strength, its effectiveness was evidenced by a 66% total gain from selection in the high direction and a 28% total decrease in stalk crushing strength in the low direction of selection, yet yield increased slightly in both directions of selection. A two-year study showed that the pith contribution to total stalk strength ranged from 20 to 50%, which was greater than previously realized. Five cycles of recurrent selection for increased stalk crushing strength resulted in increased stalk soluble solids in one of two populations under selection. Cycles 0, 4, 8, and 11 of selection for high stalk crushing strength were evaluated at yield levels of 0, ~50, and 100%, obtained by covering the ear shoot, cutting off half the ear shoot, and open pollination, respectively. As yield level increased, stalk strength declined for all check hybrids and early cycles of stalk strength selection. Crushing strength of cycle 11 of MoSQA and MoSQB, however, was unaffected by yield levels. Seeking a more cost-effective tool that could replace the stalk crushing strength technique, modified electronic and manual rind penetrometers were evaluated and compared to whole-stalk crushing strength. Coefficients of variation for the rind penetrometers were one-half of that obtained with whole-stalk crushing strength, and the electronic rind penetrometer was much simpler to use than the manual rind penetrometer. Rind penetrometer resistance was correlated with whole-stalk crushing strength at r=0.85 (P≤0.01). More recently, probing the internode below the ear attachment node has further simplified use of the rind penetrometer and made it a tool that is currently strongly recommended for commercial application.

Root strength

Improvement of root strength lagged behind that of stalk strength because of lack of suitable measurement techniques. To study roots better, a trailer-mounted root washer powered by a 1500 Watt alternator driven by a gasoline engine was developed. An application study, including four single crosses, showed differential root development and a response to planting date. Zuber first described a method of measuring root-pulling strength by using a small garden tractor and front-mounted, upright, hydraulic cylinder. Characteristics from roots grown by using sand culture were compared with field culture measurements. Total root weight, root volume, and weight of nodal roots of plants from sand culture were positively correlated with root clump weight and vertical root pulling resistance of plants from field culture. To determine the development of root strength over time, a set of 20 hybrids and six inbreds were pulled at six dates ranging from 42 to 126 days after planting. With flowering occurring about 77 days after planting, maximum root strength developed from three to four weeks afterwards. Pulling date x hybrid interaction was minimal from one week before flowering to three weeks after flowering, indicating a suitable time period to make pulling strength measurements. A great increase in speed of root pulling has been realized by using a small skid-steer loader as the power source. The skid-steer loader and an experienced field crew can pull a row of 10 alternate, competitive plants in 30 seconds. Currently, a load cell is used to replace the maximum reading scale, and a flexible-wire pulling grip is used to reduce stalk breakage as compared to that which occurs when using a clamp. A clamp is currently being used to investigate improvement of root and stalk strength simultaneously.

Corn rootworm, corn earworm, and European corn borer resistance: A split-plot method of assessing resistance to the corn rootworm (*Diabrotica* spp.) was suggested after studying the relationships among several root characteristics. This included a paired-row plot where one row of the pair was protected with insecticide and the other was not. Desired was a root pulling measurement from the untreated row that equaled that of the treated row. Procedures in use today compare a rootworm-infested row vs. a noninfested row.

Corn earworm resistance factors in Zapalote Chico were found to be easily transmitted to a series of topcross progenies, and Zapalote Chico was recommended as a source of earworm resistance for breeders. Following up on this earlier study, a flavone glycoside (maysin), which retarded the growth of the corn earworm, was found in the silks of Zapalote Chico. Using a topcross of (Mo22 x T8) x Zapalote Chico, broadsense heritability of maysin was estimated to be 0.8 and gain predicted from using S_1 recurrent selection was 25.7%.

The effect of whorl leaf, leaf sheath, and stalk pith pH, and feeding damage by European corn borer was studied in collaboration with A. J. Keaster, a University of Missouri entomologist. It was determined that these measurements were not sufficiently associated with European corn borer feeding damage to be of value in selection. Working closely with B. Dean Barry, USDA-ARS entomologist on the maize improvement project, Zuber collaborated in the release of Mo-2ECB(S1)C5 and Mo-2ECB-2, which are breeding populations with resistance to the second-generation European corn borer. These populations continue to serve as valuable sources of European corn borer resistance genes for transfer into elite breeding germplasm.

Pipe corn improvement

Zuber undertook to improve cobs used for cob pipe manufacturing by working with companies in Washington, Missouri. Specialty pipes were made from selected cobs of Mo Open-pollinated Pipe corn. Zuber developed three pipe corn hybrids that were used for production-run pipes: Mo Pipe 4 about 1969, Mo Pipe 12 in 1973, and Mo Pipe 65 in 1976. Cob quality was affected by location and plant density, leading to recommendations for contract growing at low plant densities and with irrigation if possible. The pipe corn work has engendered numerous public-interest newspaper stories that have brought widespread attention to the approaches and strategies of maize improvement.

Food-maize performance and quality testing

Beginning in 1977, Dr. Zuber coordinated a food-maize research group that has been actively supported by The Quaker Oats Company, American Corn Millers' Federation and, later, by The Snack Food Association, in addition to other private companies. Under Zuber's leadership, the research group has contributed to hybrid development and regional performance testing of new, superior white and yellow food-corn hybrids. Tests have been expanded to include quality parameters of test weight, kernel weight, kernel size, thins, density, percentage horneous endosperm, and pericarp removal. The test forum initiated by Dr. Zuber allows smaller seed companies to benefit by having their products tested over a much wider area than could be accomplished by the company itself.

Endosperm mutants

Breeding programs introduced homozygous opaque-2 (*o2*) mutant endosperm genotypes into elite germplasm following recognition of potential nutritional benefits. Distinguishing the mutant phenotype from normal can be done by using backlighting with a frosted glass surface; however, identification in some background genotypes and in genotypes with kernel endosperm modifiers is difficult. To assist in distinguishing these phenotypes, a light transmission measuring system was developed that increased the reliability of such determinations. Studies that followed determined the inheritance of test weight components of normal, *o2*, and floury-2 (*fl2*) endosperm types. Lysine and lysine content of protein were studied in sets of S_2 progeny lines differing for *o2* and *fl2* with genetically equivalent background. Differences were found among hybrids of progeny lines and between endosperm types. The association of zein body classification with lysine content in *o2* line conversions was studied by Choe, *et al.*, and a negative correlation of -0.65 was found between zein body classification and lysine content.

Protein content of Coroico, a landrace from South America with two to six layers of aleurone vs. the typical one layer, was determined as another means to increase the nutritive value of maize for human consumption. Total protein in Coroico aleurone was found to be 35% to 38% compared to 22% in yellow dent maize.

Endosperm dosage effects of dull (*du*) and waxy (*wx*) loci on amylose content were studied in inbred line B37 by crossing and selfing in the following stocks: *DuWx*, *duWx*, *Duwx*, and *duwx*. Addition of *wx* alleles reduced amylose content while addition of du alleles increased amylose content of the endosperm. Test weight of amylose extender (*ae*) endosperm mutant strains of several inbred lines was studied by Helm, *et al*. Diallel analysis of sets of crosses showed a large number of significant, additive genetic effects and suggested that test weight of high amylose maize could be improved by breeding methods that utilize accumulation of additive gene effects.

Pericarp thickness

Helm and Zuber measured pericarp thickness of 33 dent maize inbred lines over four crop years using similar cultural practices. Average thicknesses ranged from 160 μ for inbred B37 down to 62 μ for inbred H30, and environment did not significantly affect pericarp thickness. Pericarp thickness is related to tenderness of sweetcorn and contributes to popping expansion in popcorn. Sweetcorn inbred lines tend to have thin pericarps, popcorn inbreds have thicker pericarps, and dent maize pericarps are intermediate in thickness.

Stalk-rot resistance

Stalk lodging in the field is affected not only by characteristics of the stalk per se, but also by susceptibility to stalk-rotting fungi such as *Diplodia* and *Gibberella*. Zuber studied interrelations among several stalk characters and susceptibility to *Diplodia* and *Gibberella* in six selected inbred lines and four F_1 crosses/backcrosses. In the set of genotypes studied, stalk lodging was not associated with high incidence of *Diplodia* and *Gibberella* stalk rot. This study used the toothpick method of inoculation to provide a uniform amount of inoculum and allow a large number of plants to be inoculated rapidly. Because rind splitting was often associated with the method when an ice pick was used to make the insertion hole, Calvert and Zuber developed use of a cordless electric drill to make the insertion hole. This allowed subsequent harvest and crushing of the 5.1-cm section centered on the inoculation point to determine the fungus' effect on stalk strength.

Reduction of aflatoxin

Marcus Zuber was the *defacto* group leader of an international, interdisciplinary team screening for resistance to aflatoxin *and Aspergillus flavus* (Link:Fr.). He was Principal Investigator on a five-year grant (AID/ta-C-1451, July 1977– August 1982) from the U.S. Department of State, Agency for International Development, entitled "Reduction of aflatoxin in maize (*Zea mays* L.) through breeding.

Early studies compared normal and *o2* endosperm counterparts that were grown in several states, inoculated with *A. flavus*, and harvested 15, 30, 45, and 70 days after inoculation. Almost all aflatoxin production occurred during the first 30 days after inoculation and no significant difference was found between endosperm types.

The origin of inoculum responsible for preharvest aflatoxin was determined in a multi-state study by Fennell, *et al.* Silk samples and insects were collected from test ears. Over one-half of the *A. flavus* isolates collected from silks produced aflatoxin and 32% of those collected from insects produced aflatoxin.

Fungal competition between *A. flavus* and *A. parasiticus* (Speare) in ratios of 0/100, 25/75, 50/50, 75/25, and 100/0 was studied on two F_1 crosses, one of which had a thick pericarp and one a thin pericarp. *A. flavus* severely limited the ability of *A. parasiticus* to compete for growth substrate. Production of aflatoxin on the thin pericarp cross was significantly greater than on the thick pericarp cross. Pinboard injury of opened ears produced greater amounts of aflatoxin than did other inoculation methods.

Effects of hybrids and locations on levels of aflatoxin in 12 hybrids were studied in a uniform test grown in Florida, Georgia, Illinois, Iowa, Kansas, Mississippi, Missouri, North Carolina, Ohio, South Carolina, Tennessee, and Texas. No hybrid was resistant, but no overall pattern of susceptibility could be discerned. Over 90% of the samples from Georgia, Mississippi, Missouri, North Carolina, South Carolina, and Tennessee had detectable aflatoxin. Generally, bright, greenish-yellow fluorescence in kernels and high levels of aflatoxin were observed, but several samples with low levels of aflatoxin did not show fluorescence.

Genetic control of aflatoxin production was studied in a diallel set of crosses derived from eight inbred lines or a subset of seven of those lines. Combining-ability analysis found general effects to be significant and specific effects nonsignificant in the 1978 study using pinboard inoculation and harvesting both wounded and nonwounded kernels for aflatoxin analysis. Using this procedure, the ratio of wounded to nonwounded kernels was affected by ear size. Gardner, *et al.* used seven of the eight original inbreds, generating 21 crosses that were analyzed in single-hill plots and 20 replications. In the single-location study by Gardner, *et al.*, only wounded kernels were sampled for aflatoxin analysis. Mean aflatoxin levels reported were 5 to 10 times greater than found in the 1978 study. In this study, specific combining ability effects accounted for about 65% of the genotype sum of squares and rankings of crosses and parents were in good agreement with the 1978 study. The final study in the series used the same seven-line diallel and modified-natural inoculation (silks sprayed with an *A. flavus* spore suspension 5 d after full extrusion) with whole ears harvested. Data from environments in Georgia, Mississippi, North Carolina, South Carolina, and Tennessee showed significant general combining ability effects, but nonsignificant specific combining ability effects. The most important finding, however, was that results from modified-natural inoculation did not agree with those obtained earlier by using pinboard wounding and inoculation with *A. flavus*. Use of the earlier wounding technique bypassed natural plant protection mechanisms, such as the husk and pericarp.

Inbred Lines and Germplasm Released

Forty-three white and yellow dent inbred lines were released by Dr. Zuber. Among these 43 was Mo17, released January 24, 1964. The "Notice to plant breeders and seed producers relative to release of a yellow dent corn inbred line" included the following description:

"Mo17 was derived from the single cross C103 x CI187-2. The relative maturity of this line is about the same as for the inbred C103. The grain type is considerably better than C103 and it is much easier to maintain under Missouri growing conditions. Leaf blight ratings have been good but not superior to C103. Stalk quality and stalk lodging appear to be equal to C103. The single cross Mo17 x C103 exhibits some hybrid vigor which indicates this cross might be a suitable replacement for C103, especially when used as a pollen parent."

In a survey of the 1970 U.S. maize germplasm base, Sprague reported 4,749,700 kg of Mo17 were used in seed production, representing approximately 1.8% of the total need. Zuber, in a similar survey done in 1975, reported 30,678,800 kg of Mo17 (7.0% of the total need) were used. With release of B73 from the Iowa Agricultural Experiment Station in 1972 and industry promotion of the wide adaptability and high yield of the B73 x Mo17 hybrid, use of Mo17 continued to increase. In the 1979 survey, 61,364,900 kg (12.2% of the total need) of Mo17 were used in U.S. seed production and this was second only to B73. Use of Mo17 had peaked prior to 1984 when only 3,501,000 kg were used (1.5% of the total need) as newer inbreds replaced 1960s vintage lines.

Releases of specialty maize lines have included 25 *ae* inbreds, four *wx* inbreds, four white endosperm *o2* inbreds, and 10 white gametophyte-factor (*Ga*s) inbreds (Table 1). In addition, 11 open-pollinated, synthetic, or composite populations improved for specialty characters, stalk quality, or insect resistance have been developed and released to breeders.

Challenges for Maize Breeders

In 1972, Zuber was asked to summarize what he thought were the challenges for breeders to increase production for the future for a presentation at the American Seed Trade Association meeting in Chicago, Illinois. His challenges, largely relevant today, are quoted:

1. Improve yield testing techniques to increase efficiency of selection.
2. Determine the cause of low observed frequency of recovering superior inbred lines from advanced cycles of populations improved by recurrent selection.
3. Conduct studies to elucidate the physiological aspects of grain yield, stalk and root lodging, and disease and insect resistance.
4. Develop strains that are adapted for minimum and no-tillage systems.
5. Determine how maize kernels lose moisture.
6. Develop techniques to help the breeder select genotypes with rapid field dry-down capabilities.
7. Develop synthetic and composite populations for future cyclic improvement with at least one component that has resistance to the diseases and insects that are of economic importance in the U.S.
8. Obtain information on the components of root systems and their relationship with root lodging resistance.
9. Study the interrelationship of "stay green," stalk sugar accumulation, premature cell death in stalks and their effect on stalk rots and stalk lodging.
10. Develop methods of ear inoculation, along with methods and times of rating for kernel rots.

11. Wherever necessary make use of exotic germplasm for insect and disease resistance. Evaluate the B-A translocation approach of sampling exotic germplasm for suitable genes.
12. Monitor the development of new findings in genetic engineering for their possible use in conventional maize breeding programs.
13. Improve grain quality by selecting for resistance to endosperm-stress fracture due to rapid, high temperature drying.
14. Develop hybrids with early maturity that will fit into a double-cropping system when grown in the southern Corn Belt and in the southern U.S.

Students

Dr. Zuber made a significant contribution to graduate education at the University of Missouri while maintaining a diverse, productive research program. From 1948 to 1982, he served as major advisor for nine M.Sc. and 22 Ph.D. students. In addition, he co-advised six M.Sc. and four Ph.D. students with L. L. Darrah. Almost 25% of the 31 students he advised came from other countries. A sampling of his Ph.D. students who continued in maize research includes C. O. Grogan (1950), V. L. Fergason (1964), J. P. Thomas (1968), J. L. Helm (1969), A. V. Paez (1970), H. Z. Cross (1971), F. D. Cloninger (1973), K. C. Wrede (1973), C. F. Stark (1975), T. R. Colbert (1976), M. H. Blanco (1977), M. S. Kang (1977), and C. A. C. Gardner (1982). Dr. Zuber served as a member of many other students' graduate committees.

Honors

Marcus S. Zuber has been recognized by his peers for his many contributions to maize research and the maize industry. Zuber is a Fellow of the American Association for the Advancement of Science (1956), the American Society of Agronomy (1963), and the Crop Science Society of America (1963). He has received the Gamma Sigma Delta Senior-Faculty Award of Merit (1969) and the Faculty-Alumni Award from the University of Missouri (1972), and was named Outstanding Educator in America in 1973. In 1976, Zuber received the Crop Science Award from the Crop Science Society of America. The National Agri-Marketing Association presented Zuber with its National Award for Agricultural Excellence, and South Dakota State University awarded him an Honorary Doctor of Science following retirement in 1976. In 1986, Zuber was selected to receive the National Council of Commercial Plant Breeders Genetics and Plant Breeding Award.

Zuber—The Man

Before development of precision plot planters, a hand jabber was "the tool of the day" for planting maize research plots. Marcus was one of the first to pick up on the jab planter and get it to work. He enjoyed this activity even after his second retirement. Research ideas, experimental designs, and sources of testing material were always being suggested even as we planted. We would routinely be locating small parcels of land to plant "one more experiment" and this usually continued for several experiments until the magic date of June 8.

Dr. Zuber was anxious to get started as soon as it was time to pollinate the nursery and stayed with the crew until the task was done. A memorable occasion was to see Dr. L. A. Tatum, then an Area Director for ARS (and former maize breeder), join Marcus and his crew doing weekend pollinations.

When Dr. Zuber was back in the office (6–7 days a week), the telephone was like a piece of his body. He would be answering questions, giving advice, and garnering information to keep track of people or the maize industry itself. He was in "the know."

Colleagues benefited greatly from Zuber's "open door policy" and availability for discussion. For many associated with him, you did not do a formal literature review; you just talked to Marcus Zuber. His near-photographic memory would recall subject matter details and authors of 20-year-old journal articles. The synthesis of all that knowledge enables him to define areas of needed research and to recommend approaches most likely to succeed. Discussing research with Marcus was always a pleasure because he knew all facets of maize production, insect biology and ecology, and maize disease symptoms and organisms, and willingly shared that knowledge.

Table 1. Missouri lines and breeding populations released by M. S. Zuber.

Inbred line/Population	Pedigree/Description
	Normal endosperm lines
Mo1W	Mo22 x WF9^2
MoB2	Reid Yellow Dent
Mo2RF	MoB2 x 33-16^3
Mo3	38-11 x R136
Mo5	(K55 x N6) K55^2
Mo6	Midland x T8
Mo7	Midland x T8
Mo8W	Cob pipe open pollinated (pipe corn)
Mo9W	Cob pipe open pollinated (pipe corn)
Mo10	Reselected K201 sublines
Mo11	Mo22 x WF9^2
Mo12	WF9 x Mo22^2
Mo13	Missouri hybrid 1100.3 x T8^2
Mo14W	Mo22 x WF9
Mo15W	Open-pollinated Pipe Corn (pipe corn)
Mo16W	Open-pollinated Pipe Corn (pipe corn)
Mo17	C1187-2 x C103
Mo18W	WF9 x Mo22^2
Mo20W	N6 x Mo21A^5
Mo21R	H22 x Mo21A^5
Mo22	Laguna
Mo23W	(K10 x Ky49/Mammoth White Pearl) (pipe corn)
Mo24W	(K10 x K49/Ziler Hi-Cob) (pipe corn)
Mo25W	(K10 x K49/Ziler Hi-Cob) (pipe corn)

Table 1. (Continued)

Inbred line/Population	Pedigree/Description
Mo26W	MoSQA(S$_0$)C$_4$
Mo27W	MoSQA(S$_0$)C$_4$
Mo28W	MoSQA(S$_0$)C$_4$
Mo29W	MoSQA(S$_0$)C$_4$
Mo30W	MoSQA(S$_0$)C$_4$
Mo31W	MoSQA(S$_0$)C$_4$
Mo32W	MoSQA(S$_0$)C$_4$
Mo33W	MoSQA(S$_0$)C$_4$
Mo34	MoSQB(S$_0$)C$_4$
Mo35	MoSQB(S$_0$)C$_4$
Mo36	MoSQB(S$_0$)C$_4$
Mo37	MoSQB(S$_0$)C$_4$
Mo38	MoSQB(S$_0$)C$_4$
Mo39	MoSQB(S$_0$)C$_4$
Mo40	Mo17 x C103 3rd cycle
Mo41	Mo17 x C103 3rd cycle
Mo42	Mo17 x C103 3rd cycle
Mo43	Mo17 x C103 3rd cycle
Mo44[#]	Mo22 x Pioneer Mexican Syn. 17

Amylose extender (ae) endosperm lines

Mo305 *ae*	(Mo3 x *ae*K)Mo3^4 selected
Mo306 *ae*	(WF9 x *ae*K)WF9^4 selected
Mo307 *ae*	(C103 x *ae*K)C103^4 selected

[#] Joint release by Missouri, North Carolina, and Nebraska; seed available from Nebraska.

Table 1. (Continued)

Inbred line/Population	Pedigree/Description
Mo308 ae	(C17 x aeK)C17^3 selected
Mo309 ae	(MoB2 x aeK)MoB2^4 selected
Mo310 ae	(Mo10 x aeK)Mo10^5 selected
Mo311 ae	(Mo8W x aeK)Mo8W^2 selected
Mo312 ae	(T202 x aeK)T202^3 selected
Mo313 ae	(C142A x aeK)C142A^4 selected
Mo314 ae	(C138B x aeK)C138B^5 selected
Mo315 ae	(Mo12144 x aeK)Mo12144^2 selected
Mo316 ae	Bolivia 561 selected (multiple aleurone)
Mo317 ae	(Mo1W x aeK)Mo1W^5 selected
Mo318 ae	(Mo9W x aeK)Mo9W^5 selected
Mo319 ae	(H28 x aeK)H28^3 selected
Mo320 ae	(Ky122 x aeK)Ky122^5 selected
Mo322 ae	(Mp303 x aeK)Mp303^5 selected
Mo323 ae	(aeK x Cuzco)Cuzco^4 selected
Mo324 ae	(Ga153 x aeK)Ga153^5 selected
Mo325 ae	(K41 x aeK)K41^5 selected

White endosperm opaque-2 (o2) lines

Mo2RF$o2$	Mo2RF
Mo20W$o2$	Mo20W
K41$o2$	K41
K55$o2$	K55

Waxy (wx) endosperm lines

(B37 x wx)	B37^5 selected

Table 1. (Continued)

Inbred line/Population	Pedigree/Description
(B41 x *wx*)	B41^6 selected
(380-11 x *wx*)	Mo3^3 selected

White endosperm, Gametophyte sterile (Ga^s) lines

Mo401	(B37 x *wx*)B37^5 selected
Mo402	(B41 x *wx*)B41^6 selected
Mo403	(380-11 x *wx*) Mo3^3 selected

White endosperm, Gametophyte sterile (Ga^s) lines

Mo501W	(K55 x *Ga^s*)K55$^{5-7}$
Mo502W	(Ky49 x *Ga^s*)Ky49$^{5-7}$
Mo503W	(C1127 x *Ga^s*)C1127$^{5-7}$
Mo504W	(C164 x *Ga^s*)C164$^{5-7}$
Mo505W	(H25 x *Ga^s*)H25$^{5-7}$
Mo506W	(K6 x *Ga^s*)K6$^{5-7}$
Mo507W	(T115 x *Ga^s*)T115$^{5-7}$
Mo508W	(H30x *Ga^s*)H30$^{5-7}$
Mo509W	(T111 x *Ga^s*)T111$^{5-7}$
Mo510W	(N72 x *Ga^s*)N72$^{5-7}$
Mo511W	(A509 x N72^2)S
Mo512AW	(W64A x K6*Ga^s*)S$_9$
Mo512BW	(W64A x K6*Ga^s*)S$_9^{10}$

Breeding populations

PR-Mo2	Ibadan B selected
MoSQA	White endosperm, selected for stalk quality
MoSQB	Yellow endosperm, selected for stalk quality

Table 1. (Continued)

Inbred line/Population	Pedigree/Description
MoCRW-1	Late maturity corn rootworm composite
MoA100A	High amylose composite, medium maturity
MoA100B	High amylose composite, medium-late maturity
MoA100C	High amylose composite, late maturity
Mo *sh2*	Shrunken-2 population selected for heavy kernels
Mo Open-pollinated Pipe	Cob pipe maize selected for resistance to cob splitting and cob and grain yield
Mo17 White Composite	White endosperm composite of Mo17 inbred line conversions
MoSCSSS	Yellow endosperm synthetic of 14 released Stiff Stalk Synthetic line

41 David B. Walden

A Distinguished Professor and Visionary Scientist*

An Appreciation by V.R. Bommineni[1] and C.L. Baszczynski[2]

We would like to acknowledge Dick Greyson for his assistance and sharing his experiences with Dave at UWO. Additionally, we thank Prem P. Jauhar, Norm D. Williams and Terrance S. Peterson for their comments and suggestions and to Peter A. Peterson for making this volume of MAYDICA possible.

It is with great pleasure, enthusiasm, and sincere appreciation that we, colleagues, friends and former students, dedicate this volume of *MAYDICA* to David Burton Walden on the occasion of his retirement after more than three decades of teaching and significant research centered around genetics. All of us, including the maize genetics community at large, benefited greatly from his vision, creativity, practical advice and tireless enthusiasm. His excellent leadership and dedication to maize genetics and developmental biology influenced many of his students, who currently occupy eminent positions in and outside of science. He has set an outstanding example of selfless devotion and sacrifice in research. His exceptional vision in research ranged from classical to molecular genetics and biotechnology and serves as an endless source of guidance to present and future researchers in these disciplines.

* *MAYDICA* 1997, Vol. 42, No. 2
[1] USDA-ARS, Northern Crop Science Laboratory, State University Station, Fargo, ND 58105-5677.
[2] Pioneer Hi-Bred International, Inc., Research Center, P.O. Box 1004, Johnston, IA 50131-1004.

David B. Walden introduces himself as Dave and is known by several names: "DB" and "Dave" at school, "Burt" at home and all three by the corn colleagues. Dave is the elder child of David C. Walden and Fannie M. Walden, born March 29, 1932 at New Haven, Connecticut. His father died when Dave was 14. His mother celebrated her 91st birthday in 1995. Dave spent most of his childhood in Mt. Carmel, Connecticut; and worked for a market gardener and as a farm hand on a local horse farm until he started as a field worker in the Genetics Department (1948) at the Connecticut Agricultural Experiment Station. The early experiences in field genetics during his high school days were deeply ingrained in his mind. With consistent effort and dedicated hard work he matured into a recognized maize geneticist and thus, honored by his students and colleagues today.

After graduating from Hamden High School in 1949, Dave pursued his undergraduate studies at Wesleyan University, Middleton, Connecticut. During the next few summers (until 1953) he assisted the well known corn geneticists, D. F. Jones, H. L. Everett, W. C. Galinat and H. T. Stinson (Jr.) in their corn nurseries at the Connecticut Agricultural Experiment Station. He was associated with the work which, in 1952, resulted in the report of genic restoration of cytoplasmic male sterility in maize. While at Wesleyan, he worked as an assistant in the laboratory of E. W. Caspari and E. Wright. During this period Dave interacted closely with Caspari and V. W. Cochran both of whom influenced his future academic endeavors. After graduation from Wesleyan in 1954, Dave continued his education as a Ph.D. candidate in the Department of Plant Breeding at Cornell University under the supervision of H. L. Everett focusing on in vitro pollen viability. It was in his senior year at Wesleyan that Dave met his future life partner, Carol A. Isherwood, who at that time was attending Boston University School of Nursing, and they were married in 1956. They have a son (David John, born in 1958) and a daughter (Karen Ruth, born in 1961) both of whom are married. Grandchildren, Alex Walden was born in 1994 and Chantelle d'Ailly was born in 1996.

In 1959, Dave accepted a post-doctoral fellowship to work with Marcus M. Rhoades and Ellen Dempsey in the Department of Botany at Indiana University. It was during this time when maize and *Drosophila* cytogenetics were at the forefront. He was actively involved in many discussions on maize genetics and gradually developed his life time interest in "genome organization." Several of his earlier ideas in classical cytogenetics that were developed in

Marcus Rhoades' laboratory influenced research projects until the end of the 70s. Dave was markedly influenced during graduate school by the teaching skills of A. M. Srb, F. C. Steward, and H. L. Everett at Cornell, and of M. M. Rhoades, T. Sonneborn and C. Hagen at Indiana. In 1961, Dave accepted the position of assistant professor in the Department of Plant Sciences (at that time, Department of Botany) at the University of Western Ontario (UWO), London, Canada, where he continued the rest of his teaching and research career.

Mission Accomplished at Western

Basically, two challenging factors influenced his decision to accept the position at the University of Western Ontario: 1) to design and start an undergraduate Genetics program; and 2) to establish the first corn genetics research program at Western and in fact, in Canada. (See following paragraphs for recognition of Dave's contributions and services at Western.) Even today, he generates and maintains special genetic stocks of maize in summer nurseries for research and teaching purposes. Within only a few years after his arrival, Dave successfully accomplished these goals and then focused intensely on an array of research-related activities. In 1964 he became an associate professor and through persistent contributions, was appointed full professor in 1971.

Richard (Dick) Greyson, who was his close associate in the Department, remembers that Dave's commitment to undergraduate Genetics program went far beyond the development and administration of academic programs, but extended to his superb and extensive commitment to teaching assignments and to the personal attention to the welfare of the individual students. His door was always open (lineups were frequent) for undergraduates who wanted some further exploration of a point made in a lecture, or for someone who needed some encouragement or to share some personal problem. This dedication to undergraduate teaching was recognized in his being awarded one of the first UWO Awards for Excellence in Teaching in 1982. The Genetics program has become one of the most successful undergraduate honors programs in the Faculty of Science in the last 30 years. What exists currently at Western is a great tribute to Dave's foresight, dedication and hard work.

In addition to his efforts with the undergraduate Genetics program, Dave's dedication in support of research and the training of graduate students have also been a high priority. The topics that he chooses for his own research and makes available to his students are always fundamental, topical and utilize the most up-to-date technologies. His expectations of originality and effort from his students are high. For all his students he has been a demanding supervisor, setting high performance standards. But his students have grown to respect these standards and many have become lifelong friends and colleagues.

To his colleagues and students, Dave is constantly a source of new ideas and suggestions. Many who passed through the Department of Plant Sciences will remember the famous Tuesday night ("corn night") seminars in which participation was a requirement for graduate students. The theme was "current topics" and beyond that anything was game. Sometimes, topic suggestions arose from an article in Science or were stimulated by some other seminar that Dave or a colleague had just attended. Frequently, Dave's suggestions involved ways in which maize could be used to solve various developmental or molecular questions. Indeed, these weekly "corn nights" were stimulating opportunities for students to extend their background and hone their skills for various future national and international presentations. For Dave they were a wonderful mechanism to suggest new research directions and explore the usefulness of maize in biological research.

Since 1962 Dave has directed and supervised the research of 20 graduate students, trained post-doctoral fellows, and hosted several visiting scientists. In addition, he influenced the career development of many undergraduate and graduate students. He attracted continuous funding from national, international and private sources. For example, he as PI with colleagues was one of the first recipients of a grant from the Human Frontier Science Program (1991, the first in Canada and one of the first world wide in plant molecular biology). Dave has published more than 50 research papers in refereed journals that provided guidance to other researchers.

From the beginning, Dick Greyson recalls, the development and maintenance of a maize field in the summer has been an important and necessary adjunct to Dave's overall laboratory-based research objectives. For much of the time this was a lonely responsibility, for few graduate students really appreciated how the quality and quantity of the corn seed on which their theses rested, was a direct result of Dave's attention, effort and breeding knowledge. In some cases the very availability of selected crosses in sufficient amounts was due to the careful planning and thought expanded years in advance of the project or the student. The maintenance of his laboratory on campus was made possible by his ability to maintain and enlarge his corn stocks. Certainly, only a few of his departmental colleagues recognized how central the field was to his total program. Still fewer are aware of how he has enjoyed the solitude and the intellectual stimulation that those late afternoon pollinations provided. Many of his best research projects were 'hatched' during those very private sessions. His field program, as well, over the years produced large quantities of seed for other researchers, both here and abroad.

Throughout his earlier days at Western, Dave was associated with and actively participated in several science-related tasks. He takes his commitment to genetics, and to good science on the whole very seriously. He has provided special lectures on corn, genetics and science to schools, county councils and other public forums. More direct were his contributions to on-campus committees and a variety of contributions to national science. The following lists some of his more obvious work, contributions and services. In 1962 with M. L. Barr, he established the annual Great Lakes Chromosome Conference and later, from 1965–72 Dave served as a Consultant in Cytogenetics for Case-Western Reserve University, Cleveland, OH. Dave served in several organizations and committees outside the UWO campus. During the 70s, he served: three terms on the UWO senate; in 1972–75 as a chairman in the International Maize Symposium and in 1978, as editor of the proceedings of the International Maize Symposium. Dave also served as: President, the Biological Council of Canada (1974–76), and the Genetics Society of Canada (1981–83); Chairperson of the Steering Committee, Canadian Congress of Biology (1983–85); Secretary-General, XI International Congress of Genetics (1985–88); President, the International Genetics Federation (1988–93); and Chairperson of the Local Organizing Committee for the Annual Meeting of the Genetics Society of Canada (1996–97).

Over his career, Dave's research contributions and services have been recognized and he has received a number of prestigious awards. These include the University of Western Ontario Award of Excellence in Teaching (1982); Distinguished Research Professor (1987); Stadler Lecturer, University of Missouri (1994); Sigma Xi Distinguished Canadian Research Lecturer, University of Guelph (1995) and Award of Excellence, Genetics Society of Canada (1996). He is the only member of the Faculty at UWO ever to have received both the Teaching and Research Awards.

Significant Research Contributions

Dave's dedication and commitment to maize research emerged during his Wesleyan and Cornell days. Having made that choice, he cultivated and fine tuned his ideas during his stay at Indiana, and developed the ability to use them successfully. He continued to use maize to probe current questions and research thrusts—whether they were questions about the structure or organization of the nucleus or genome, breeding strategies which explore gene action, gene mapping; genetic transformation, biology of meristems and more recently, the isolation and sequencing of genes. From time to time, sabbatical leaves provided him an opportunity for intensive interactions with corn folks and other colleagues (such as J. R. Laughnan, E. B. Patterson, M. Bennett, P. Heslop-Harrison and J. G. Parker) which helped him to expand research directions.

From very early on, Dave took an interest in how the nucleus is organized in terms of chromosomal behavior through the cell cycle. With his laboratory's research efforts focusing on maize cytogenetics in the 70s, many studies were done looking at morphology and behavior of chromosomes in normal and aberrant stocks of maize, as well as following perturbation of the normal processes by chemical and physical treatments of maize seedlings. Thoughts about the nucleus being an organized entity, with chromosomes having defined behavior and organization during the cell cycle, have been borne out in work of others demonstrating chromatin domains as functional units, the importance of the nuclear scaffold/matrix in anchoring regions of chromatin and impacting gene expression, three-dimensional reconstructions of nuclei and simulations of chromosome movements, role of cell cycle genes in cellular processes, etc. Quantification and statistical analysis were always a key component of data collection and manipulation, whether cytogenetic or molecular. Early studies in Dave's laboratory showing that temperature regimes influence the rate of cell divisions (mitotic indices) led to a whole new era of stress-induced gene expression studies in maize, initially with temperature (heat-shock) and later with other stresses. In the transition, a strong association was established between the laboratories of Dick Greyson (who retired from the Department of Plant Sciences in 1992) and Burr Atkinson of the Department of Zoology. This union, which crossed departmental boundaries, facilitated increased progress through pooled and shared resources, with the common thread being a love of genetics and developmental biology, and the realization of the powerful utility of the maize genetics system to address many problems. A large number of publications resulted from this association.

Dave's group was one of the first to demonstrate the molecular impact of different temperature shift regimes on gene expression in plants. As we know, maize is a dynamic plant experiencing many environmental stimuli during its life cycle, temperature variation being one. Additionally, heat unit accumulation is an important contributor to biomass and grain yield. Attempting to understand what impact temperature stresses have on maize lead to the identification, and molecular and immunochemical characterization of several novel classes of heat-shock proteins (HSPs) following high temperature stress or other thermal shifts. Further characterization of proteins and genes continued. Meanwhile, different laboratories around the world cloned and established sequence homology among several of the plant and animal heat-shock proteins. In subsequent studies, for the first time, it was implicated that the highly conserved ubiquitins are heat-shock proteins in plants and in follow-up studies the genes were characterized and mapped. In the following years, Dave's laboratory utilized the cloned 18kd hsp genes and studied their regulation by in situ anti-sense RNA hybridization experiments.

As a person of vision, Dave directed his research projects into other areas of developmental and applied biology, which he believed would lead to a better understanding of genome organization in maize. Some of these studies included regulation of proteins in developing flower organs and embryos of maize, and regeneration of plants through protoplasts and shoot apical meristem culture. Organization of cells within the apical dome after microsurgery lead to an estimate of the number of cells comprising a functional dome. The future applications of this information remain to be seen through studies that follow cell lineages and regulation of apical meristem related genes. Dave's group was among the first to conclude that the plant genome was exposed to more stress during reorganizational processes associated with plantlets from callus, in comparison to development of plants from organized meristems. Although Dave collaborated with many colleagues within and outside of the Department and corn community, it is noteworthy that Dave and Dick Greyson's collaboration resulted in one of the few laboratories in the world focused on the biology of maize meristems.

According to Dick Greyson, a serious collaboration with Dave did not begin until 1966 when Dave produced the first ABPHYL plant in the greenhouse. He further adds that this "opposite-leaved" plant lead to a series of discussions, observations and experiments resulting in a number of poster demonstrations, wax models of the meristems and publications. Although this work seemed so difficult to interpret at that time, meristem organization and leaf/node placement are now being reinvestigated and studied by other laboratories. Through their continuous independent research, discussions and interactions, later research lead to observations on genic male sterility, root growth, morphology of flower organs and practical aspects of tissue culture and genetic transformation. It should be noted that important research projects emerged from the Greyson-Walden collaborative tissue culture studies: 1) micropropagation by shoot tip; 2) in vitro fertilization either by cultured gametophytes or pollen obtained from cultured tassel inflorescences; and 3) regulation of flower development through cultured ear inflorescences. Clonal micropropagation, important for asexual preservation of maize germplasm, demonstrated that more than 2600 normal fertile plants can be produced from a single original explant within 22 months of culture, and can be maintained for an indefinite number of transfer generations. In addition to germplasm preservation, this propagation procedure was shown to be useful to recover transgenic chimeric segments in other laboratories.

Clearly Dave's contributions, whether directly or by way of "planting the seed" of an idea into others' thoughts, resulted in a plethora of contributions not only to maize genetics and development, but also to science in general.

A Man of Hospitality

On behalf of all of us who have benefited from Dave and Carol's kindness and hospitality over the years, we recognize the very generous way in which their home was open to undergraduate, graduate and faculty parties, corn barbecues and many social occasions. This generosity and hospitality coupled with the strong support Carol always provided Dave, is a direct measure of not only their dedication to science, but also of their interest and concern for people, both collectively and individually.

42 James D. Smith

A Tribute from His Students and Colleagues*

By David Hole[1]

James D. Smith has been known to students and colleagues simply as JD for as long as any of us can remember. JD was born on December 14, 1927 to Stuart N. and Ann Smith on a Paullina, Iowa farm. He began his scientific career under the tutelage of his father in the cornfield when he began pollinating corn for ten cents an hour at the age of 11. His father obtained his Ph.D. with Dr. E. W. Lindstrom in the Genetics Dept. at Iowa State College. Soon after, he became a breeder of sweet corn and owned a sweet corn firm which was later acquired by Asgrow. JD's father was also mayor in Ames, Iowa, and the town

showed its esteem for him by naming one of the city parks, the Stuart Smith Park, after him. JD graduated from Ames High in 1946 and played guard on the Central Iowa championship football team—a feat made more remarkable because he couldn't see without his glasses, and the goggles he wore kept fogging up.

JD was mentored in his graduate work by Professors Peter A. Peterson and Joseph O'Mara at Iowa State College, later, Iowa State University. His Ph.D. thesis included a luteus mutant (chromosome 4) that originated in the Bikini-A-Bomb tests and was uncovered in the screening of this material by Dr. E. G. Anderson at the California Institute of Technology. His studies were interrupted by the Korean War, and JD eventually separated from the U.S. Army with the rank of Captain.

He began his career at Texas A & M University in 1959. JD, in turn, has mentored over 65 graduate students at the M.S. and Ph.D. level. His masters

* *MAYDICA* 1998, Vol. 43, No. 1
[1] Plants, Soils, and Biometeorology, Utah State University, Logan, UT 84322

and doctoral students have cast to the winds ending up as academicians, private plant breeders, attorneys, medical doctors, pharmacists and leaders of the plant biotech industries. The breadth of knowledge JD imparted to his students, from biochemical to statistical and quantitative genetics, is made apparent by the careers many have followed. While names such as Ann Blakey and Chris Carson remain familiar to those in the maize community, many have found other paths to success. Morgan Chiang, JD's first Ph.D. student, spent his career in Canada investigating quantitative traits in maize and cabbage.

James Fuchs is another case in point. Here was a young student who arrived at Texas A & M having experienced a severe drought on the family cotton and sorghum farm in West-Central Texas. Becoming excited about genetics in his junior year, he sought out JD and continued there for both of his advanced degrees. Jim went on to several laboratories in Europe and finally to Minnesota where he has pursued bacterial regulation and genetics en route to becoming a molecular biologist in the Biochemistry Department at the University of Minnesota. In 1984, he co-authored a DNA techniques textbook. His brother Roy, a senior research scientist at Monsanto also in recombinant DNA work, was instrumental not only in increasing Bt toxin expression in transgenic cotton to effective levels, but also in clearing the path through EPA regulations for commercial release. These accomplishments from a drought-stricken farm in Texas through the mentoring of JD.

Even further afield are Dr. Richard Stockert, who works with glycoprotein receptors in the Marion Bessin Liver Research Center of the Albert Einstein College of Medicine, Dr. D-G Bai at Chonnam National University in Korea who is using the expertise he gained from studying the role of ABA in maize to examine the effects of hormones on reproduction in frogs and Dr. Bryan Bailey, who is a USDA-ARS pathologist in Beltsville Maryland working on biocontrol of Erythroxylum coca.

JD also served on hundreds of committees (at least 592). Although he was chronically late to committee meetings, he would occasionally show up a week early for a meeting. In a statistical sense, that averaged out all of his late arrivals. In the lab, we always kept the clock running a few minutes fast to try to keep him on time to committee meetings, but he would invariably get used to the increase and factor it in to his departure time so we would have to move it up a few minutes more. We believe that the clock was, in fact, running about two days ahead by 1996. He was still late for committee meetings. One of his colleagues has mentioned that for many of the students, on whose committee he served, he was more instrumental in their success than their major professor. One of these students, Dr. Helen Belefant-Miller recently wrote.

"As a graduate student at TAMU (Texas A & M University), I often played 'hunt the professor.' My major professor was not JD but I could sometimes run mine to the ground in JD's lab. Eventually I realized why my professor would be there—it was fun! JD and a continually changing host of students would be holding forth on sports, anthocyanins, history, corn varieties, farming, TAMU, noted scientists, and home improvements, in that order. Inducing and altering this flow were people coming in with questions on classes, statistics, computer programming, planting, carotenoid purification, and HPLC repair. JD was the keystone. By being of good humor, a curious mind, and gregarious nature, he provided all of us a pleasant place of interest and information."

Another of his students, Dr. Ann Blakey adds these comments: "Two fundamental concepts always come to mind when thinking about JD, these are: the value of history and the importance of integrity in scientific interactions. The value of the history of genetics, and in particular maize genetics, to the development of science was never omitted from discussions on any topic. Whether he was reminiscing about the Minnesota Mafia or the Wisconsin Gang of Five, the impact, the richness, the sheer diversity in the nature of maize genetics was always a key element of discussions with JD. The 'Evening at JD's' seminars were the scene of many of these discourses. It is here that students would present information on topics chosen for the semester, and a historical development was always included. JD would often add anecdotes that would bring his students closer to the science as a realm of human interaction that is all too often ignored in the lecture halls of today."

These human interactions bring up the second area that JD seemed to have a tremendous impact on his students. JD's openness, his sharing of time, knowledge, experience, research materials and ideas were all a part of the education of his students. JD led by example in this field where science was an endeavor that involved the participation of many and not just the one. Integrity and respect were keys elements in these interactions. And although some might fail, in their humanness, to reciprocate, JD's belief in the open sharing of ideas remained undaunted.

Time with JD was always time well spent. The stories he told breathed life into the science which we all came know under his tutelage. He instilled his students with integrity, respect, humor, and a sharp eye for any place where he might have laid his glasses since his last cup of coffee.

While his children were young, JD had established a routine of going home at 5:00 p.m. to spend time with them, returning to the lab after their bedtime, and frequently working in the lab until 2:00 a.m. This routine continued long after his children were grown. Many of his colleagues and students would find this late night discussion time fruitful for hearing his theories of ABA biosynthesis and dormancy induction in maize and for bouncing their own ideas off him.

The other great draw in his lab was the coffee maker. It attracted geneticists like moths to a flame. Such sessions were frequently brainstorming sessions about whatever eclectic topic had arisen first. Graduate students in the lab learned more from these drop-in visitors and participating in these discussions that they catalyzed with JD than they ever learned in their courses. Jim Dunlap, one colleague who was influenced by JD early in his career recently wrote: "I can remember my first contact with Dr. J. D. Smith. It was a graduate course in genetics. JD went quickly through about ten different subjects and associated concepts within the 60 minutes allotted for the class. At the end of the class, JD was still expanding on the already vast array of ideas presented to this group of students. At that time, I knew that JD was a person totally immersed in his intellectual pursuits of answers to biological questions. I was both fascinated and frustrated. JD had communicated the excitement of searching for hidden truths but I had failed to take even one note for the entire lecture. The rest of the semester was the closest I ever come to an academic race as I attempted to anticipate important issues and sort out the many possibilities that JD had presented. It was JD that initiated my interest and emotional involvement with maize. This was long before the 'Field of Dreams' but certainly set the scene. JD was the mentor on a special project related to the regulation of precocious germination in maize. On the first day, he took me to the field and introduced his large 'family' of maize genetic material. We looked at endosperm types and pericarp color until I was utterly confused. Over the next few months, JD worked daily with me and his many other student colleagues as we collected pollen and made crosses. It was an experience with a man and his crop that I will cherish the rest of my life. I still see JD once or twice a year. He is still able to deliver ideas faster than I can take notes. It is not often that people are able to nudge others in life-changing ways. JD is one of those exceptional people. He certainly has made an impact on my life that will always be special."

JD's great belief in liberal causes coupled with frequently conservative graduate students often led to lively discussions of politics. While no one ever changed JD's mind, he succeeded in changing a few with his impassioned beliefs and his willingness to take unpopular stances when he believed in the cause. His passions for science and love of all things maize were contagious to his students. The other source of many discussions was the crossing block. JD led a band of undergraduate and graduate students as well as other professors into the crossing block each day. As we split up to work our way through the various ranges, ideas seem to spring from the fertile Brazos bottom soil. Such discussions usually involved shouting back and forth between the various ranges of stocks. Slogging in the mud and dodging fire ants also took our concentration. Most of the students frequently wilted some in the Texas July heat, but JD, fortified with his thermos of coffee, always seemed somehow comfortable. The most anticipated event each year was the corn party, held at JD's house after nearly all of the crossing was done, but before the Kandy Korn, Mainliner, and Silver Queen border rows got too old.

For a number of years, a colloquium was also held at his home one evening a week during the semester. Graduate students, and other professors took turns giving seminars about a mutually agreed upon topic. One year it was particularly enjoyable as JD's wife, Edna, was serving as editor of a cookbook and was testing every recipe in the book, using the colloquium participants as an eager testing population. Edna and JD also frequently had a large menagerie of animals around the house. Cats ruled the roost, but a pet raccoon and other assorted animals were also around. JD did a number of backcrosses trying to separate coat color and eye color genes in cats but gave up after a large number of backcrosses failed to produce a blue-eyed black cat. During those years, graduate students were required to adopt kittens in addition to other incidentals, such as passing their prelims prior to graduation.

JD's research on ABA biosynthesis was pioneering. Utilizing various mutant stocks in a genetic dissection, he elucidated the biochemical pathway that is now accepted as the most likely source of ABA in maize. Along the way his research interests turned to dormancy induction and the flavonoid pathways.

JD allowed his students free reign and did his best to support their research whether it dealt with ABA biosynthesis, or genes for resistance to southern rust. As long as his students were interested in maize genetics, he always encouraged their explorations. While his critique of research interpretations was intense, and manuscripts frequently came back from his office dripping red, there was a pervasive feeling that he respected the students' efforts even as he sought to sharpen our mental processes.

PART C

RECENT DEVELOPMENTS IN THE GENETICS AND MOLECULAR BIOLOGY OF MAIZE MUTANTS

by M. Motto[1], A. Bianchi[2] and F. Salamini[3]

[1] Istituto Sperimentale Cerealicoltura, 000191 Rome, Italy.
[2] Experimental Institute of Cereal Research—Maize Section, Via Stezzano, 24-24126 Bergamo, Italy.
[3] Max-Planck Institüt for Zuchtungsforschung, Carl-von-Linne-Weg 10, D-50829 Cologne 30, Germany.

Introduction

Perhaps the most intriguing aspect of this book on the scientific career and achievements of several important maize geneticists can be summarized in a simple question: what motivated them? The answer is, without any doubt equally simple: their feeling and respect for the maize plant. This plant is fascinating, diverse and aesthetically attractive, with a vegetative and reproductive morphology simple enough to make it a natural model system for genetic studies. But perhaps even more important is the fact that the plant itself is deeply rooted in the culture and economy of the Americas. This concept is the *leitmotif* of the book "Corn and Its Early Fathers" by H. A. Wallace and W. Brown (1956), which is dedicated to maize and to the scientists who established its success as a crop species.

Maize, as a wild plant, has existed for a long time, and since the time of domestication has followed man from the Americas to Asia and Europe over millennia. The maize plant has adapted successfully to hostile environments, a characteristic which geneticists term "genetic flexibility."

In their book, Wallace and Brown profess a deep respect for the American Indians who selected and used the maize plant, often adopting it as the dominant symbol of their civilization. On the other hand, these same authors take pride in describing how a small group of pioneers achieved one of the most important applications of plant biology: the exploitation of hybrid heterosis. In this book the theme is again the same: Man, the dominating figure who makes decisions, suffers, produces results and is occasionally happy.

The synthesis by H.A. Wallace and W. Brown, together with the introductory chapters (Part A) by P. A. Peterson and the biographies (Part B) that represent the backbone of this book, outline a segment of the history of science that demonstrates that successful maize breeding has always been accompanied by genetic discoveries. "Corn And Its Early Fathers" also gives credit to C. Darwin, G. Mendel, G. H. Shull and E. M. East as pioneers in the development of maize hybrids, together with the 17th and 18th century agronomists who noted the heterotic effects of intervarietal crosses. Needless to say, these four scientists were motivated in their studies more by the challenge of discovering basic biological principles than by the practical potential of their discoveries.* Indeed, for the first several decades of this century, the contribution of genetics to maize breeding and to the agronomic success of the crop was cultural, rather than practical: the outstanding maize geneticists of that era had the role of training those maize breeders who would later lay the foundations for maize improvement programs carried out by private companies and by public institutions.**

Figure 24 in Part A of this book shows that the modern schools of maize breeding and genetics can all be traced back to E. M. East. Hence the extraordinary success of maize as a field crop is due to the work of scientists motivated by practical considerations, but who had their origins in academic maize genetics. Genetics and breeding studies in this species have nevertheless followed different pathways, with genetics representing a noble soul that has remained separate from practical breeding. The biographies in Part B also reflect the clear-cut division between genetics and breeding: one series presents the achievements of maize geneticists, while a second group illustrates the contributions of breeding-oriented scientists to the improvement of the crop. Briefly, we wish to show here that the investigations carried out on maize in the first seven decades of this century, encompassing studies on cytogenetics, formal and biochemical genetics, mutagenesis and transposon mutagenesis, genetic dissection of metabolic pathways and developmental genetics, generated a body of genetic knowledge. These features became the basis for the continuous, efficient and fruitful development of genetic improvement programs. At the same time it has often been difficult for many basic research programs to predict what practical content or applicative scope they might have.

---

*In 1904, C.B. Davenport, at that time director of the Station of Cold Spring Harbor, asked G. Shull to grow a few maize plants to demonstrate Mendel's Laws to visitors. For this purpose Shull chose sweet and dent genotypes of different origin to be crossed and self-fertilized. In this way Shull began the experiments that led him to the discovery of the effects of outbreeding and inbreeding.
**Although we assume that this conclusion is correct, as always it admits large and relevant exceptions. These concern particularly the scientific and methodological contributions given to maize breeding from the schools of quantitative genetics at North Carolina State University in Raleigh and at Iowa State University in Ames (Hallauer and Miranda, 1988). A second important exception was the development of systems of cytoplasmic male sterility and of pollen fertility restoration (Duvick, 1965). These genetical contributions were rapidly and efficiently exploited by the seed companies for producing maize hybrids.

It is our feeling, however, that genetics and breeding of maize have more recently fused into a common discipline. This conclusion emerges from a consideration of the results that molecular biology and advanced maize genetics are now making available for the future improvement of this crop. The final goal of these molecular studies, in fact, is to assign a function to all maize genes. Knowledge of the function of a gene is the prerequisite for its possible practical manipulation based on recombinant DNA techniques. This is a realistic goal, thanks to the development of techniques for controlling gene expression and for genetic transformation of maize. The study at the molecular level of the function of a gene and of its metabolic relevance, on the other hand, itself relies on the availability of mutants. These mutants are the result of much work by maize geneticists in collecting, classifying, and mapping them, and in genetically dissecting physiological and developmental pathways. Finally, the investigations on transposable elements pioneered by B. McClintock, P. A. Peterson, and M. M. Rhoades—studies considered in the 50s and 60s at best esoteric—are now of tremendous importance for applied genetics. The use of transposons is now a routine approach to the induction of unstable mutations—the starting material for the cloning of agriculturally important genes and understanding their metabolic function. The central role of this technique is demonstrated by the fact that one of the most successful maize seed companies has automated the procedure and can now assign functions to genes for which the DNA sequence is known (Meeley and Briggs, 1995).

In conclusion, it is now evident that the role played by basic genetical research has been a dominant force in agricultural progress. This underlines the wisdom of the many institutions and private groups who have financially supported the Maize Genetics Cooperative and, more generally, basic studies of maize biology. The object of Part C of this book is to highlight the nature and variety of scientific discoveries that are emerging from the work of the maize scholars honored in *MAYDICA* monographs. These modern contributions represent a synthesis between the maize genetics tradition and molecular biology. In fact, a majority of the scientists who will be directly mentioned or cited in this chapter have pedigrees springing from the tree shown in Part A, Chapter 16 of this book, an obvious case of cultural inheritance in genetic studies.

43 Gene Cloning

The Role of Endosperm Genetics

The maize kernel provides a particularly tractable genetic system. The conspicuous endosperm and embryo can easily be monitored, as several hundred kernels are available in a single ear (Coe and Neuffer, 1978). At the end of the 1970s, the wealth of studies on endosperm genetics supplied the major rationale for cloning and characterizing maize genes. The cloning of genes responsible for zein synthesis, the primary storage protein of maize endosperm, is particularly noteworthy. The genes encoding these proteins were among the very first to be cloned and sequenced in plants (Geraghty *et al.*, 1981; Burr *et al.*, 1982; Pedersen *et al.*, 1982), mainly because the deposition of zeins had previously been investigated in great detail (Nelson, 1980). In addition, within the endosperm, transcripts encoded by storage protein genes were abundant and easily accessible for molecular analysis.

Oliver Nelson (University of Wisconsin) was particularly active in establishing the association between some endosperm mutations and their effects on storage product accumulation or on enzymatic deficiencies, with the objective of genetically dissecting specific biosynthetic pathways. Genes encoding zeins belong to multigene families that vary in size, organization and chromosomal localization (reviewed in Motto *et al.*, 1989a). Several mutants for genes regulating the production of maize storage proteins during endosperm development have been studied in some detail, for example *opaque-2 (o2)*, *floury-2 (fl2)*, *opaque-7 (o7)*, *opaque-6 (o6)*, *floury-3 (fl3)*, *Mucronate (Mc)*, and *De-B30*. The *O2* gene has been cloned (Schmidt *et al.*, 1987; Motto *et al.*, 1988) and shown to encode a DNA-binding protein belonging to the b-ZIP class of eukaryotic transcriptional activators (Hartings *et al.*, 1989; Schmidt *et al.*, 1990).

The groups presently active in the study of the genetic control of zein deposition are those of B. Burr (Brookhaven National Laboratory, Upton), B.

Larkins (Arizona University), R. Boston (North Carolina State University, Raleigh), M. Motto (Bergamo), F. Salamini and R. Thompson (Max-Planck Institüt, Köln).

The organization of zein genes has been studied by J. Messing (Rutgers University, Piscataway), I. Rubenstein (University of Minnesota, St. Paul), A. Viotti (Milano, Italy), G. Feix (Freiburg, Germany). The use of the recessive *o2* allele in maize breeding was considered by K. S. McWhirter (University of Sydney, Canberra), S. K. S. Vasal (CIMMYT, Mexico City), and M. Motto (Bergamo). Recent studies have been devoted to the functional characterization of the *O2* transcriptional activator (Aukerman *et al.*, 1991; Lohmer *et al.*, 1991, 1993); the study of the ribosome-inactivating protein b-32, which under the control of *O2*, protects the developing endosperm against pathogens (Maddaloni *et al.*, 1991, 1997; Bass *et al.*, 1995; Mehta *et al.*, 1995); the characterization of other genes under the control of *O2* (Maddaloni *et al.*, 1996); the nature of the regulatory gene *fl2* (Coleman *et al.*, 1995) and of other dominant regulators (Marocco *et al.*, 1991; Boston *et al.*, 1995); the organization of zein genes in clusters (Liu and Rubenstein, 1993; Chaudhuri and Messing, 1994); the effect of nitrogen nutrients on endosperm gene expression (Balconi *et al.*, 1993).

A second research area which has grown out of endosperm genetics is the study of starch biosynthesis. A cardinal role in these investigations was also played by O. E. Nelson (reviewed in Nelson and Pan, 1995). Several mutants are known in this pathway and the enzyme defects associated with individual mutant genes are clear in many cases. Progress has also been stimulated by the advent of gene cloning technologies and by the commercial interest in genetically modified starches. Hannah (1997) has proposed, in a recent review, a scheme of starch-carbohydrate metabolism in maize endosperm based on the correlation between enzyme activities and corresponding mutants. Gene-enzyme relationships are known for the following loci: *Mn1* (*Miniature1*) and invertase, *Sh1* (*Shrunken1*) and sucrose synthase, *Sh2* (*Shrunken2*) and *Bt2* (*Brittle2*) and ADP-glucose pyrophosphorylase, *Bt1* (*Brittle1*) and a membrane-bound metabolite transporter, *Wx* (*Waxy*) and granule-bound UDP-glucose transferase, *Ae* (*Amylose-extender*) and starch branching enzyme, *Su1* (*Sugary1*) and starch debranching enzyme. The association of *mn1* with invertase activity was reported by Miller and Chourey (1992) and Taliercio *et al.* (1995). That sucrose degradation was the most important role of sucrose synthase was revealed by Chourey and Nelson (1979) while the cloning of the gene encoding this enzyme was achieved by Werr *et al.* (1985). The second sucrose synthase gene *Sus1* was cloned by McCarty *et al.* (1996). The genes *Sh2* and *Bt2* were cloned by Bhave *et al.* (1990) and BAE *et al.* (1990), and later shown not to be expressed in the embryo or leaf tissue. The *Bt1* locus was cloned by transposon tagging (Sullivan *et al.*, 1991) and found to be related to a membrane-bound metabolite transporter. The *Wx* gene was cloned independently in the laboratories of H. Saedler and N. Fedoroff (Fedoroff *et al.*, 1983a; Klösgen *et al.*, 1986). *Amylose extender* was cloned by Fisher *et al.* (1993) and Stinard *et al.* (1993). James *et al.* (1995) have cloned the *Su1* gene using the transposable element *Mu*. They also confirmed the previous observation of Pan and Nelson (1984) that the gene is involved in the debranching of starch, showing that starch debranching is necessary for starch synthesis. Hannah (1997) lists as attractive areas in the study of starch synthesis the following topics: genes involved in signal transduction; phosphorylation of starch biosynthetic enzymes; the role of isoforms of several enzymes; the role of the *Mn1* gene in regulating invertase activity; limiting steps in starch biosynthesis, in addition to the one catalyzed by adenosine diphosphate glucose-pyrophosphorylase; the coordination of starch biosynthetic pathways in the endosperm and the existence of accessory pathways also leading to starch synthesis. Structural features of starch synthetic genes and of their promoters have also been revealed as important for the understanding of regulative pathways (Maas *et al.*, 1990; Lugert and Werr, 1994). Among the active groups studying starch synthesis in the maize endosperm, particularly at the molecular level, are those of C. Hannah (University of Florida, Gainesville), P. Schnable (Iowa State University, Ames), W. Werr (University of Köln, Germany), P. Chourey (University of Florida, Gainesville), and M. Guiltinan (Pennsylvania State University).

The anthocyanin biosynthetic pathway has a long and rich scientific history, beginning with Mendel who studied the segregation of dark-red kernels on the maize ear. In maize, the genetic control of anthocyanin biosynthesis has been the subject of intensive research and more than 20 structural and regulatory genes have been identified (Coe *et al.*, 1988a). These studies have also lead to the discovery of such important biological phenomena as genetic transposition, paramutation, chromosomal breakage-fusion-bridge (BFB) cycles and the use of visible markers for maize transformation. Several genes encoding enzymatic activities of the pathway itself have been identified and cloned via transposon tagging. These include the structural genes for dihydrogenetic reductase (*A1*) (O'Reilly *et al.*, 1985), *A2* (Menssen *et al.*, 1990), chalcone synthase (*C2*, colored 2 and *Whp*, white pollen) (Wienand *et al.*, 1986; Franken *et al.*, 1991), UDP glucose-flavonol glucosyl transferase (bronze, *Bz1*) (Fedoroff *et al.*, 1984; Dooner *et al.*, 1985), and glutathione-S-transferase (bronze, *Bz2*) (McLaughlin and Walbot, 1987; Theres *et al.*, 1987; Marrs *et al.*, 1995). The structural genes are coordinately regulated by other genetic loci, which condition the tissue-specific production of the anthocyanin pigment. At least two classes of regulatory genes exist: *Pl*-like and *C1*-like (Paz-Ares *et al.*, 1987; Cone and Burr, 1989) and members of the R and B families (Chandler *et al.*, 1989; Ludwig *et al.*, 1989). These genes have been cloned in several laboratories. The predicted *Pl*-encoded and *C1*-encoded proteins are homologous to each other and to the product of the animal *myb* proto-oncogene family (Paz-Ares *et al.*, 1987; Cone *et al.*, 1993) and contain an acidic transcriptional activator function. R and B genes are functionally equivalent (Goff *et al.*, 1992). Molecular studies (Ludwig *et al.*, 1989; Perrot and Cone, 1989; Radicella *et al.*, 1991) revealed that the R family encodes Myc-like proteins which contain a basic-helix-loop-helix motif found in other eukaryotic transcriptional activators such as Max and MyoD1 (Davis *et al.*, 1987; DePinho *et al.*, 1987). The *Intensifier* (*In*) gene, which regulates the expression of *Whp*, shows sequence homologies to the R gene (Burr *et al.*, 1996). A characteristic feature of *In* is that most of the gene-specific transcripts are differentially spliced. The P locus, involved in the synthesis of a red pigment in the pericarp, cob and other floral tissues, was cloned and characterized by Lechelt *et al.* (1989) and Grotewold *et al.* (1991).

Coe (University of Missouri) and Doebley (University of Minnesota) investigating the role of regulatory and enzymatic loci in the maize anthocyanin pathway, suggested that purple kernels resulted from changes in *cis* regulatory elements at regulatory loci and not changes in either regulatory protein function nor the enzymatic loci (Hanson *et al.*, 1996).

Laboratories still active in elucidating the mechanisms affecting the accumulation and tissue-specific expression of anthocyanins are those of J. L. Kermicle (University of Wisconsin, Madison), K. C. Cone (University of Missouri, Columbia), S. L. Dellaporta (Yale University, New Haven), F. A. and B. Burr (Brookhaven National Laboratory, Upton), E. H. Coe, (University of Missouri, Columbia), V. Walbot (Stanford University, Stanford), V. Chandler

(University of Oregon, Eugene), G. Gavazzi and C. Tonelli (University of Milan), U. Wienand (University of Hamburg), T. Peterson (Cold Spring Harbor Laboratory) and H. Saedler (Max-Planck Institüt, Cologne).

One feature of the anthocyanin studies has been the focus on the phenomenon known as paramutation. Paramutation, an interallelic interaction, leads, at relatively high frequency, to a heritable change in one of the two alleles participating in a cross. In maize, paramutation has been studied at some loci that control anthocyanin production. Two of the systems considered by R. A. Brink and J. Kermicle involve primarily the *R* and *B* loci of the *R* gene family. The structures of the paramutagenic allele *R-st* (Eggleston *et al.*, 1995; Kermicle *et al.*, 1995) and the paramutable allele *R-r* (Walker *et al.*, 1995) have now been determined. The paramutagenic *R-st* allele has four *r* genes repeated in tandem and separated by several kilobases. The first centromere-proximal gene is interrupted by a transposable element (I-R), but neither this gene alone nor the I-R transposon is responsible for the paramutation (Kermicle *et al.*, 1995). Instead, the other three *R* genes or the sequences that lie between them confer paramutagenicity (Eggleston *et al.*, 1995; Kermicle *et al.*, 1995). These three genes are expressed at a very low level, resulting in a near-colorless phenotype. Derivative alleles, generated by unequal crossing-over, have fewer repeats, are less paramutagenic and confer higher levels of seed pigmentation. The upstream regions of the component *R* genes are extensively methylated in the most paramutagenic derivatives (Eggleston *et al.*, 1995). In another example, the *R-sc* allele, which has only a single functional *R* gene, may interact with a particular allele of the *R* homologue *Sn*. This apparent non-allelic silencing results in extensive methylation, and can be reversed, to some extent, by treatment with 5-azacytidine (Ronchi *et al.*, 1995). G. Patterson (Cold Spring Harbor Laboratory) and V. Chandler (University of Oregon) have mapped the differences between two alleles at the B locus, *B'*, the dominant "changer," and *B-I*, the recipient, to the 5' end of the locus between the transcription start site and a position 0.1 map unit upstream of it (Patterson *et al.*, 1993). The promoter region is responsible for paramutation (Patterson *et al.*, 1995), but the precise nature of the rearrangements that distinguish these alleles from others that are non-paramutational has not yet been determined.

While paramutation is known to result in drastically reduced levels of transcription of the allele involved, a strong inverse correlation has been established between the degree of methylation and the expression of some genes like the zein genes (Bianchi and Viotti, 1988; Lund *et al.*, 1995a). In addition, DNA methylation has been implicated in the control of a number of cellular processes in maize, including transcription (Bianchi and Viotti, 1988; Walbot and Warren, 1990; Lund *et al.*, 1995b; Rossi *et al.*, 1997), imprinting (Chaudhuri and Messing, 1994; Lund *et al.*, 1995a), and transposition (Schläppi *et al.*, 1994; Wang *et al.*, 1996).

In general, the establishment of an association between a specific endosperm mutation and a particular step in a pathway leading to the synthesis of either storage products or to anthocyanin deposition was readily accomplished. Recently, there has been interest in understanding other seed-specific developmental processes, based on the isolation of a class of mutations having simultaneous effects on both endosperm and embryo development. Sheridan and Neuffer (1980) and Neuffer and Sheridan (1980), for instance, analyzed 855 *defective kernel* (*dek*) mutants representing about 285 different loci. Similarly, Scanlon *et al.* (1994) isolated, in an active *Mutator* (*Mu*) stock, 63 mutations affecting the development of the maize kernel. These probably result from insertion of the *Mu* transposon within a *dek* gene; thus, the affected loci are accessible to molecular cloning via transposon-tagging. Some *dek* mutants have also been used to clarify the relationship between mitotic activity and DNA endoreduplication in endosperm cells (Knowles *et al.*, 1990, 1992).

A recent investigation has, moreover, suggested that in maize endosperm the endoreduplication of DNA is a result of two events: inhibition of factors that promote entry into the M-phase of the cell cycle, and induction of S phase-related protein kinases (Grafi and Larkins, 1995). Recently, cDNA markers for the development of basal transfer cells were isolated (Hueros *et al.*, 1995), which may provide more information about the role of this layer in endosperm formation. The transfer layer is affected in a number of *defective kernel* mutants including *mn-1*, which suggests that it may carry out rate-limiting process in nutrient uptake.

44 Visiting the Genome

Transposition

In the late 40s, Barbara McClintock, working at the Cold Spring Harbor Laboratory, began to call attention to the novel behavior of maize genetic factors which she called "controlling elements." These genetic elements, first noticed because they inhibited the expression of other maize genes, did not have a fixed chromosomal location: they seemed to move, inserting and excising from other genes. After excision, the function of a previously repressed gene was often restored. Thus, the genes associated with controlling elements were rendered somatically unstable. Genetic transposition represents the most significant contribution of maize genetics to general genetics. For this discovery B. McClintock received in 1983 the Nobel Prize in Physiology or Medicine (Lewin, 1983). The discovery of this complex phenomenon was possible, because by the '30s, maize genetics had already made available a number of endosperm markers, particularly the linked series Wx, Bz, Sh, and C, located in the short arm of chromosome 9. In the years preceding the discovery, B. McClintock had already developed a phenotypic color code that permitted her to decipher genetical events, such as chromosomal cycles of breakage-fusion-bridge (BFB), based on the pattern of somatic variegation of the endosperm. At this time, other scientists who had a significant role in developing maize genetics, such as M. M. Rhoades and P. A. Peterson, were also attracted by the phenomena of somatic instability. P.A. Peterson, like B. McClintock, recognized the theoretical importance of the phenotype of a *pale green* mutable allele expressed in the seedling, which later contributed to clarification of several aspects of gene transposition (Peterson, 1953).

Transposable elements today represent a basic tool in the study of the genetics of several crop plants. The cardinal feature of these elements, the ability to transpose, has been used by molecular biologists to detect and clone interesting genes not only in maize but also in species into which these mobile elements have been transferred via genetic engineering. The schools active in the molecular investigation of maize transposable elements originated particularly with B. McClintock, but also P. A. Peterson, M. M. Rhoades, O. E. Nelson, N. Fedoroff, D. Schwartz, and from the Köln groups led by P. Starlinger and H. Saedler, whose interest in plant transposable elements derived from their experience with IS elements in bacteria. In addition, the genetic materials from the laboratories of B. McClintock and P. A. Peterson were made available to them.

Activator (*Ac*) is a maize transposon that was first identified genetically by McClintock (1945) because of its ability to promote chromosome breakage at a specific chromosomal site designated the *Dissociation* (*Ds*) locus. The *Ac* family includes a large and structurally heterogeneous group of elements collectively designated *Ds* elements, which can transpose only if provided with the *Ac*-encoded transposase function. The genetic behavior of *Ac* indicates that it transposes by a non-replicative mechanism and that transposition can be coupled to the replication process, as recently demonstrated by S. Dellaporta and coworkers (Moreno *et al.*, 1992). The molecular characterization of the *Ac* and *Ds* elements, investigated by the Fedoroff and Starlinger groups (Fedoroff *et al.*, 1983b; Behrens *et al.*, 1984; Müller-Neumann *et al.*, 1984; Pohlman *et al.*, 1984), has led to the cloning of a number of maize genes that have been insertionally inactivated by members of this family. The *Ac* element is transcribed into a 3.5 kb mRNA (Kunze *et al.*, 1987; Finnegan *et al.*, 1988) which codes for a transposase. The *Ac* Tpase is a DNA-binding protein which recognizes subterminally located, reiterated sequences of the same element (Kunze and Starlinger, 1989). Somatic and germinal activities of *Ac* Tpase have been studied in transgenic tobacco by Kunze *et al.* (1993). In addition, the utility of *Ac* for gene tagging (Fedoroff *et al.*, 1984) and cell lineage analysis (Finnegan *et al.*, 1989; Dawe and Freeling, 1990, Dellaporta *et al.*, 1991) has been well documented. *Ac* has been shown to be active in *Nicotiana tabacum* (Baker *et al.*, 1986) and in numerous other dicotyledon plant species. The laboratories active in research at the molecular level on the *Ac/Ds* system are numerous. Here we mention those led by J. Jones and K. Theres for transposition in tomato (Carroll *et al.*, 1995; Nussaume *et al.*, 1995), J. L. J. Kermicle for position dependence of insertion element activity (Alleman and Kermicle, 1993), S. L. Dellaporta for methylation and *Ac* activity (Brutnell and Dellaporta, 1994), B. Burr and S. Wessler for allelic change in state due to *Ds* transposition (Weil *et al.*, 1992), and for the mode of excision of *Ac* (Baran *et al.*, 1992), H. K. Dooner for transpositional sites of *Ac* (Dooner *et al.*, 1994), and for frequency and timing of transposition in tobacco and *Arabidopsis* (Keller *et al.*, 1993a, b), R. Kunze for binding motifs for Ac transposase required for excision of *Ds* elements (Bravo-Angel *et al.*, 1995), and *Ds* transposition elements in transgenic tobacco (Firek *et al.*, 1996).

The *En/Spm* system of transposable elements was isolated independently by P. A. Peterson (1953; *En*) and B. McClintock (1954; *Spm*) and genetically analyzed by these two scientists since its discovery (for review, see Peterson, 1987). The complete element *En-1* was cloned (Pereira *et al.*, 1986) and the DNA sequence of the 8.3 kb element was determined (Pereira *et al.*, 1986). An *Spm* element was also sequenced (Masson *et al.*, 1987) and shown to differ from the *En-1* sequence in only eight positions. This homology was expected from genetic comparisons of *En* and *Spm* (Peterson, 1965). The element encodes alternatively spliced transcripts (Gierl *et al.*, 1988; Masson *et al.*, 1989). A combination of two of these gene products is necessary and sufficient to promote transposition of the element, as well as of transposition-defective derivatives, designated *dSpm* elements (Frey *et al.*, 1989). The Fedoroff group has shown that *Spm* is genetically inactivated by C-methylation near its transposition start site (Schläppi *et al.*, 1994). The *En/Spm* element has been a useful tool for isolating several structural and regulatory maize genes (Gierl and Saedler, 1992). Initial studies in tobacco and potato showed that the autonomous *En/Spm* elements were mobile in these species (Frey *et al.*, 1989; Masson and Fedoroff, 1989; Pereira and Saedler, 1989). The low excision frequency, however, seemed to preclude the use of *En/Spm* for transposon tagging in heterologous hosts. More recently, the Saedler laboratory, by studying the mobility of the *En/Spm* element in *A. thaliana,* have shown that the rate of germinal excision of *En/Spm* is high enough for transposon tagging in *Arabidopsis* (Cardon *et al.*, 1993). The groups still active in the study of *En/Spm* elements are those of P. A. Peterson—loss of the element (Dash and Peterson, 1994), chromosome localization (Chang and Peterson, 1994), role of the element in creating genetic variability (Peterson, 1993), role of *En*-transposase level on transcription (Peterson, 1995); N. Fedoroff: interaction between *Spm*-encoded proteins (Schläppi *et al.*, 1995), interaction of *Spm* transposons with *cis*-elements (Raina *et al.*, 1995), epigenetic regulation (Fedoroff *et al.*, 1995); O. E. Nelson: suppressibility of *Spm* (Bunkers *et al.*, 1993); of H. P. Döring, M. Motto and F. Salamini: gene tagging with *En/ Spm* (Michel *et al.*, 1995; Tacke *et al.*, 1995); H. Saedler on the biological relevance of *En/Spm* in heterologous hosts.

The *Mutator* (*Mu*) transposable element system of maize was discovered by Robertson (1978) on the basis of its ability to induce mutations at an extremely high rate usually 50- to 100-fold higher than spontaneous mutation rates. Molecular analysis of *Mu*-induced mutations led to the discovery of a family of at least 8 related transposable elements. *Mu1* was the first member of the family to be cloned and was recovered from an unstable *Adh1* allele (Bennetzen *et al.*, 1984). It is approximately 1.4 kb long and has inverted repeats of ~200 bp (Barker *et al.*, 1984). Subsequently, *Mu1.7* (*Mu2*), *Mu3*, *rcy:Mu7* and *Mu8* were also found as insertions in *Mu*-induced mutants (Taylor and Walbot, 1987; Oishi and Freeling, 1988; Schnable *et al.*, 1989; Fleenor *et al.*, 1990). Two other *Mu* elements, *Mu4* and *Mu5,* were cloned from non-*Mutator* stocks via their homology to the inverted repeats of the *Mu1* element (Talbert *et al.*, 1989), and a further *Mu* element, has recently been found to correspond to the autonomous *Cy* transposon (Hsia and Schnable, 1996). The system is regulated by the *MuDR* class of element. This regulatory *Mu* element was independently cloned in several laboratories (Qin and Ellingboe, 1990; Chomet *et al.,* 1991; Hershberger *et al.*, 1991; Qin *et al.,* 1991; Chandler and Hardeman, 1992) and different groups have initiated the examination of the molecular mechanisms of *Mu* transposition and regulation. *MuDR* is 4.9 kb in length, and two *MuDR*-homologous transcripts are present in active *Mutator* stocks (Chomet *et al.,* 1991; Hershberger *et al.*, 1991; Qin *et al.*, 1991). The groups of V. Walbot (Stanford University, Stanford), M. Freeling (University of California, Berkeley), J. L. Bennetzen (Purdue University), V. L. Chandler (University of Oregon) and P. Schnable (Iowa State University) continue to actively study the basic aspects of *Mu* biology. The Freeling group (Lisch *et al.*, 1995) have shown that *Mu* transposons can transpose duplicatively, that *MuDR1* suffers frequently deletions, and that the MURB protein is localized in a tissue-specific manner which may explain the pattern of *Mu* transposition *in planta* (Donlin *et al.*, 1995). The group of V. Walbot is contributing relevant data on the characterization of the major transcript of *MUDR* (Hershberger *et al.*, 1995), on promoter activity of terminal sequences of *Mudr* (Benito and Walbot, 1994), on the similarity of *Mu* to bacterial insertion sequences (Eisen *et al.*, 1994). J. L. Bennetzen studied the basis for preferential insertion of *Mu* elements into genomic DNA (Cresse *et al.*, 1995) and has developed models of the origins of *Mu* subfamilies (Bennetzen and Springer, 1994). V. L. Chandler has provided data on the preferential targeting of different classes of *Mu* elements (Hardeman and Chandler, 1993) and the association of *Mu* insertion with large tandem duplications (Harris *et al.*, 1994). P. Schnable has demonstrated the sequence identity between the autonomous transposon *Cy* and *MuDr* element (Hsia and Schnable, 1996). The amplification of *Mu* elements in active *Mutator* crosses, together with their tendency to insert in transcribed regions, implies not only a high probability of identifying a *Mu* insertion allele of a gene of interest, but also that it should be possible to identify *Mu* insertion alleles of genes of known sequence whose mutant phenotypes are not known. Such alleles could be identified by utilizing the PCR protocol on pooled genomic DNA, using as primers a sequence within the *Mu* TIR and a second one within the gene of interest. This approach, the "Mu gene machine" which was developed by Meeley and Briggs of Pioneer Hi-Bred International (1995) is proving to be extremely powerful in the study of gene function and fitness. Also maize breeding is being improved by the use of the Meeley and Briggs procedure which allows, among other applications, facilitates the discovery of genes limiting maize performance.

Several other maize transposons have been genetically characterized (reviewed by Peterson, 1993). Some of them have now been cloned. The maize transposable element *Bs1* is similar to retrotransposons, a class of mobile DNA sequences in eukaryotes that transpose through a reverse-transcribed RNA intermediate. *Bs1* was identified as an inactivating insertion of the *alcohol dehydrogenase (Ad1)* gene (Mottinger *et al.*, 1984). *Bs1*-like elements were found to be present at a copy number of one to five per haploid genome (Johns *et al.*, 1985). Studies by Jin and Bennetzen (1994) have indicated that *Bs1* may represent a defective version of a plant retrovirus which has incorporated a fragment of the maize plasma membrane ATPase gene. Fuerstenberg and Johns (1990) have studied the distribution of *Bs1* retrotransposons in *Zea* and related genera concluding that *Bs1* is not highly mobile. Sequences coding for a reverse transcriptase but lacking LTRs were also identified in the maize element *cin4* (Schwarz-Sommer *et al.*, 1987).

The *Bg-rbg* system of maize was originally reported to be responsible for the instability of the mutable allele *o2m(r)* at the *Opaque-2 (O2)* locus (Salamini, 1980, 1981). This system consists of the complete *Bg* elements and defective copies incapable of transposing (*rbg* members). Chromosomal and genetic instabilities originating from revertant derivatives of the *o2m(r)* mutable alleles and the tranpositional capability of the *Bg-rbg* system have been reported (Montanelli *et al.*, 1984; Motto *et al.*, 1987, 1989b). The autonomous and receptor elements of the *Bg-rbg* system were isolated and characterized at the molecular level (Motto *et al.*, 1989b; Hartings *et al.*, 1991a, b). The receptor element resembles the autonomous element in size, and both elements share extensive sequence homology (Hartings *et al.*, 1991a). Some 10 to 12 copies of *Bg*-homologous sequences are present in the maize genome (Motto *et al.*, 1989b). The distribution of *Bg*-like sequences in maize and in related species has been investigated (Hartings *et al.*, 1996). *Bg*-homologous sequences are present in similar numbers in all *Zea* taxa.

S. Wessler (University of Georgia), by studying mutations at the *Wx* locus has detected a new family of transposable elements designated *Tourist* (Bureau and Wessler, 1992). *Tourist* elements are, on average, 133 bp in length and are characterized by 13- to 14-bp terminal repeats, a 3-bp target site, and the potential to form DNA secondary structures. *Tourist* exists in at least 20 different sequences. These elements may transpose preferentially to AT-rich sequences, which tend to be non-coding DNA regions. Although several indications of element insertion have been reported, no evidence exists for element excision.

Other maize transposable elements are *Cin1*, *Cin2*, and *Cin3* (Blumberg Vel Spalve *et al.*, 1990; Shepherd *et al.*, 1984), *Stoner* (M. Marillonnet, University of Georgia, cited in Stapleton and Phillip, 1993) and the PREM-2 (Turcich *et al.*, 1996) and the *Zeon-1* families of retrotransposons (Hu *et al.*, 1995), the *Dotted* element (*Dt*) (Brown *et al.*, 1989) and the *a1-mrh* element (Shepherd *et al.*, 1989). More recently, Bennetzen and coworkers (San Miguel *et al.*, 1996), investigating the relative organization of genes and repetitive DNAs in the maize genome, have found that a 280 kbp region containing the maize genes *Adh1-F* and *w22* is composed primarily of retrotransposons inserted within each other. Because of their intermingled nature, high copy number and location—flanking maize genes—they suggested that such nested retroelement structures might represent the standard organization of intergenic regions.

Although maize transposable elements are proving their mettle in gene tagging campaigns, not all genes have a mutant allele with a known insertion.

Moreover, some eukaryotic sequences are simply not stable in prokaryotes. Yeast artificial chromosome (YAC) libraries have the potential to solve both of these problems, making chromosome walking possible and allowing "difficult" sequences to be maintained stably. K. Edwards and his colleagues at Zeneca Seeds have constructed a YAC library that contains 79,000 clones with an average insert size of 145 kb (Edwards *et al.*, 1992). The entire library can be screened at once by high-density colony hybridization, or pools of clones can be screened, with positive pools being progressively subdivided. This will be a step forward for maize gene isolation.

45 Genome Dissection and Fingerprinting

The Age of Molecular Markers

Modern contributions of maize cytology

The cytogenetics of maize is well developed and has played a pivotal role in defining many aspects of plant chromosome behavior, in addition to equipping maize geneticists with useful tools for addressing genetic problems (Carlson, 1988a). Two major figures dominated maize cytology in the past: M. M. Rhoades and B. McClintock. Their remarkable contributions are nowadays continued by the studies of W. Carlson on the control on chromosome B non-disjunction (Carlson, 1986, 1988b), control of meiotic loss of B chromosomes (Carlson and Roseman, 1992), systems for stable duplication of A chromosome segments (Carlson and Curtis, 1986; Carlson and Roseman, 1991); of J. A. Birchler, on compound B-A translocations (Birchler, 1994), marker systems for r-x1 deletions (Birchler and Coe, 1994); dosage effects on gene expression in maize ploidy series (Guo *et al.*, 1996); of J. B. Beckett, on the use of B-A translocations for locating recessive genes (Beckett, 1994), and on segregation products of B-A translocations (Kindiger *et al.*, 1991).

The recent adoption of molecular markers has opened new horizons in maize cytogenetic studies. Although still in its initial phase, the new strategy has already produced relevant results on the molecular characterization of B chromosome centric sequences (Alfenito and Birchler, 1993), on a recombination hot spot present in the *a1* gene (Xu *et al.*, 1995), on the structure of maize telomeres (Burr *et al.*, 1992), on DNA regions with high affinity for the nuclear matrix (Avramova *et al.*, 1995), on the relationships between genetic and physical distances (Civardi *et al.*, 1994) and between active genes and repetitive DNA sequences (Bennetzen *et al.*, 1994; Springer *et al.*, 1994), on physical and genetic mapping of morphological variants, with respect to RFLP markers (Chao *et al.*, 1996), and characterization of meiotic crossover identified by RFLP method (Timmermans *et al.*, 1996).

Using three-dimensional light microscopy, Dawe *et al.* (1994) have been able to produce a detailed analysis of changes in chromosome structure during meiotic prophase I. Chromosomes were found to condense completely before they pair, and pairing first becomes evident at the telomeres, at least those of chromosome 9S, which in the inbred line used—the very line used by Creighton and McClintock (1931) to demonstrate the relationship between crossing over and genetic exchange—are marked by readily distinguished knobs.

Molecular maps: a new approach to quantitative variability

The conflict between the Mendelian theory of particulate inheritance and the observation of continuous variation in nature for most traits was resolved in the early 1900s by the concept that quantitative traits can result from segregation of multiple genes.

Breeding studies reported by E. M. East on ear-length of maize confirmed the predictions of this theory.

This concept of quantitative variation was used by a number of maize geneticists including R. Comstock, H. Robinson, C. Cockerham and R. Moll, working at North Carolina State University, A. Hallauer, S. Eberhart, W. Russell at Iowa State University, and C.O. Gardner at the University of Nebraska, in developing biometrical methods designed to yield quantitative estimates of genetic and nongenetic contributions to measured characters. Quantitative genetic theory has been useful to breeders in characterizing the genetic variability of populations and in providing a basis for interpreting the phenomena of heterosis and response to selection. The spectacular progress in the last 60 years in improving the performance of maize hybrids is, in large part, attributable to these studies. The B73 inbred line, derived from long-term selection experiments by W. Russell conducted at Iowa State University, has been widely used throughout the world in commercial hybrid production, and represents the most visible practical result of these studies. A book which has become the basic reference text for modern maize breeders summarizing maize quantitative genetics has been published by A. Hallauer and F. Miranda (Hallauer and Miranda F., 1988). The American contribution to maize quantitative genetics was of paramount importance. In Europe also, schools developed, with contributions from Schnell and coworkers in Hohenheim (Germany), A. Gallais and coworkers in Gif sur Yvette (France) and E. Ottaviano in Milan (Italy). In Brazil E. Paterniani and coworkers in Piracicaba should also be mentioned.

Maize has a well-developed genetic map with over 600 genetic loci localized on the 10 chromosomes, many of which are expressed as seedling or kernel factors that are easily diagnosed. The identification and use of molecular markers has helped to characterize the maize genome further. This approach proved successful, at least in part, because maize DNA is highly polymorphic (Evola *et al.*, 1986). Helentjaris *et al.* (1986) constructed the first RFLP map for maize. Weber and Helentjaris (1989) used B-A translocations and markers to characterize chromosome segments around the centromeres. Helentjaris *et al.* (1988) observed that 29% of their cloned maize sequences hybridized to mappable, duplicated nucleotide sequences. Burr *et al.* (1988) developed an RFLP map based on segregation data in two recombinant inbred populations. Further RFLP maps have been developed by the Agrigenetics Company (Murray *et al.* 1988; Shoemaker *et al.*, 1992), Pioneer Hi-Bred International (Beavis and Grant, 1991) and by Hoisington *et al.* (1988), Gardiner *et al.* (1993). The RFLP procedure has been widely used for assigning maize inbreds to heterotic groups and for detecting pedigree relationships among lines and populations (Melchinger *et al.*, 1991; Smith and Smith, 1991; Livini *et al.*, 1992; Stuber *et al.*, 1992; Boppenmaier *et al.*, 1993; Williams *et al.*, 1995).

Until recently, investigations of quantitative variation were based on biometrical procedures, which characterize, *en masse*, the genetic factors involved in the expression of this type of variation. The use of mapped genetic markers provides a new and powerful approach to quantitative variation. Molecular markers, i.e. isozymes, RFLPs, RAPDs, AFLPs (Burr *et al.*, 1983; Helentjaris *et al.*, 1985; Evola *et al.*, 1986), are far superior to morphological markers for such applications as the tagging of major genes of agricultural importance and the genetic dissection of complex traits with a quantitative mode of inheritance (quantitative trait loci, QTLs) (Stuber *et al.*, 1992). Molecular markers facilitate breeding programs (Ragot *et al.*, 1995) and provide better estimates of types and magnitude of gene effects attributed to QTLs. Their use may lead to a better understanding of phenomena such as epistasis, pleiotropy and heterosis. This approach to quantitative genetics contributes relevant new methods to maize improvement (reviewed by Stuber, 1995). In addition, the use of molecular markers hybridizing to DNA of different species has opened new approaches to the study of chromosomal evolution (Helentjaris 1993; Paterson *et al.,* 1995) and to the study of meiotic recombination (Williams *et al.,* 1995).

The groups currently developing and using molecular markers at present are those of M. Lee (Iowa State University), A. Paterson (Texas A & M University), T. Helentjaris (Johnston, Iowa), M. Sari (University of Milan), C. Stuber (North Carolina State University), B. Burr (Brookaven National Laboratory, Upton), D. Hoisington (CIMMYT, Mexico), J. Dudley (University of Illinois), A. Melchinger (University of Hohenheim), J. Bennetzen (Purdue University, Indiana), A. Gallais (Institute Nationale de la Recherche Agronomique, Gif sur Yvette), M. Motto (Cereal Crop Institute, Bergamo), and several seed companies such as Pioneer Hi-Bred International (Johnston), CIBA Seeds (Basel), Zeneca Seeds (Bracknell, Berkshire), Limagrain Genetics (Champaign), DuPont de Nemours (Wilmington), Key Gene (Wageningen).

Partial sequencing and mapping of randomly selected cDNAs

In maize, the wealth of mutants, many of which have already been mapped and extensively studied, increases the chances of correlating map position of cDNA clones with mutant loci. This allows the identification of gene function and the use of single genes in biotechnological applications. Recent studies have also considered the analysis of expressed DNA sequences in order to characterize and classify in gene banks the genetic information present in plant species. The sequencing of anonymous cDNAs producing expressed sequence tags (ESTs), has shown great promise in the identification of putative clone functions through similarity with entries already present in genome databases. Helentjaris and coworkers (Shen *et al.,* 1994) and Keith *et al.* (1993) have reported partial sequencing and mapping of a large number of cDNAs from maize cDNA libraries. Sequencing and mapping data are useful in addressing basic questions of genomic structure and organization, such as gene clustering, gene duplication and synteny among species.

46 Biotechnology

New Methods for Engineering
Varieties with Superior Characters

Maize breeders have made significant progress in the improvement of maize over the last 70 years. About 50% of the increase in the yield is attributable to the use of improved varieties: the remainder derives from greater and more efficient use of fertilizers and crop management (Russell, 1991). The progress achieved by conventional maize breeding is summarized by the reviews of Crosbie (1982), Duvick (1984) and Russell (1984, 1991), to which the interested reader is directed for information on recent developments. Here we will focus on the contribution of recombinant DNA technology, which provides new opportunities for manipulating the maize genome.

In applied molecular biology common strategies can often be used for a large spectrum of crop species. A particular crop, however, may also require the development of specific technologies. Maize molecular breeding is supported by several agro-chemical companies, such as Monsanto, Ciba Geigy, DuPont de Nemours, Sandoz, Zeneca. Several schools of genetics and molecular biology have also significantly contributed to this development which can all be traced back to the tradition of maize genetics. We have already mentioned the key role played by maize geneticists in developing gene-tagging systems, reverse genetic screens, recombinant inbred lines, molecular maps, gene expression methods, and promoter characterization. Methodological contributions to DNA manipulation and the specificity of directed gene expression by the group of S. Dellaporta include a DNA miniprep protocol (Dellaporta et al., 1983) and the development of a visible marker for maize transformation (Ludwig et al., 1990).

Tissue culture

Maize has been more difficult to manipulate in tissue culture than most other species. Laboratories which have contributed in this area are those of B.G. Gengenbach, R. Phillips and C. Green (University of Minnesota), I. K. Vasil (University of Florida, Gainesville), J. A. Widholm (University of Illinois, Urbana) I. K. Hodges (Purdue University), H. Lörz (University of Hamburg, Germany) and M. Beckert (INRA, Clermont-Ferrand, France). Progress has been made over the last 20 years, and it is now possible to regenerate different genotypes from calli derived from immature embryos (Green and Phillips, 1975; Edallo *et al.*, 1981). The laboratory of Hodges has provided data on the inheritance of the tissue culture response (Willman *et al.*, 1989; Armstrong *et al.*, 1992). Maize genotypes can be associated with compact and organized callus cultures (Edallo *et al.*, 1981) and with callus which is friable and highly embryogenic, as with A188 and hybrids between A188 and B73 (Armstrong and Green, 1985). Embryogenic suspension cultures derived from friable calli are of particular interest as a source of regenerable protoplasts (Vasil and Vasil, 1987).

Maize is a good model system for studying the process of fertilization (Dumas and Mogensen, 1993) and embryogenesis (Meinke, 1991). Protocols for the isolation of male and female gametes have been developed initially by Dupius *et al.* (1987) and Wagner *et al.* (1989). Artificial zygotes that are capable of developing into plants have been produced by electrofusion of male and female gametes (Kranz and Lörz , 1993). The course of this 'electro-syngamy' has been studied by transmission electron microscopy (Faure *et al.*, 1993). An *in vitro* system for adhesion and fusion of maize gametes has been reported by the C. Dumas group (University of Lyon; Faure *et al.*, 1994). Zygotes produced *in vitro* and isolated after *in vivo* fertilization can now be regenerated into fertile plants.

A wide range of genetic changes has been shown to occur during tissue culture of maize (Edallo *et al.*, 1981; Zehr *et al.*, 1987; Armstrong and Phillips, 1988; Fluminhan and Kameya, 1996). Selection experiments carried out *in vitro* have yielded disease- and herbicide-resistant plants (Gengenbach *et al.*, 1977, Shaner and Anderson, 1985). An increase of free threonine in the grain has been obtained by tissue culture selection in the presence of the amino acids lysine and threonine (Hibberd *et al.*, 1980; Hibberd and Green, 1982; Miao *et al.*, 1986). More recently, lysine accumulation in cell cultures transformed with a lysine-insensitive form of the maize gene encoding the dehydropicolinate synthase enzyme has been reported by the laboratory of B. G. Gengenbach (Bittel *et al.*, 1996; Shaver *et al.*, 1996).

Protoplasts are the starting point for several genetic manipulations. These include genetic modification by incorporation of foreign DNA (direct gene transfer by PEG, electroporation, and by other methods), intra- and interspecific protoplast fusion and generation of somaclonal variants. There have been several reports of the regeneration of fertile plants from maize protoplasts (Prioli and Söndahl, 1989; Shillito *et al.*, 1989).

Transformation

Genetic transformation is a fundamental requirement when tailoring cells to express desirable characteristics. This requirement has been fulfilled by the development of systems for the delivery and integration of foreign genes into the plant genome. Fertile transgenic maize plants have been produced through bombardment of embryo suspension cultures with gold particles covered by a layer of DNA solution containing the gene to be delivered (Fromm *et al.*, 1990; Gordon-Kamm *et al.*, 1990), or bombarding immature zygotic embryos (Koziel *et al.*, 1993). Wounding of such embryos followed by electroporation has also been used successfully to transform maize (D'Halluin *et al.*, 1992). Production of fertile transgene maize plants has been also achieved using embryogenic maize suspension cultures that were transformed using silicon carbide whiskers to deliver plasmid DNA (Frame *et al.*, 1994). Methods based on the capacity of the *Agrobacterium* to transfer DNA to plant cells have been also the focus of considerable attention. This system is the most thoroughly characterized available for dicot plants, but successful application of such vectors to the manipulation of the maize genome has been reported only recently by Ishida and coworkers (1996).

Resistance to pests

The European Corn Borer (ECB) is a major pest of maize in North America and Europe. Yield loss of 3% to 7% per borer per plant can result from ECB feeding at various stages of plant growth. Chemical control methods are available but inadequate. Natural host plant resistance factors exist, but provide inadequate control. Resistance to leaf feeding by the first generation of the borer has been correlated with high concentrations of particular chemical metabolites. Hybrids with good resistance to the first-generation leaf feeding damage have been developed, but current hybrids are still susceptible to sheath and collar feeding by the later generations of borers. Quantitative resistance factors for later-generation feeding damage have been identified through molecular mapping techniques (Schön *et al.*, 1993). Progress has also been made in developing hybrids which will stand and retain ears despite heavy infestations of borers. Recently, the *Bx1* gene, which controls DIMBOA formation, was isolated by A. Gierl and coworkers (University of München; personal communication). The *Bx1* gene does not encode P450 monooxygenases that are specifically expressed in young maize plants and match the expression pattern of DIMBOA (Frey *et al.*, 1995).

Today, molecular biology offers plant breeders a new approach to the development of insect-resistant plants. The most outstanding success has been the incorporation into the maize genome of a synthetic gene encoding an insecticidal protein derived from *Bacillus thuringiensis,* as reported by the firms Ciba-Geigy and Monsanto (Koziel *et al.*, 1993; Armstrong *et al.*, 1995). This is a new and powerful tool which is already being used in integrated pest management strategies directed against insect attack.

Resistance to pathogens

Groups active in the area of maize-pathogen interaction are those of J. Bennetzen (Purdue University), T. Pryor (CSIRO, Canberra) and S. Briggs (Pioneer Hi-Bred International, Johnston). In maize, cloning of a gene mediating a gene-to-gene relationship between the plant and the pathogen (reviewed by Staskawicz *et al.*, 1995), has not yet been reported; a number of laboratories are however close to accomplishing this task. Rust resistance caused by *Puccinia sorghi* has been particularly considered. The present understanding of the genetics of rust resistance in maize originates with the work of Hooker and his students at the University of Illinois and has been reviewed several times (Hooker, 1985). Major *Rp* genes specifying resistance against *Puccinia sorghi* were identified. The genes have been mapped to four chromosomal locations (Lee *et al.*, 1963; Saxena and Hooker, 1968): *Rp3* with six alleles, *Rp4* with two alleles, *Rp5* and *Rp6* with one allele each. Some 14 distinguishable resistance specificities were mapped to the *Rp1* gene. They were initially considered as alleles of the locus and designated *Rp1*-A to *Rp1*-M. Subsequent work indicated that these specificities represent closely linked loci (Hulbert and Bennetzen, 1991). DNA markers tightly linked to *Rp1* and *Rp3* were used to resolve the genetic fine structure of these loci (Hulbert and Bennetzen, 1991; Sudupak *et al.*, 1993). *Rp1* is a complex locus comprised of multiple *Rp* genes. The germinal instability of this locus is associated with unequal crossing over between tandemly repeated DNA sequences in the *Rp1* region (Pryor, 1987; Bennetzen *et al.*, 1988b; Hong *et al.*, 1993). More recently Richter *et al.* (1995) have confirmed that new rust resistance specificities are associated with recombination in the *Rp1* complex. By contrast, *Rp3* appears to be a relatively simple and stable locus (Sanz-Alferez *et al.*, 1995). Molecular tagging of the *Rp1* gene has been attempted (Bennetzen *et al.*, 1988a). Further results (Bennetzen *et al.*, 1988b) indicate that the hypersensitive response elicited by *Puccinia sorghi* requires a cell-autonomous, non-diffusible factor specified by the *Rp1* locus.

Leaf spot of maize, caused by *Cochliobolus carbonum*, is emerging as one of the better understood fungal diseases of plants. The interaction between *C. carbonum* race 1 and maize has been studied by A. J. Ullstrup and coworkers (Nelson and Ullstrup, 1964); the data indicate that the specificity of this disease is determined by a cyclic tetrapeptide toxin produced by the fungus. Studies on the structure and biochemistry of the toxin have been reported by the J. D. Walton laboratory active at the Michigan State University (Walton, 1991; Walton and Holden, 1988; Meeley and Walton, 1991; Meeley *et al.*, 1992; Brosch *et al.* 1995). Fungal isolates producing the toxin are exceptionally virulent on plants that are homozygous recessive at the *Hm* locus. Maize that is heterozygous or homozygous dominant at *Hm1* is 100 times less sensitive to the toxin. Johal and Briggs (1992) from Pioneer Hi-Bred International generated and cloned several *Hm1* alleles by transposon-induced mutagenesis. The wild-type gene shares homology with dehydroflavonol-4-reductase genes of Petunia and snapdragon, suggesting that *Hm1* encodes a reductase that inactivates the *Cochliobolus* toxin, a mode of action that may be typical of other natural resistance genes.

A gene coding for a pathogenesis-related protein was cloned by Casacuberta *et al.* (1991). A close relationship exists between the expression of this gene and infection by the fungus *Fusarium moniliforme*, a common pathogen of maize that induces stalk and ear rot (Kuchareck and Kommedahl, 1966). The increased expression of the gene may be part of a defense mechanism against pathogens active during seed germination.

Plant inhibitors of gene translation, called ribosome-inactivating proteins, are active against fungal and viral growth (Stirpe *et al.*, 1992). By sequence analysis and *in vitro* assay Maddaloni *et al.* (1991), Bass *et al.* (1992), Walsh *et al.* (1991) have reported that the maize *b-32* protein, a 32-kDa endospermic albumin under the control of the gene *O2*, is a ribosome-inactivating protein. This protein may play a role in defense against pathogenesis, as is implied by the increased susceptibility of *o2* mutant kernels to pathogen attack. Maddaloni *et al.* (1997) have reported that transgenic tobacco plants expressing the b-32 protein show increased tolerance to infection by the soil-borne fungal pathogen *Rhizoctonia solani*.

Plant viruses have a negative impact on crop production throughout the world. Murry *et al.* (1993) from Sandoz Agro Inc., Palo Alto, California, have reported the first example of cp-mediated protection of maize transformed with the coat protein (*cp*) gene of the maize dwarf mosaic virus strain B (MDMV-B). Fertile plants were regenerated which express high levels of MDMV-cp and these were resistant to MDMV strains A and B and to mixed inoculations with MDMV and maize chlorotic mottle virus.

Stress resistance

Early studies on the effects of stress conditions in maize concern the response of maize roots to oxygen deprivation developed by the Freeling laboratory. This research can be traced back to a group having its origin with D. Schwartz, a student of M. M. Rhoades. This group collaborated with several investigators who are currently active in studies on oxygen deprivation (Peschke and Sachs, 1994; Sachs *et al.*, 1996), regulatory pathways in hypoxia (Devetten and Ferl, 1995) and dehydration stress response elements (Mao *et al.*, 1995). The laboratory of J. Peacock (Canberra) has also provided relevant contributions to the study of hypoxia (Olive *et al.*, 1990, 1991). Curiously, J. Peacock also cooperated with M. M. Rhoades, and therefore has a scientific ancestry with M. Freeling. Major polypeptides synthesized by plant roots in anaerobiosis are the alcohol dehydrogenase (ADH) proteins (Ferl *et al.*, 1979; Sachs *et al.*, 1980). Sequences located upstream of the maize *Adh1* gene are capable of conferring hypoxia- or anaerobiosis-dependent expression. Analyses of these upstream sequences have led to the definition of the anaerobic responsive element required for anaerobic inducibility, as well as other cis-acting elements present in the *Adh1* gene promoter (Ferl and Nick, 1987; Olive *et al.*, 1990;

Paul and Ferl, 1991; Devetten *et al.*, 1992; Devetten and Ferl, 1994; Kyozuka *et al.*, 1994). In addition, characterization of chromosomal mapping and expression of different polyubiquitin genes in tissues from control and heat-shocked maize seedlings were reported by the Walden group (Liu *et al.*, 1995).

Heat shock proteins (Hsps) have also been studied in maize by the Walden (University of Western Ontario) and Ho laboratories (Washington University, St. Louis). Studies of the Hs response have addressed the issues of induction temperature, tissue specificity, Hsp profiles, and intracellular localization (Baszczynski and Walden, 1982; Cooper and Ho, 1983; Cooper *et al.*, 1984, Atkinson *et al.*, 1989). Among these proteins, Hsp70s, the most important group studied, are encoded by multiple genes which are closely related (Bates *et al.*, 1994). Similarly, it has been shown that the heat shock proteins of low molecular weight are the products of different but related genes (Goping *et al.*, 1991). Greyson *et al.* (1996) have reported that maize seedlings show cell-specific responses to heat shock as revealed by expression of RNA and proteins encoded by 18 kDa heat-shock protein genes, while Gagliardi *et al.* (1995) have studied the expression of heat shock factor and heat shock protein 70 genes during maize pollen development. The study of low-temperature stress in maize has seen a significant contribution of the group of V. Walbot (Stanford University) who considered the activation of the general phenylpropanoid and anthocyanin pathways under low-temperature conditions (Christie *et al.*, 1994).

Water deficit also results in the activation of several genes, as studied particularly by the group of F. J. Close (University of California, Riverside) which has focused its research activity particularly on a class of proteins called dehydrins (Close *et al.*, 1994; Asghar *et al.*, 1994).

The effect of water stress has also been approached by using a class of maize mutants known as *viviparous* kernels (Robertson, 1955). In several of these mutants the biosynthesis of the stress-related hormone abscisic acid (ABA) is altered. The results available support a direct role for ABA in regulation of dehydrin expression (Paiva and Kriz, 1994; Mao *et al.*, 1995). The *viviparous-1* (*Vp1*) gene encodes a transcriptional activator required for ABA induction of maturation-specific genes correlated with the acquisition of desiccation tolerance. The gene was isolated by transposon tagging by D. R. McCarty and coworkers (1989a, b; 1991). The Vp1 protein functions as a transcriptional activator (McCarty *et al.*, 1991) and is still being actively studied in the context of ABA-mediated drought tolerance in plants (Hoecker *et al.*, 1995; Vasil *et al.*, 1995). Abscisic acid and drought have also been suggested to affect the differential transcription of two maize ferritin genes, *ZmFer1* and *ZmFer2* (Fobis-Loisy *et al.*, 1995).

A frequently observed consequence of environmental adversity is the phenomenon of oxidative stress. By perturbing cellular metabolism, particularly photosynthetic processes, many stress factors induce the production of reactive oxygen species and cause oxidative damage within the cell. Protection against oxidative stress is complex and includes both enzymic and non-enzymic components. Superoxide oxidoreductases (SODs) catalyze the first step in active oxygen scavenging, while catalase provides protection against reactive oxygen toxicity by dismutating hydrogen peroxide to water and oxygen. In maize, SOD and catalase isozymes are coded for by nonallelic nuclear genes (Redinbaugh *et al.*, 1990; Zhu and Scandalios, 1994). J. Scandalios and coworkers (North Carolina State University) have produced exhaustive data on isoenzymatic forms and induction of SOD proteins (Zhu and Scandalios, 1994, 1995; Guan and Scandalios, 1995), on the resolution and characterization of the Sod4 and SodA genes (Kernodle and Scandalios, 1996), and on the isolation, characterization and expression of the maize catalase genes (Guan and Scandalios, 1993; Abler and Scandalios, 1994; Guan *et al.*, 1996).

Glutathione-S-transferases (GST) catalyze the conjugation of glutathione, or in some cases homoglutathione, to a wide variety of electrophilic, lipophilic substrates (Mannervik and Danielson, 1988). The addition of glutathione, via the sulphydryl group of the cysteine, to a range of toxic xenobiotics, can lead to the production of water-soluble conjugates with reduced toxicity. These enzymes are of agronomic importance, since glutathione conjugation has been implicated in the detoxification of fungicides, organophosphorous insecticides and a range of widely used herbicides (Lamoureux and Rusness, 1989). Much of the early work on plant GST was focused on herbicide metabolism in crop and weed species. Safeners are commonly used to prevent injury from pre-emergency herbicides, such as chloroacetanilides. They act primarily by elevating herbicide metabolism via increased GST activity. In maize, these enzymes comprise 1–2% of soluble protein (Mozer *et al.*, 1983; Holt *et al.*, 1995) and exist in several isoforms (Dean *et al.*, 1991). Addition of herbicide safeners causes the appearance of two additional isoforms and increases the expression of constitutive isoforms (Wiegand *et al.,* 1986). The ability of GST isoforms to conjugate glutathione to a range of substrates, including chloroacetanilide herbicides (alachlor) has been reported (Holt *et al.,* 1995). Work by Jepson *et al.* (1994) reports the isolation of cDNA clones encoding safener-induced subunits present in several GST isoforms. These clones are being used as molecular probes to study mRNA induction upon safener treatment. Work by Rossini *et al.* (1995) indicated that the *Gst1* gene controls the expression of the two GST isoforms expressed in roots. They also showed by RFLP analysis that *Gst1* was localized on the long arm of chromosome 8, while two putative *Gst II* loci were mapped to chromosome 8 and 10, respectively. Marrs *et al.* (1995) have shown that the *Bz2* locus encodes a GST involved in vacuolar transfer of anthocyanins. These results provide a biochemical function for *Bz2*, and suggest a common mechanism for the ability of plants to sequester structurally similar but functionally diverse molecules in the vacuole.

Male sterility

The genetic study of male sterility has almost always been concerned with practical applications. This is a story of success (the use of cytoplasmic male sterility in hybrid seed production) and of failure (the susceptibility of maize Texas male-sterile cytoplasm, CMS-T, to the fungal pathogen *Helminthosporium maydis*). Mitochondrial DNA carries genetic determinants which condition male fertility and each of the CMS cytoplasms can be distinguished by restriction endonuclease fragment analysis of their respective mitochondrial DNAs (Levings and Pring, 1976; Pring and Levings, 1978). The T cytoplasm is a model for cytoplasmic male sterility and fertility restoration. Restoration requires the action of the dominant alleles of two nuclear restorer genes, *Rfl1* and *Rfl2* (Laughnan and Gabay-Laughnan, 1983). T-cytoplasm maize is sensitive to the pathotoxin HmT, produced by *Cochliobolus heterostrophus* Drechsler race T (Hooker *et al.,* 1970). S. Levings and coworkers at the North Carolina State University have shown that the mitochondrial gene *T-urf13* present in T cytoplasm and absent in N cytoplasm, is responsible for male sterility and toxin sensitivity (Dewey *et al.,* 1986, 1988). *T-urf13* encodes a 13-kDa mitochondrial polypeptide (Dewey *et al.,* 1988; Wise *et al.* 1987; Korth *et al.,* 1991). The *Rf2* gene, required for fertility restoration in male-sterile T-cytoplasm maize, was recently cloned. The protein predicted by the *Rf2* sequence is a putative aldehyde dehydrogenase, which suggests several mechanisms that might explain *Rf2*-mediated fertility restoration in CMS-T maize (Cui *et al.*, 1996). Groups still actively involved in maize male sterility and related subjects are those led by S. Dellaporta: molecular studies of tassel seed genes (Delong *et al.*, 1993; Dellaporta and Calderon-Urrea, 1994); C.S. Levings, III and J. N. Siedow: characterization of mitochondrial proteins (Korth and Levings III, 1993; Rhoads *et al.*, 1994); L. P. Taylor: conditional male sterility (Pollak *et al.*, 1995); P. Schnable: dissecting molecular basis of fertility restoration (Wise *et al.*, 1996); C. D. Chase: restoration-of-fertility alleles (*Rf3*) in diploid pollen.

A novel genetic method for inducing pollen sterility, which promises to be applicable to many plant species, maize included, has been reported by Mariani *et al.* (1990), a group working at Plant Genetic Systems (Gent, Belgium). The genetically engineered hybridization system comprises two components: a male sterility component (*barnase* gene) and a fertility restoration component (*barstar* gene). Crossing one parent expressing the male sterility component with a second line carrying the fertility restoration components results in the production of fully fertile hybrid seed.

47 Maize Development

Understanding the Rules Governing the Differentiation of Tissues and Organs

The advanced state of its genetics and the availability of well-characterized transposable elements have made maize a convenient organism for the study of molecular biology. The history of genetics shows that progress in understanding the function and functioning of genes is dependent on the formulation of the right questions and on the use of the appropriate organism to answer them. Geneticists moved from *Drosophila*, through *Neurospora* to bacteria and bacteriophages in search of systems in which genetic analysis could be carried out faster and with finer higher power resolution. Similarly, plant geneticists have exploited species with special features, like maize, and now *Arabidopsis thaliana*. *Arabidopsis* has many advantages. It can be selfed from which large progenies can be obtained. Its generation time of five to six weeks makes it possible to grow several generations in one year. The plant is physically small, which allows the handling of many individuals in a single experiment, an important attribute for large-scale mutagenesis projects. In this species, cloning and sequencing of genes is proceeding rapidly because *Arabidopsis* has only 10^8 base pairs of DNA in its haploid genome.

In spite of the obvious arguments favoring the use of *Arabidopsis* as an organism for studying plant genetics, basic research carried out with maize is not yet suffering from neglect. Several maize genes involved in the process of differentiation and morphogenesis have already been cloned, either because insertion mutants were available or because they were easily generated when necessary. A second reason for choosing maize as a model system is the relative size of its organs, which makes them suitable for detailed developmental studies. This is, for instance, the case for endosperm development: the tissue that can be sampled in sufficient quantities as early as 8 days after pollination. Moreover, maize, like all grasses, has a plant structure with bilateral symmetry and the vegetative part of the plant body is divided into well defined developmental units, the phytomers (Galinat, 1994). These attributes make maize suitable for studying clonal relationships in organogenesis, and for assigning roles to genes controlling cell differentiation and morphogenesis.

Origin of the species

The origin of *Zea mays* L. spp. *mays* has been the subject of intense debate. Three major hypotheses have been advanced to explain the emergence of maize and teosinte, but various individual interpretations of each exist: the "Common origin and divergent evolution" hypothesis by Randolph (1976) and Weatherwax (1955); the "Descent from teosinte" hypothesis by Galinat (1971), Beadle (1978), Iltis (1983) and Doebley (1983); and the "Maize as an ancestor of teosinte" hypothesis by Mangelsdorf (1974). In recent years, a large body of evidence has accumulated in support of the hypothesis that maize is a domesticated form of teosinte (Goodman, 1988). J. Doebley (University of Minnesota) in cooperation with H. Iltis (University of Wisconsin), has reorganized the relationships among taxa of the tribe *Andropogoneae*, and this work was later confirmed by molecular studies (Doebley, 1990). Moreover, a new species was recently discovered (*Zea diploperennis*) which helps to elucidate the evolutionary relationships in the genus (Iltis *et al.*, 1979).

The Goodman-Stuber-Doebley group has provided a large body of isozyme-based data supporting a close relation between maize and Balsas teosinte (*Z. mays* ssp. *parviglumis*) again implying an origin of maize from teosinte (Doebley *et al.*, 1988). J. Doebley (University of Minnesota) is currently investigating the genetic changes that converted teosinte to modern-day maize. His data suggest that the morphological differences between maize and teosinte, although complex, can be restricted to five key traits (Doebley and Stec, 1991). In addition, Doebley provides evidence that a single gene, *teosinte glume architecture-1* (*tga1*), converts the hard triangular case covering the teosinte kernel into a soft, short maize-type glume (Dorweiler *et al.*, 1993). The *tga1* mutation may have been the key event that made maize attractive for domestication, because the seeds became easily accessible for harvest and propagation. A second genomic region, detected by QTL analysis, is responsible for the conversion of teosinte-type long lateral branches tipped by ears. Interestingly, this region maps near the *teosinte branched* (*tb1*) locus described several years ago by C. Burnham. Molecular data (Doebley *et al.*, 1995) indicate that a few regions of the genome can account for the dramatic differences in inflorescence morphology between maize and teosinte. More recently, Doebley *et al.* (1997) have cloned *tb1* by transposon tagging and show that it encodes a protein with homology to the *cycloidea* gene of snapdragon (Luo *et al.*, 1995). Similarly, a QTL mapping strategy has been employed to determine the number of QTLs that control the difference in kernel weight in two maize-teosinte hybrid populations (Doebley *et al.*, 1994). The small number of QTLs detected (4 to 6) and the large magnitude of their effects suggest that differences in kernel weight between maize and teosinte should be described as oligogenic rather then polygenic.

At present, two hypotheses exist on how the maize ear could have arisen, assuming that teosinte is the ancestor of maize. The older theory, presented in great detail by Galinat (1992), proposed that the maize pistillate inflorescence (or ear) originated from the pistillate inflorescence of teosinte through a series of single gene mutations. Iltis (1983) has a very different view of the evolution of the maize ear. He proposed that the terminal spike of the teosinte lateral branch, was transformed into the maize ear by a process described as sexual transmutation.

Clonal analysis of plant development

Cell lineage analysis can be used in understanding the cellular dynamics of morphogenetic processes. In plants, patterns of cell lineage have been investigated by using chimeras, spontaneous sectoring, sectoring induced at specific stages of development and by sector boundary analysis (Dawe and Freeling, 1991; Spena and Salamini, 1995). Although an organism may be a genetic mosaic without this being manifest in the phenotype, in clonal analysis a cell or its clonal progeny are tagged by a convenient visible, cell-autonomous trait that differentiates them from the surrounding background. Mutations affecting chlorophyll or flavonoid pigments have most often been used to mark genetic mosaics, but change in cell size due to alteration in ploidy level, shape of epidermal hairs, amylose content and morphology of epicuticular waxes have also been used.

Historically, plant chimeras were the first mosaics used in clonal analysis (reviewed in Tilney-Basset, 1986). Chimeras can be reproduced by stem cuttings but are lost during sexual reproduction. For this reason, annual plants reproducing via seeds—like maize—have not been used in chimera studies. Maize has, nevertheless, contributed significant methods and concepts to cell lineage studies. Spontaneous sectors were first used in maize as clonal markers by B. McClintock. She analyzed in 1931 the L. J. Stadler's collection of maize plants derived by pollinating homozygous recessive mutant lines with irradiated pollen carrying wildtype alleles of the same genes (McClintock, 1984). Some of the variegated plants had a ring chromosome with fused broken ends. Ring chromosomes are unstable and their loss results in variegation (McClintock, 1932). Thus, they can be used in clonal analysis.

Endosperm development has been followed by studying sectors generated by transposon excision at the *Wx* and *A* loci (McClintock, 1978). Cell lineages proliferate in the endosperm as radially expanding clones (Coe and Neuffer, 1978). To evaluate the contribution of meristematic layers to the formation of the maize carpel, the female organ of the flower, Dellaporta *et al.* (1991) have followed the excision of the transposon *Ac* from the *P* locus and concluded that the gynoecium derives from two genetically independent layers of the meristem. The role of early cell divisions in leaf development has been studied by following the excision of a transposable element from the *Gl1* locus (Cerioli *et al.*, 1994). This study concluded that a clonal type of development exists early during the formation of the leaf epidermis. Spontaneous sectors have also been used to show that intermediate and small veins of the maize leaves—developing in the internal layer of the leaf mesophyll with a basipetal orientation—represent the fusion of two transversal cell lineages (Langdale *et al.*, 1989). Based on the observation that maize mosaics with 50% of the leaf surface taken up by a sector had a coincidence of sector boundary and midrib, Steffensen (1968) suggested that the maize shoot originates from two precisely orientated cells. However, Bossinger *et al.* (1992) found some plants in which the border between wildtype and mutant tissues was positioned in a plane normal to the leaf midrib. This supports the possibility that the left-right plane of the maize seedling either coincides with the plane along which the two shoot founder cells divide, or lies perpendicular to it (see comments in Spena and Salamini, 1995). The Freeling Group (University of California, Berkeley) has performed analysis of maize leaves mosaic for the *liguleless-1* (*lg1*) mutation. They showed that the wildtype allele of *Lg1* locus is necessary in the mesophyll for normal ligule development and auricle epidermal cell differentiation (Becraft *et al.*, 1996) while the expression of the *Lg3* phenotype, originally thought to result only in a leave-to-sheath transformation, can also transform blade to auricle (Fowler *et al.*, 1996). By using the *Ac* transposable element as a molecular tag, a novel *lg1* allele, *lg1-m1*, was isolated and cloned by Moreno *et al.* (1997). Analysis of somatic revertant sectors confirmed that the *Lg1* gene product in a cell-autonomous fashion. cDNA cloning as well as RT-PCR analysis of the *Lg1* mRNA indicate that the *lg1* gene is expressed at very low levels in the ligular region of developing maize leaf primordia. Cellular localization studies in a heterologous system indicate that the *Lg1* product localizes exclusively to the nucleus. The predicted amino acid sequence of the *Lg1* protein is largely novel, but contains an internal domain of 77 amino acids with significant similarity to a domain present in two recently identified Squamosa Promoter-Binding, proteins 1 and 2 (SBP1 and SPB2) in *Antirhinum majus* (Klein *et al.*, 1996).

Clonal analysis based on sectors induced at a specific stage of development was adopted for maize by L. J. Stadler (1930), who first noted that in the adult plant the size of an induced sector, representing the lineage of one mutated cell, decreased if the X-ray irradiation was delayed. The method can be used to obtain fate maps of the cells of the embryo at the moment of the mutagenic treatment, or maps of other meristematic tissues, and to estimate the apparent cell number present in a primordium at the given stage. E. Coe (University of Missouri, Columbia) has been the pioneer of these studies, and the method proposed by him is now being used in several other plant species (Coe and Neuffer, 1978). After irradiation, appearance and size of the somatic sectors and their position on the plant are recorded and used to develop the fate map. Fate maps of plant meristems are represented by concentric rings of cells, with the most external ring responsible for the organization of the basal part of the plant. Maize fate maps based on induced sectors have been produced by Coe and Neuffer (1978), McDaniel and Poethig (1988); the clonal origin of the maize ear has been considered by Coe and Neuffer (1978), and Uhrig *et al.* (1997); number and position of leaf founder cells was studied by Steffensen (1968), Poethig (1984); Poethig and Szymkowiak (1995).

Sector boundary analysis was first proposed by Sturtevant (1929) in *Drosophila* and used for the first time in plants only in 1986 to describe the development of the cotton embryo (Christianson, 1986). In maize, the method was used by Dawe and Freeling (1992) who irradiated germinating seedlings heterozygous for genes inducing anthocyanin pigmentation. The borders of large sectors were used to study the development of anther primordia and the anther was shown to originate from 4 founder cells surrounded by a ring of 8 cells. The orientation of the 4 founder cells predicts the orientation of the microsporangia. Sector boundary analysis has been also exploited by Uhrig *et al.* (1997) to establish that during the inception of the lateral meristem leading to the formation of the ear shoot, a few apical cells are clonally destined to originate the reproductive part of the ear. An intercalary type of growth, starting from about 200 cells, is, moreover, responsible for the formation of the ear peduncle and connected lateral organs.

The cultural role of the schools of B. McClintock and L. J. Stadler in maize clonal development has been highlighted by the work of D. M. Steffensen and of E. H. Coe (University of Missouri, Columbia). Present groups active in the field are those led by R. S. Poethig (University of Pennsylvania), M. Freeling (University of California, Berkeley) S. Dellaporta (Yale University, New Haven), and F. Salamini (Max-Planck Institüt, Köln). Neuhaus and coworkers (Lusardi *et al.*, 1994) have developed an approach towards genetically engineered cell fate mapping in maize using the *Lc* gene as a visible marker. The aim of this approach is to manipulate single cells of the shoot apical meristem by microinjection of genes encoding visible markers, and follow the fate of the marker during development.

Genes establishing developmental patterns and clonal analysis

In all flowering plants, cells that are potentially capable of further differentiation are supplied by a group of dividing cells known as meristem. The meristem may consist of one or more functionally distinct layers of cells (two in maize) which continually self regenerate and produce lateral organs. Our understanding of the layered structure of the meristem is based on clonal analysis. More recently, major genes which participate both in meristem formation and in the determination of early morphogenetic patterns have been considered. Once again the problem was approached genetically. The laboratories of M. Freeling (University of California, Berkeley) and S. Hake

(USDA-Plant Genetic Expression Center, Albany) have long been interested in leaf development, cloning genes identified by mutant phenotypes.

One of first such leaf development genes analyzed was the maize locus *Knotted-1* (*Kn1*), a dominant mutation which causes abnormal leaf development. *Kn1* was cloned and sequenced (Vollbrecht *et al.*, 1991), and was the first gene in plants shown to contain a homeobox, a sequence motif found in genes involved in pattern formation in many other eukaryotes. Further studies have now clarified that many homeobox genes exist in plants. Those similar to *knotted-1* can be divided into two classes based on sequence analysis and expression pattern (Kerstetter *et al.*, 1994). Class 1 genes are highly expressed in meristem-enriched tissues, such as the vegetative meristem and ear primordia, but not in leaves. Class 2 genes are transcribed in all tissues but most abundantly in roots. Although originally discovered during the study of leaf development, the role of maize homeobox genes, or at least of some of them, seems more closely related to meristem activity. Recently, for instance, *knotted 1* was found to be a very precise marker of meristem formation during embryogenesis (Smith *et al.*, 1995). Two other maize homeobox genes, were identified by W. Werr and coworkers (University of Köln, Germany; W. Werr is a former student of P. Starlinger). Both homeobox genes are transcribed simultaneously in meristematic and proliferating cells of the maize plant (Klinge and Werr, 1995; Klinge *et al.*, 1996). The ectopic expression of these genes causes an alteration in vegetative and floral development in transgenic tobacco (Ueberlacker *et al.*, 1996).

Other maize mutations are currently being mapped to homeobox genes (Becraft and Freeling, 1994; Schneeberger *et al.*, 1995). This raised the possibility that early developmental pattern in plants can be considered to be determined by homeobox genes and mediated by the specificity of their expression in given positions in the meristem. For instance, Jackson *et al.* (1994) have noted that the expression of three homeobox genes predicts patterns of morphogenesis in the vegetative shoot.

MADS box genes discovered first in *Antirrhinum* and *Arabidopsis,* are involved in the transition of the meristem from the vegetative to the reproductive phase and have a major role in establishing organ identity in the flower (Schwarz-Sommer *et al.,* 1990; Yanofsky *et al.*, 1990). Similar genes have also been isolated in maize (Mena *et al.*, 1995; Theissen *et al.*, 1995; Fisher *et al.*, 1995) and their role in inflorescence and flower development is being established. Of special interest in maize is the genetic control leading floral meristems to develop only male (the terminal staminate inflorescence or tassel) or female (the axillary pistillate inflorescence, or ear) flowers. Available data (summarized by Dellaporta and Calderon-Urrea, 1994) show that flower primordia are basically bisexual (Peterson, 1976) and have the potential to develop organs of either sex. Sexual fate is determined at a critical developmental stage. The search for mutants blocked in this process that would allow one to dissect genetically the developmental pathway leading to male and female maize inflorescences has focused particularly on the tassel seed phenotypes. S. Dellaporta and coworkers (Yale University, New Haven) have cloned the *tassel seed 2* (*ts2*) gene which encodes an alcohol dehydrogenase-like protein with similarities to sterol dehydrogenases (Delong *et al.*, 1993).

Studies on leaf epidermal differentiation have recently been approached by Becraft *et al.* (1996) using the *crinkly4* (*cr4*) mutation. This mutation affects leaf epidermis differentiation such that cell size and morphology are altered, and surface functions are compromised, allowing graft-like fusions between organs. In the seed, loss of *Cr4* inhibits aleurone formation in a pattern that reflects the normal progression of differentiation over the developing endosperm surface. The *Cr4* gene was isolated by transposon tagging and found to encode a putative kinase receptor. The extracellular domain contains a cysteine-rich region similar to the ligand binding domain in mammalian tumor necrosis factor receptors, and seven copies of a previously unknown 39-amino acid repeat. The results suggest a role for *Cr4* in a differentiation signal.

The laboratory of Sheridan (University of North Dakota) has recently isolated a recessive pleiotropic mutation, termed *multiple archesporial cells1* (*mac1*), controlling the commitment of the meiotic pathway in maize (Sheridan *et al.*, 1996). Its cytological phenotype suggests that this locus plays an important role in the switch of the hypodermal cells from the vegetative to the meiotic (sporogenous) pathway in maize ovules. The sporophytic expression of this gene is also important for normal female gametophyte development. This mutation is putatively tagged with *Mu* element and Sheridan and coworkers have initiated an effort to clone this gene.

In flowering plants, the polarity of the developing embryo sac is evident in the megasporocyte at early meiosis and is displayed along the micropylar-chalazal axis of the female gametophyte throughout its development. The polar distribution of the DNA-containing organelles is believed to be a key factor in the ultimate determination of cell fate during the selection of the functional megaspore and the formation of the seven-celled embryo sac (Huang and Sheridan, 1994). The elucidation of cytoskeletal organization and modification during embryo sac formation promises to be helpful in understanding the role of genes in the normal developmental process. Huang and Sheridan (1996) have characterized nuclear behavior in maize and changes in the organellar DNA and microtubular cytoskeleton during embryo sac development in both normal and *indeterminate gametophyte 1* (*ig1*) mutant. This mutation was first reported by Kermicle (1969) as a mutation that conditions the production of haploid embryos of paternal origin and inhibits development of the female gametophyte (Lin, 1978, 1981). By looking in detail at the division patterns and cytoskeletal organization of the developing embryo sac, Sheridan and Huang have found that the nuclei of *Ig* mutants divide asynchronously. In addition, the placement of the spindle poles is often abnormal, as is the orientation of the division plane. These results indicate that in addition to acting primarily in controlling nuclear divisions, the *ig* gene acts secondarily in regulating microtubule behavior. This cytoskeletal activity most probably controls the polarization and nuclear migration underlying the formation and fate of the cells of the normal embryo sac.

Current groups actively studying major genes determining maize developmental patterns in addition to those already mentioned, include M. Freeling (University of California, Berkeley), S. Hake (USDA Plant Gene Expression Center, Albany), R. Schmidt (University of California, San Diego), W. Werr (University of Cologne), H. Saedler (Max-Planck Institüt, Cologne), S. Dellaporta (Yale University, New Haven), P. W. Becraft (Iowa State University, Ames), P. S. Stinard (USDA-ARS, Urbana), and W. Sheridan (University of North Dakota).

Genetics of plant hormones

A massive and complex literature exists on auxin physiology, but the molecular mechanisms of auxin action remain largely unknown. The maize endosperm, grown *in vivo* and *in vitro* is widely used to study the metabolism and synthesis of indolacetic acid (reviewed by Jensen and Bandurski, 1994). Progress has been made in the isolation of maize genes encoding auxin binding protein— possible molecular components of the auxin signal transduction pathways

(Yu and Lazarus, 1991; Schwob *et al.*, 1993). These genes encode proteins that are closely related in size and sequence. Relevant breakthroughs in this area, however, have been difficult to achieve, possibly because mutant phenotypes corresponding to evident defects in IAA metabolism have only rarely been reported. One of the few cases is the association of a defective endosperm mutant with a marked reduction in accumulation of free and bound IAA (Torti *et al.*, 1986).

The *orange pericarp* mutant represents a second interesting phenotype. A. Wright and M. G. Neuffer (University of Missouri, Columbia) investigated this mutant which has a visible phenotype only when the recessive alleles of two independent genes are homozygous (Wright *et al.*, 1991). The two loci map in chromosomes 4 and 10. The orange pigment was analyzed and an excess of indole was found, which is consistent with a defect in the last step of the tryptophan pathway, catalyzed by tryptophan synthase. In collaboration with K. Cone (University of Missouri, Columbia) this research group used a probe from *Arabidopsis* to clone two tryptophan synthase maize genes. RFLP mapping confirmed that the 2 genes considered in the *orange pericarp* study are structural genes for the b subunit of tryptophan synthase (Wright *et al.*, 1992). Subsequent analysis revealed that in mutant seedlings grown on media containing precursors labeled with stable isotopes, IAA was more abundant than tryptophan. No incorporation of label from tryptophan into IAA could be detected. These results established that in maize IAA can be produced *de novo* without needing tryptophan as an intermediate (Wright *et al.*, 1991).

Gibberellin (GA) deficiency varies among plant genera and is associated with morphological consequences that include reduced cell elongation and aberrant floral development (Reid, 1986). The phenotype of GA-responsive mutants of maize includes reduced plant stature, shorter and broader leaves, reduced number of branches in the tassels, presence of anthers on flowers that are only pistillate in the wildtype. The relationship between dwarfism and gibberellin metabolism in maize was first established by the observation that a wildtype phenotype could be restored by treating homozygous recessive dwarf mutants with GA (Phinney, 1956; Phinney and West, 1960). Evidence obtained from bioassay of GA intermediates has indicated that the loci *D1*, *D2*, *D3*, *D5* and *An1* control specific and different steps in the early 13-hydroxylation pathway leading to *Ga1* (Phinney and Spray, 1982; Spray *et al.*, 1984). Furthermore, Spray *et al.* (1996) presented evidence that the *d1* mutation blocks three steps in the GA biosynthetic pathway. Using Mu elements as molecular probes, the *Anther ear1* (*An1*) gene was cloned by Bensen *et al.* (1995). The *An1* gene product is involved in the synthesis of ent-kaurene, the first tetracyclic intermediate in the gibberellin biosynthetic pathway. The same approach was used by Winkler and Helentjaris (1995) to clone the *Dw3* gene, which was shown to encode a cytochrome P450-mediated, early step in gibberellin biosynthesis. It is noteworthy that the same group of T. Helentjaris (active at Pioneer Hi-Bred International, Johnston) was able to show that in a cross between tall and short inbred maize lines, a QTL was segregating which mapped to the *Dw3* locus on chromosome 9 near the centromere (Touzet *et al.*, 1995).

The interested reader can find information on the genetics of abscisic acid metabolism in maize in the section of this chapter devoted to environmental stresses; the review of McCarty (1995) is also helpful.

Groups still active in hormone genetics in maize are those of M. G. Neuffer and K. Cone (University of Missouri, Columbia), T. Helentjaris and S. Briggs (Pioneer Hi-Bred, Johnston) and K. Palme (Köln, Germany).

Looking at the maize seedling

When grown in large quantities, maize seedlings can easily be screened for mutant phenotypes. In the review of Coe *et al.* (1988b), 117 genetic loci are reported which affect pigmentation of the leaf. Only a few of these genes have been considered in detail with respect to their indirect or direct effect on the leaf photosynthetic apparatus. High-chlorophyll-fluorescence mutants have been characterized (Miles, 1994) and are believed to affect directly the system of thylakoid membrane proteins which make possible the conversion of light into chemical energy. Some of these mutants are important in clarifying the role of specific proteins in photosynthetic electron transport.

Leaf mutations which are being actively worked on are those disrupting the C4 syndrome. Leaves of maize display Kranz anatomy with veins surrounded by an inner layer of photosynthetic bundle sheath cells (BS) and an outer layer of photosynthetic mesophyll (M) cells. Each cell provides a unique subset of the enzymatic activities required for C4 photosynthesis, due mainly to cell-specific expression of the genes encoding C4 pathway enzymes (Laetsch, 1974; Nelson and Langdale, 1992, 1993). A large number of chlorophyll-photosynthetic mutants, in addition to affecting chlorophyll biosynthesis, may be defective in chloroplast and/or cell maturation in both photosynthetic cell types, and are thus likely to define genes that function well after BS and M cell identity has been established. However, a small number of mutations exhibit BS cell-specific phenotypes (Langdale *et al.*, 1995). These are of interest because they may define genes that determine BS/M cell identity or act later in BS/M cell photosynthetic maturation. Langdale and Kidner (1994) have characterized a mutation, *bundle sheath defective 2-mutable1* (*bsd2-m1*), that disrupts the coordinated differentiation of BS and M cells. What is particularly interesting about this mutation is that it affects the accumulation of BS-specific C4 gene products from a very early stage, before BS and M cells can be distinguished, but after the Kranz anatomy has been established.

In other maize mutants, including *ij* (*iojap*), *j1* (*japonica 1*), *j2* (*japonica 2*), *sr1* (*striate 1*) and *sr2* (*striate 2*), albino sectors are not distributed randomly over the leaves, but form stripes, confined to the leaf and sheet margins. This pattern of striping is in contrast to the clonal and random distribution that is expected from transposon insertion mutations. Of the striping mutants of maize, the recessive *ij* mutant has been extensively characterized since it was first described. The phenotypic defects of *ij* are most prominent in the margins of leaves. A series of genetic and biochemical studies indicates that the *ij* gene has a significant role in plastid biogenesis and plant development (Siemenroth *et al.*, 1980). A transposon tagging experiment using Robertson's *Mutator* (*Mu*) has generated an insertion allele of *Ij*. The gene was cloned and the structure of the gene and its DNA sequence determined. The *Ij* gene encodes a 24.8 kDa protein which has no significant similarity with proteins listed in databases (Han *et al.*, 1992).

A recent paper on a similar subject concerns a *leaf permease 1-mutable 1* (*Lpe1-m1*) mutant of maize with disrupted chloroplast ultrastructure, preferentially affecting BS chloroplasts under low light. Despite the disrupted ultrastructure, the metabolic cooperation of BS and M cells in C4 photosynthesis remains intact in the mutant. The *Ac* transposon-tagged allele was cloned by T. Nelson (Yale University, New Haven) and coworkers. The *Lpe1* gene encodes a polypeptide with significant similarity to microbial pyrimidine and purine transport proteins (Schultes *et al.*, 1996).

Nuclear genes essential for the biogenesis of the chloroplast cytochrome *b-6f* complex were identified by Voelker and Barkan (1995) using mutations that cause the specific loss of the complex. The four mutations define two nuclear genes and block the biogenesis of the cytochrome *b-6f* complex at a post-transcriptional step. The two nuclear genes identified by these mutations may encode previously unknown subunits, be involved in prosthetic group synthesis or attachment, or facilitate assembly of the complex.

Studies on the timing of leaf senescence and characterization of senescence-related cDNAs have been undertaken by the Greaves group (Smart *et al.*, 1995).

In all photosynthesic organs carotenoids are lipid-soluble pigments which, among other functions act to protect chlorophyll from photo-oxidation. Maize contains carotenoids in the endosperm also, but the function of carotenoids in this tissue is still unknown. A large number of white endosperm mutants of maize have been analyzed at the genetic level. Many of these give rise to albino plants (Robertson, 1975). Some of these white endosperm-albino seedling mutants are devoid of b-carotene, but can accumulate its precursors in leaves and endosperms. The *Y1* gene is responsible for the production of b-carotene in the endosperm. Buckner *et al.* (1990) have cloned a Robertson's *Mutator*-tagged allele of the *Y1* gene. More recently, Buckner *et al.* (1996) have shown that the *Y1* gene codes for phytoene synthase. Similarly, Li *et al.* (1996) have cloned and characterized a maize cDNA encoding phytoene desaturase (PDS), another enzyme of the carotenoid biosynthetic pathway. Their results suggested that PDS may be encoded by the *Vp5* locus. An understanding of how the *Y1* and *Vp5* alleles activate the carotenoid biosynthetic pathway in the endosperm of maize may suggest strategies for activating this pathway in the endosperm of other important grains such as rice or wheat.

Other maize mutants that can be scored at the seedling stage are those affecting the morphology of the leaf. Recent reviews of leaf development based on this class of mutations are those of Freeling (1992), Sinha *et al.* (1993) and Harper and Freeling (1996). As previously discussed, homeobox gene-related mutants have been instrumental in discovering that *Knox* genes create patterns of leaf/non-leaf boundaries in the meristem (Jackson *et al.*, 1994). Other mutations, like *lax, midrib, hairy sheath*, and *frayed rolled* have been associated with the functional definition of leaf domains (Freeling and

Hake, 1985; Becraft and Freeling, 1994; Schneeberger *et al.*, 1995; Fowler and Freeling, 1996). The mutant alleles of the *narrow leaf* gene show a precise deletion of the margin domain of the leaf (Scanlon *et al.*, 1996).

Moreover, studies of cell lineage indicate that the maize unit of vegetative growth, the phytomer—which consists of internode, node, leaf and axillary bud—may be induced as a very specific group of founder cells (Poethig and Symkowiak, 1995) and that the tissues derived from the internal layer (LII) of the meristem induce the epidermal cell fate (Langdale *et al.*, 1989; Cerioli *et al.*, 1994). These studies indicate an integrated developmental model of leaf growth.

Maize leaf development is at present of major interest in the group of M. Freeling (University of California, Berkeley), S. Hake (USDA-ARS PGEC, Albany), R. S. Poethig (University of Pennsylvania), M. Motto–F. Salamini (Bergamo-Köln), T. Nelson (Yale University, New Haven) and J. Langdale (University of Oxford, Oxford).

In maize, mutants which can be easily identified at the seedling stage have the so-called *glossy* phenotype. This can be easily scored by spraying water on seedlings: waxless mutants appear green and wildtype seedlings silver. The mutants produce a reduced amount of epicuticular waxes. Maize waxes are a mixture of very long chain fatty acids, aldehydes, alcohols, and hydrocarbons. A large fraction of the alcohols and fatty acids are used to form the corresponding esters. The biosynthesis and deposition of epicuticular waxes in the first five or six leaves of young maize plants is governed by at least 18 genes. The groups which have been particularly concerned with the genetics of this system are those of Sprague, formerly at Iowa State University, and Bianchi and Salamini in Italy. It was hypothesized that two pathways can give rise to different kinds of leaf waxes. One pathway responsible for wax synthesis in the first five or six juvenile leaves, whereas the second pathways produces waxes during the whole life cycle of the maize plants (Bianchi *et al.*, 1985). The biosynthesis of wax formation is poorly understood, as is the function of epicuticular leaf waxes in preventing water loss. For this reason the *glossy* genes are being actively tagged with transposons (*En/Spm* in the group of Motto-Salamini in Bergamo, Italy and of Sisco at North Carolina State University, and *Mu* at Iowa State University by P. Schnable and coworkers). Tacke *et al.* (1995), starting with the allele *gl2-m2*, and Hansen *et al.* (1997) starting with the allele *gl1-755048*, have recently been successful in cloning *Gl2* and *Gl1* genes, respectively. While for the GL2 protein no similar amino acid sequence was found in protein data banks, the sequence of the GL1 protein is similar to a CER1 protein of *Arabidopsis* (Aarts *et al.*, 1995), and to a *Kleinia odora* protein in which the transcript accumulates mainly in the epidermis. These proteins have predicted structures similar to those of a class of membrane receptor characterized in animals. The *glossy-15* mutant has also been tagged. This mutation is of particular interest to the study of the juvenile-to-adult phase transition in the leaf epidermis (Evans *et al.*, 1994; Moose and Sisco, 1994). The *Gl15* gene encodes a putative transcription factor with significant sequence similarity to the *Arabidopsis* regulatory genes *Apetala2* and *Aintegumenta*, which act primarily to regulate floral organ identity and ovule development (Moose and Sisco, 1996). This finding expands the known functions of both vegetative and reproductive lateral organ identity in plants. In addition, the results provide molecular support for the hypothesis that leaves and floral organs are related structures derived from a common growth plan. Studies like these, which allow a genetic approach to heterochrony in maize (Dudley and Poethig, 1993), are being carried out particularly in the group of R. S. Poethig (University of Pennsylvania).

References

Aarts M.G.M., C.J. Keijzer, W.J. Steikema, A. Pereira, 1995. Molecular characterization of the *CER1* gene of *Arabidopsis* involved in epicuticular wax biosynthesis and pollen fertility. Plant Cell 7: 2115-2127.

Abler M.A., J.G. Scandalios, 1994. Isolation and characterization of a genomic sequence encoding the maize *Cat3* catalase gene. Plant Mol. Biol. 22: 1031-1038.

Alfenito M.R., J.A. Birchler, 1993. Molecular characterization of a maize B chromosome centric sequence. Genetics 135: 589-597.

Alleman M., J.L. Kermicle, 1993. Somatic variegation and germinal mutability reflect the position transposable element dissociation within the maize *R* gene. Genetics 135: 189-203.

Armstrong C.L., C.E. Green, 1985. Establishment and maintenance of friable, embryogenic maize callus and the involvement of L-proline. Planta 164: 207-214.

Armstrong C.L., R.L. Phillips, 1988. Genetic and cytogenetic variation in plants regenerated from organogenic and friable, embryogenic tissue cultures of maize. Crop Sci. 28: 363-369.

Armstrong C.L., J. Romero-Severson, T.K. Hodges, 1992. Improved tissue culture response of an elite maize inbred through backcross breeding and identification of chromosomal regions important for regeneration by RFLP analysis. Theor. Appl. Genet. 84: 755-762.

Armstrong C.L., G.P. Parker, J.C. Pershing, S.M. Brown, P.R. Sanders, D.R. Duncan, T. Stone, D.A. Dean, D.L. Deboer, J.Hart, A.R. Howe, F.M. Morrish, M.E. Pajeau, W.L. Petersen, B.J. Reich, R. Rodriguez, C.G. Santino, S.J. Stato, W. Schuler, S.R. Sims, S. Stehling, L.J. Tarochone, M.E. Fromm, 1995. Field evaluation of European Corn Borer control in progeny of 173 transgenic corn events expressing an insecticidal protein from *Bacillus thuringiensis*. Crop Sci 35: 550-557.

Asghar R., R.D. Fenton, D.A. Demason, T.J. Close, 1994. Nuclear and cytoplasmic localization of maize embryo and aleurone dehydrin. Protoplasma 177: 87-94.

Atkinson B.G., L. Liu, I.N. Goping, D.B. Walden, 1989. Expression of the genes encoding hsp73, hsp18, and ubiquitin in radicles of heat-shocked maize seedlings. Genome 31: 698-704.

Aukerman M.J., R.J. Schmidt, B. Burr, F.A. Burr, 1991. An arginine to lysine substitution in the Bzip domain of an opaque-2 mutant in maize abolishes specific DNA binding. Genes & Development 5: 310-320.

Avramova Z., P. SanMiguel, E. Georgieva, J.L. Bennetzen, 1995. Matrix attachment regions and transcribed sequences within a long chromosomal continuum containing maize *Adh1*. Plant Cell 7: 1667-1680.

Bae J.M., M. Giroux, L.C. Hannah, 1990. Cloning and characterization of the *brittle-2* gene of maize. *MAYDICA* 35: 317-322.

Baker B., J. Schell, H. Lörz, N.V. Fedoroff, 1986. Transposition of the maize controlling element *Activator* in tobacco. Proc. Natl. Acad. Sci. USA 83: 4844-4848.

Balconi C., E. Rizzi, M. Motto, F. Salamini, R. Thompson, 1993. The accumulation of zein polypeptides and zein mRNA in cultured endosperms of maize is modulated by nitrogen supply. Plant J. 3: 325-334.

Baran G., C. Echt, T. Bureau, S. Wessler, 1992. Molecular analysis of the maize *wx-b3* allele indicates that precise excision of the transposable *Ac* element is rare. Genetics 130: 377-384.

Barker R.F., D.V. Thompson, D.R. Talbot, J. Swanson, J.L. Bennetzen, 1984. Nucleotide sequence of the maize transposable element *Mu1*. Nuc. Acids Res. 12: 5955-5967.

Bass H.W., G.R. O'Brian, R.S. Boston, 1995. Cloning and sequencing of a 2nd ribosome-inactivating protein gene from maize (*Zea mays* L.). Plant Physiol. 107: 661-662.

Bass H.W., C. Webster, G.R. O'Brian, J.K.M. Robers, R.S. Boston, 1992. A maize ribosome-inactivating protein is controlled by the transcriptional activator *Opaque-2*. Plant Cell 4: 225-234.

Baszczynski C.L., D.B. Walden, 1982. Regulation of gene expression in corn (*Zea mays* L.) by heat shock. Can. J. Biochem. 60: 569-579.

Bates E.E.M., P. Vergne, C. Dumas, 1994. Analysis of the cytosolic *hsp70* gene family in *Zea mays*. Plant Mol. Biol. 25: 909-916.

Beadle G.W., 1978. Teosinte and the origin of maize. In Maize Breeding and Genetics, edited by D.B. Walden. J. Wiley & Sons, New York, pp. 113.

Beavis W.D., D. Grant, 1991. A linkage map based on information from four F2 populations of maize *Zea mays* L. Theor. Appl. Genet. 82: 636-644.

Beckett J.B., 1994. Locating recessive genes to chromosome arm with B-A translocations. In The maize handbook, edited by M. Freeling, and V. Walbot. Springer-Verlag New York, pp. 315-327.

Becraft P.W., M. Freeling, 1994. Genetic analysis of *rough sheath1* developmental mutants of maize. Genetics 136: 295-311.

Becraft P.W., P.S. Stinard, D.R. McCarty, 1996. CRINKLY4: a TNFR-like receptor kinase involved in maize epidermal differentiation. Science 273: 1406-1409.

Behrens U., N. Fedoroff, A. Laird, M. Müller-Neumann, P. Starlinger, J. Yoder, 1984. Cloning of *Zea mays* controlling element *Ac* from the *wx-m7* allele. Mol. Gen. Genet. 194: 346-347.

Benito M.I., V. Walbot, 1994. The terminal, inverted repeat sequences of mudr are functionally active promoters in maize cells. *MAYDICA* 39: 255-264.

Bennetzen J.L., P.S. Springer, 1994. The generation of Mutator transposable element subfamilies in maize. Theor. Appl. Genet. 87: 657-667.

Bennetzen J.L., J. Swanson, W.C. Taylor, M. Freeling, 1984. DNA insertion in the first intron of maize *Adh1* affects message levels cloning of progenitor and mutant *Adh1* allele. Proc. Natl. Acad. Sci. USA 81: 4125-4128.

Bennetzen J.L., W.E. Blevins, A.H. Ellingboe, 1988a. Cell-autonomous recognition of the rust pathogen determines *Rp1*-specified resistance in maize. Science 241: 208-241.

Bennetzen J.L., M.M. Qin, S. Ingels, A.H. Ellingboe, 1988b. Allele-specific and *Mutator*-associated instability at the *Rp1* disease-resistance locus of maize. Nature 332: 369-370.

Bennetzen J.L., K. Schrick, P.S. Springer, W.E. Brown, P. SanMiguel, 1994. Active maize genes are unmodified and flanked by diverse classes modified, highly repetitive DNA. Genome 37: 565-576.

Bensen R.J., G.S. Joohal, V.C. Crane, J.T. Tossberg, P.S. Schnable, R.B. Meeley, S.P. Briggs, 1995. Cloning and charactrization of the maize *An1* gene. Plant Cell 7: 75-84.

Bhave M.R., S. Lawrence, C. Burton, L.C. Hannah, 1990. Identification and molecular characterization of *shrunken-2* cDNA clones of maize. The Plant Cell 2: 581-588.

Bianchi A., G. Bianchi, P. Avato, F. Salamini, 1985. Biosynthetic pathways of epicuticular wax of maize as assessed by mutation, light, plant age and inhibitor studies. *MAYDICA* 30: 179-198.

Bianchi M.W., A. Viotti, 1988. DNA methylation and tissue-specific transcription of the storage proteins of maize. Plant Mol. Biol. 11: 203-214.

Birchler J.A., 1994. Construction of compound B-A translocations. In The maize handbook, edited by M. Freeling, and V. Walbot. Springer-Verlag New York, pp. 332-334.

Birchler J.A., E.H. Coe, 1994. Marker systems of *r-x1*. In The maize handbook, edited by M. Freeling, and V. Walbot. Springer-Verlag New York, pp. 359-360.

Bittel D.C., J.M. Shaver, D.A. Somers, B.G. Gengenbach, 1996. Lysine accumulation in maize cell-cultures transformed with a lysine-insensitive form of maize dihydrodipicolinate synthase. Theor. Appl. Genet. 92: 70-77.

Blumberg Vel Spalve J., Z. Schwarz-Sommer, H. Saedler, P.A. Peterson, 1990. The Cin2 and the Cin3 insertion elements of *Zea mays* ssp. *Parviglumis*. *MAYDICA* 35: 151-156.

Boppenmaier J., A.E. Melchinger, G. Seitz, H.H. Geiger, R.G. Herrmann, 1993. Genetic diversity for rflps in european maize inbreds.3. performances of crosses within versus between heterotic groups for grain traits. Plant Breed. 111: 217-226.

Bossinger G., M. Maddaloni, M. Motto, F. Salamini, 1992. Formation and cell lineage patterns of the shoot apex of maize. The Plant J. 2: 311-320.

Boston R.S., J.W. Gillikin, R.L. Wrobel, 1995. Coordinate induction of 3 erluminal stress proteins in maize endosperm mutants. J. Cell. Biochem. Suppl. 19A: 143.

Bravo-Angel A.M., H.A. Becker, R. Kunze, B. Hohn, W.H. Shen, 1995. The binding motifs for *Ac* transposase are absolutely required for excision of *Ds1* in maize. Mol. Gen. Genet. 248: 527-534.

Brosch G., R. Ransom, T. Lechner, J.D. Waldon, P. Loidl, 1995. Inhibition of maize histone deacetylases by HC toxin, the host-selective toxin of *Cochliobolus carbonum*. The Plant Cell 7: 1941-1950.

Brown J.J., M.G. Mattes, C. O'Reilly, N.S. Shepherd, 1989. Molecular characterization of rDt, a maize transposon of the dotted controlling element system. Mol. Gen. Genet. 215: 239-244.

Brutnell T.P., S.L. Dellaporta, 1994. Somatic inactivation and reactivation of *Ac* associated with changes in cytosine methylation and transposase expression. Genetics 138: 213-225.

Buckner B., T.L. Kelson, D.S. Robertson, 1990. Cloning of the *y1* locus of maize a gene involved in the biosynthesis of carotenoids. Plant Cell 2: 867-876.

Buckner B., P. SanMiguel, D. Janick-Buckner, J.L. Bennetzen, 1996. The *y1* gene of maize codes for phytoene synthase. Genetics 143: 479-488.

Bunkers G., O.E. Nelson Jr., V. Raboy, 1993. Maize bronze 1 DSPM insertion mutations that are not fully suppressed by an active *Spm*. Genetics 134: 1211-1220.

Bureau T.E., S.R. Wessler, 1992. *Tourist*: A large family of small inverted repeat elements frequently associated with maize genes. Plant Cell 4: 1283-1294.

Burr B., F.A. Burr, T.P. St.John, M. Thomas, R.D. Davis, 1982. Zein storage protein gene family of maize. J. Mol. Biol. 154: 33-49.

Burr B., S.V. Evola, F.A. Burr, J.S. Beckman, 1983. The application of restriction fragment length polymorphism to plant breeding. In Genetic engineering. Vol. 5. Genetic engineering principles and methods, edited by J.K. Setlow, and J.A. Hollaender. Plenum Press, New York, pp. 45-59.

Burr B., F.A. Burr, K.H. Thompson, M.C. Albertsen, C.W. Stuber, 1988. Gene mapping with recombinant inbreds in maize. Genetics 188: 519-526.

Burr B., F.A. Burr, E.C. Matz, J. Romero Severson, 1992. Pinning down loose ends-mapping telomeres and factors affecting their length. Plant Cell 4: 953-960.

Burr F.A., B. Burr, B. Scheftler, M. Blewitt, U. Wienand, E.C. Matz, 1996. The maize repressor-like gene, *intensifier 1*, belongs to the R/B multigene family of transcription factors and exhibits missplicing. Plant Cell 8: 1249-1259.

Cardon C.H., M. Frey, H. Saedler, A. Gierl, 1993. Mobility of the maize transposable element *En/Spm* in *Arabidopsis thaliana*. The Plant J. 3: 773-784.

Carlson W.R., 1986. The B chromosome of maize. CRC Critical Rev. Pl. Science 3: 201-226.

Carlson W.R., 1988a. The cytogenetics of corn. Ch. 4. In Corn and Corn Improvement., edited by G.F. Sprague, and J.W. Dudley. Am. Soc. Agronomy, Crop Sci Soc. Am., Soil Sci. Soc. Am., Madison, WI, pp. 259-343.

Carlson W.R., 1988b. B chromosomes as a model system for nondisjunction. In Aneuploidy. Part B: Induction and Test Systems, edited by B. Vig and A. Sandberg. Alan Liss, NY, pp. 199-207.

Carlson W.R., C. Curtis, 1986. A new method for producing homozygous duplications in maize. Can. J. Genet. Cytol. 28: 1034-1040.

Carlson W.R., R. Roseman, 1991. Segmental duplication of distal chromosomal regions in maize. Genome 34: 537:542.

Carlson W.R., R. Roseman, 1992. A new property of the maize B chromosome. Genetics 131: 211-223.

Carroll B.J., V.I. Klimyuk, C.M. Thomas, G.J. Bishop, K. Harrison, S.R. Scofield, J.D.G. Jones, 1995. Germinal transpositions of the maize element dissociation from t-DNA loci in tomato. Genetics 139: 407-420.

Casacuberta J.M., P. Puigdomenech, B. Sansegundo, 1991. A gene coding for a basic pathogenesis related (PR-like) protein from *Zea mays*. Molecular cloning and induction by a fungus (*Fusarium moniliforme*) in germinating maize seeds. Plant Mol. Biol. 16: 527-535.

Cerioli S., A. Marocco, M. Maddaloni, M. Motto, F. Salamini, 1994. Early event in maize leaf epiderms formation as revealed by cell lineage studies. Development 120: 2113-2120.

Chandler V.L., K.J. Hardeman, 1992. The *Mu* elements of *Zea mays*. Adv. Genet. 30: 77-122.

Chandler V.L., J.P. Radicella, T.P. Robbins, J. Chen, D. Turks, 1989. Two regulatory genes of the maize anthocyanin pathway are homologous. Isolation of *B* utilizing *R* genomic sequences. Plant Cell 1: 1175-1183.

Chang R.Y., P.A. Peterson, 1994. Chromosome labeling with transposable elements in maize. Theor. Appl. Gen. 87: 650-656.

Chao S., J.M. Gardiner, S.Melia-Hancock, E.H. Coe Jr., 1996. Physical anfd genetic mapping of chromosome 9S in maize using mutations with terminal deficiencies. Genetics 143: 1785-1794.

Chaudhuri S., J. Messing, 1994. Allele-specific parental imprinting of *dzr1*, a posttranscriptional regulator of zein accumulation. Proc. Natl. Acad. Sci. USA 91: 4867-4871.

Chomet P., D. Lisch, K.J. Hardeman, V.L. Chandler, M. Freeling, 1991. Identification of a regulatory transposon that controls the *Mutator* transposable element system in maize. Genetics 129: 261-270.

Chourey P., O.E. Nelson Jr., 1979. Interallelic complementation at the *Sh* locus in maize at the enzyme level. Genetics 91: 317-325.

Christianson M.L., 1986. Fate map of the organizing shoot apex in Gossypium. Am. J. Bot. 73: 947-958.

Christie P.J., M.R. Alfenito, V. Walbot, 1994. Impact of low-temperature stress on general phenylpropanoid and anthocyanin pathways: enhancement of transcript abundance and anthocyanin pigmentation in maize seedlings. Planta 194: 541-549.

Civardi L., Y. Xia, K.J. Edwards, P.S. Schnable, B.J. Nikolau, 1994. The relationship between genetic and physical distances in the cloned *a1-sh2* interval of the *Zea mays* L. genome. Proc. Natl. Acad. Sci. USA 91: 8268-8272.

Close T.J., R.D. Fenton, F. Moonan, 1994. A view of plant dehydrins using antibodies specific to the carboxy-terminal peptide. Plant Mol. Biol. 23: 279-286.

Coe E.H., M.G. Neuffer, 1978. Embryo cells and their destinies in the corn plant. In The clonal basis of development, edited by S. Subtelny, I.M. Sussex. Academic Press, New York, pp. 113-129.

Coe E.H. Jr., D.A. Hoisington, M.G. Neuffer, 1988a. The genetics of corn. In Corn and Corn Improvements, edited by G.F. Sprague, J. Dudley. American Society of Agronomy, Madison, WI, pp. 81-258.

Coe E.H. Jr., D.L. Thompson, V. Walbot, 1988b. The phenotypes mediated by the *iojap* genotype in maize. Am. J. Bot. 75: 634-644.

Coleman C.E., M.A. Lopes, J.W. Gillikin, R.S. Boston, B.A. Larkins, 1995. A defective signal peptide in the maize high-lysine mutant *floury-2*. Proc. Natl. Acad. Sci. USA 95: 6828-6831.

Cone K.C., B. Burr, 1989. Molecular and genetic analyses of the light requirement for anthocyanin synthesis in maize. In The genetics of flavonoids, edited by D.E. Styles, G.A. Gavazzi, M.L. Racchi. Proceedings of a Post Congress Meeting of the XVI Intl. Congress of Genetics. Ed. Unicopli, Milano, Italy.

Cone K.C., S.M. Cocciolone, F.A. Burr, B. Burr, 1993. Maize anthocyanin regulatory gene *Pl* is a duplicate of *C1* that functions in the plant. Plant Cell 5: 1795-1805.

Cooper P., T.HD. Ho, 1983. Heat shock proteins in maize. Plant Physiol. 71: 215-222.

Cooper P., T.HD. Ho, R.M. Hauptmann, 1984. Tissue specificity of the heat-shock response in maize. Plant Physiol. 75: 431-441.

Creighton H.B., B. McClintock, 1931. A correlation of cytological and genetical crossing-over in *Zea mays*. Proc. Natl. Acad. Sci. USA 17: 492-497.

Cresse A.D., S.H. Hulbert, W.E. Brown, J.R. Lucas, J.L. Bennetzen, 1995. *Mu1*-related transposable elements of maize preferentially insert into low copy number DNA. Genetics 140: 315-324.

Crosbie T.M., 1982. Changes in physiological traits associated with long-term breeding efforts to improve grain yield of maize. In Proc. 37th Annu. Corn and Sorghum Industry Res Conf., edited by H.D. Loden, D. Wilkinson. Chicago, IL 5-9 Dec. Am. Seed Trade Assoc. Washington, DC, pp. 206-223.

Cui X., R.P. Wise, P.S. Schnable, 1996. The *rf2* nuclear restorer gene of male-sterile T-cytoplasm maize. Science 272: 1334-1336.

D'Halluin K., E. Bonne, M. Bossut, M.. Beuckeleer, J. Leemans, 1992. Transgenic maize plants by tissue electroporation. The Plant Cell 4: 1495-1505.

Dash S., P.A. Peterson, 1994. Frequent loss of the *En* transposable element after excision and its relation to chromosome-replication in maize (*Zea mays* L.). Genetics 136: 653-671.

Davis R., H. Weintraub, A. Lasser, 1987. Expression of a single transfected cDNA converts fibroblasts into myoblasts. Cell 51: 1061-1067.

Dawe R.K., M. Freeling, 1990. Clonal analysis of the cell lineages in the male flower of maize. Dev. Biol. 142: 233-245.

Dawe R.K., M. Freeling, 1991. Cell lineage and its consequences in higher plants. The Plant J. 1: 3-8.

Dawe R.K., M. Freeling, 1992. The role of initial cells in maize anther morphogenesis. Development 116: 1077-1085.

Dawe R.K., J.W. Sedat, D.A. Agard, W.Ze. Cande, 1994. Meiotic chromosome pairing in maize is associated with a novel chromatin organization. Cell 76: 901-912.

Dean J.V., J.W. Gronwald, M.P. Anderson, 1991. Glutathione S-transferase activity in nontreated and CGA-154281-treated maize shoots. Z. Naturforsch. 46: 850-855.

Dellaporta S.L., A. Calderon-Urrea, 1994. The sex determination process in maize. Science 266: 1501-1505.

Dellaporta S.L., J. Wood, J.B. Hicks, 1983. A plant minipreparation: version 2. Plant Mol. Biol. Rep. 1: 19-22.

Dellaporta S.L., M.A. Moreno, A. Delong, 1991. Cell lineage analysis of the gynoecium of maize using the transposable element *Ac*. Develop. suppl. 1: 141-147.

Delong A., A. Calderon-Urrea, S.L. Dellaporta, 1993. Sex determination gene tasselseed 2 of maize encodes a short-chain alcohol dehydrogenase required for stage-specific floral organ aborttion. Cell 74: 757-768.

DePinho R., K. Hatton, A. Tesfaye, G. Yancopoulos, F. Alt, 1987. The human *myc* gene family. Stucture and activity of *L-myc* and *L-myc* pseudogene. Genes & Dev. 1: 1311-1326.

Devetten N.C., R.J. Ferl, 1994. Transcriptional regulation of environmentally inducible genes in plants by an evolutionarily conserved family of G-box binding factors. Int. J. Biochem. 7: 384-391.

Devetten N.C., R.J. Ferl, 1995. Characterization of a maize G-box binding-factor that is induced by hypoxia. Plant J. 7: 589-601.

Devetten N.C., G. Lu, R.J. Ferl, 1992. A maize protein associated with the G-box binding complex has homology to brain regulatory proteins. Plant Cell 4: 1295-1307.

Dewey R.E, C.S. Levings III, D.H. Timothy, 1986. Novel recombinations in the maize mitochondrial genome produce a unique transcriptional unit in the Texas male-sterile cytoplasm. Cell 44: 439-449.

Dewey R.E., J.N. Siedow, D.H. Timothy, C.S. Levings III, 1988. A 13-kDa maize mitochondrial protein in *E. coli* confers sensitivity to *Bipolaris maydis*. Science 239: 293-295.

Doebley J.F., 1983. The taxonomy and evolution of Tripsacum and teosinte, the closest relatives of maize. In Proc. Int. Maize Virus Disease Colloquium and Workshop, edited by D.T. Gordon, J.K. Knobe, L.R. Nault, and R.M. Ritter. Ohio Agric. Res. Development Center, Wooster, Ohio, p. 15.

Doebley J.F., 1990. Molecular evidence and the evolution of maize. Econ. Bot. 44: 6-27.

Doebley J., A. Stec, 1991. Genetic analysis of the morphological differences between maize and teosinte. Genetics 129: 285-295.

Doebley J., A. Bacigalupo, A. Stec, 1994. Inheritance of kernel weight in two maize-teosinte hybrid populations: Implications for crop evolution. Heredity 85: 191-195.

Doebley J., A. Stec, C. Gustus, 1995. Teosinte branched 1 and the origin of maize: Evidence for epistasis and the evolution of dominance. Genetics 141: 333-346.

Doebley J., A. Stec, L. Hubbard, 1997. The evolution of apical dominance in maize. Nature 386: 485-488.

Doebley J.F., D.J. Wendel, J.S.C. Smth, C.W. Stuber, M.M. Goodman, 1988. The origin of Corn belt maize: the isozyme evidence. Econ. Bot. 42: 120-131.

Donlin M.J., D. Lisch, M. Freeling, 1995. Tissue-specific accumulation of murb, a protein encoded by mudr, the autonomous regulator of the mutator transposable element family. Plant Cell 7: 1989-2000.

Dooner H.K., E. Weck, S. Adams, E. Ralston, M. Favreau, 1985. A molecular genetic analysis of insertions in the *bronze* locus in maize. Mol. Gen. Genet. 200: 240-246.

Dooner H.K., A. Belachew, D. Burgess, S. Harding, M. Ralston, E. Ralston, 1994. Distribution of unlinked receptor sites for transposed *Ac* elements from the *bz-m2 (Ac)* allele in maize. Genetics 136: 261-279.

Dorweiler J., A. Stec, J. Kermicle, J. Doebley, 1993. Teosinte glum architecture 1: A genetic locus controlling a key step in maize evolution. Science 262: 233-235.

Dudley M., R.S. Poethig, 1993. The heterochronic Teopod1 and Teopod2 mutations of maize are expressed non-cell-autonomously. Genetics 133: 389-399.

Dumas C., H.L. Mogensen, 1993. Gametes and fertilization: maize as a model system for experimental embryogenesis in flowering plants. Plant Cell 5: 1337-1348.

Dupuis I., P. Roeckel, E. Matthys-Rochon, C. Dumas, 1987. Procedure to isolate viable sperm cells from the corn (*Zea mays*) pollen grain. Plant Physiol. 85: 876-878.

Duvick D.N., 1965. Cytoplasmic pollen sterility in corn. Adv. Genet. 13: 1-56.

Duvick D.N., 1984. Genetic contributions to yield gains of U.S. hybrid maize, 1930 to 1980. pp. 15-47 In: W.R. Fehr (Ed.) Genetic contributions to yield gains of five major crop plants. CSSA Spec. Publ. 7. CSSA, Madison, WI.

Edallo S., C. Zucchinali, M. Perenzin, F. Salamini, 1981. Chromosomal variation and frequency of spontaneous mutation associated with *in vitro* culture and plant regeneration in maize. *MAYDICA* 26: 39-56.

Edwards K.J., H. Thompson, D. Edwards, A. De Saizieu, C. Sparks, J.A. Thompson, A.J. Greenland, M. Eyers, W. Schuch, 1992. Construction and characterization of a yeast artificial chromosome library containing three haploid maize genome equivalents. Plant Mol. Biol. 19: 299-308.

Eggleston W.B., M. Alleman, J.L. Kermicle, 1995. Molecular organization and germinal instability of *R-stippled* in maize. Genetics 141: 347-360.

Eisen J.A., M.I. Benito, V. Walbot, 1994. Sequence similarity of putative transposases links the maize mutator autonomous element and a group of bacterial insertion sequences. Nucl. Acids Res. 22: 2634-2636.

Evans M.M.S., H.J. Passas, R.S. Poethig, 1994. Heterochronic effects of *glossy15* mutations on epidermal cell identity in maize. Development 120: 1971-1981.

Evola S.V., F.A. Burr, B. Burr, 1986. The suitability of restriction fragment length polymorphisms as genetic markers in maize. Theor. Appl. Genet. 71: 765-771.

Faure J.E., C. Digonnet, C. Dumas, 1994. An *in vitro* system for adhesion and fusion of maize gametes. Science 263: 1568-1600.

Faure J.E., H.L. Mogensen, C. Dumas, H. Lörz, E. Kranz, 1993. Karyogamy after electrofusion of single egg and sperm cell protoplasts from maize: cytological evidence and time course. Plant Cell 5: 747-755.

Fedoroff N.V., J. Mauvais, D. Chaleff, 1983a. Molecular studies on mutations at the *Skrunken* locus in maize strains with mutations caused by the controlling element *Ds*. J. Mol. Appl. Gen. 2: 11-29.

Fedoroff N., S. Wessler, M. Shure, 1983b. Isolation of the transposable maize controlling elements *Ac* and *Ds*. Cell 35: 235-242.

Fedoroff N.V., D.B. Furtek, O.E. Nelson Jr., 1984. Cloning of the *bronze* locus in maize by a simple and generalizable procedure using the transposable controlling element *Activator (Ac)*. Proc. Natl. Acad. Sci. USA 81: 3825-3829.

Fedoroff N., M. Schläppi, R. Raina, 1995. Epigenetic regulation of the maize spm transposon. Bioessays 17: 291-297.

Ferl R.J., H.S. Nick, 1987. *In vivo* detection of regulatory factor binding sites in the 5' flanking region of the maize *Adh1*. J. Biol. Chem. 262: 7947-7950.

Ferl R.J., S.R. Dlouhy, D. Schwartz, 1979. Analysis of maize alcohol dehydrogenease by native-SDS two dimentional electrophoresis and autoradiography. Mol. Gen. Genet. 169: 7-12.

Finnegan E.J., B.H. Taylor, E.S. Dennis, W.J. Peacock, 1988. Transcription of the maize transposable element *Ac* in maize seedlings and in transgenic tobacco. Mol. Gen. Genet. 212: 505-509.

Finnegan E.J., B.H. Taylor, S. Craig, E.S. Dennis, 1989. Transposable elements can be used to study cell lineages in transgenic plants. Plant Cell 1: 757-764.

Firek S., D.J. Martin, M.R. Roberts, F. Sturgess, R. Scott, J. Draper, 1996. Gametophyte-specific transposon of the maize *Ds* element in transgenic tobacco. The Plant J. 10: 569-578.

Fisher D.K., C.D. Boyer, L.C. Hannah, 1993. Starch branching enzyme II from maize endosperm. Plant Physiol. 102: 1045-046.

Fisher A., N. Baum, H. Saedler, G. Thiessen, 1995. Chromosomal mapping of the mads-box multigene family in *Zea mays* reveals dispersed distribution of allelic genes as well as transported copies. Nucl. Acids Res. 23: 1901-1911.

Fleenor D., M. Spell, D.S. Robertson, S.R. Wessler, 1990. Nucleotide sequence of the maize *Mutator* element, *Mu8*. Nuc. Acids Res. 18: 6725.

Fluminhan A., T. Kameya, 1996. Behavior of chromosomes in anaphase cells in embryogenic callus cultures of maize (*Zea mays* L.). Theor. Appl. Genet. 92: 982-990.

Fobis-Loisy I., K. Loridon, S. Lobreaux, M. Lebrun, J.F. Briat, 1995. Structure and differential expression of two maize ferritin genes in response to iron and abscisic acid. Eur. J. Biochem. 231: 609-619.

Fowler J.E., M. Freeling, 1996. Genetic analysis of mutations that alter cell fates in maize leaves: dominant *Liguleless* mutations. Dev. Genet. 17: 349:357.

Fowler J.E., G.J. Muehlbauer, M. Freeling, 1996. Mosaic analysis of the *liguleless3* mutant phenotype in maize by coordinate suppression of *Mutator*-insertion alleles. Genetics 143: 489-503.

Frame B.R., P.R. Drayton, S.V. Bagnali, C.J. Lewnau, W.P. Bullock, H.M. Wilson, J.M. Dunwell, J.A. Thompson, K. Wang, 1994. Production of fertile transgenic maize plants by silicon carbide whisker-mediate transformation. The Plant J. 6: 941-948.

Franken P., U. Niesbach-Klösgen, U. Weydemann, I. Marechal-Drouard, H. Saedler, U. Wienand, 1991. The duplicated chalcone synthase genes *C2* and *Whp* (white pollen) of *Zea mays* are independently regulated: evidence for translational control of Whp expression by the anthocyanin intensifying gene. EMBO J. 10: 2605-2612.

Freeling M., 1992. A conceptual framework for maize leaf development. Dev. Biol. 153: 44-58.

Freeling M., S. Hake, 1985. Developmental genetics of mutants that specify knotted leaves in maize. Genetics 111: 617-634.

Frey M., S. Tavantzis, H. Saedler, 1989. The maize *En-1/Spm* element transposes in potato. Mol. Gen. Genet. 217: 172-177.

Frey M., R. Kleim, H. Saedler, A. Gierl, 1995. Expression of a cytochrome P450 gene family in maize. Mol. Gen. Genet. 246: 100-109.

Fromm M.E. F. Morrish, C. Armstrong, R. Williams, J. Thomas, T.M. Klein, 1990. Inheritance and expression of chimeric genes in the progeny of transgenic maize plants. Bio/Technology 8: 833-839.

Fuerstenberg S.I., M.A. Johns, 1990. Distribution of *Bs1* retrotransposons in *Zea* and related genera. Theor. Appl. Genet. 80: 680-686.

Gagliardi D., C. Breton, A. Chaboud, P. Vergne, C. Dumas, 1995. Expression of heat shock factor and heat shock protein 70 genes during maize pollen development. Plant Mol. Biol. 29: 841-856.

Galinat W.C., 1971. The origin of maize. Annu. Rev. Genet. 5: 447-478.

Galinat W.C., 1992. Evolution of corn. Adv. Agron. 47: 203-231.

Galinat W.C., 1994. The patterns of plant structures in maize. In The maize handbook, edited by M. Freeling, V. Walbot. Springer-Verlag New York, Berlin, pp. 61-65.

Gardiner J.M., E.H. Coe, S. Melia-Hancock, D.A. Hoisington, S. Chao, 1993. Development of a core RFLP map in maize using an immortalized F-2 population. Genetics 134: 917-930.

Gengenbach B.G., C.E. Green, C.M. Donovan, 1977. Inhertiance of selected pathotoxin resistance in maize plants regenerated from cell cultures. Proc. Natl. Acad. Sci. USA 74: 5113-5117.

Geraghty D., M.A. Peifer, I. Rubenstein, J. Messing, 1981. The primary structure of a plant storage protein: zein. Nucl. Ac. Res. 9: 5163-5174.

Gierl A., H. Saedler, 1992. Plant transposable elements and gene tagging. Plant Mol. Biol. 19: 39-49.

Gierl A., S. Lutticke, H. Saedler, 1988. *TnpA* product encoded by the transposable element *En-1* of *Zea mays* is a DNA binding protein. EMBO J. 7: 4045-4053.

Goff S., K.C. Cone, V.L. Chandler, 1992. Functional analysis of the transcriptional activator encoded by the maize *B* gene-evidence for a direct functional interaction between two classes of regulatory proteins. Genes Dev. 6: 864-875.

Goodman M.M., 1988. The history and evolution of maize. Critical Rev. Plant Sci. 7: 197-220.

Goping I.S., J.R. Frappier, D.B. Walden, B.G. Atkinson, 1991. Sequence identification and characterization of cDNA encoding two different members of the 18 kDa heat shock family of *Zea mays* L. Plant Mol. Biol. 16: 699-711.

Gordon-Kamm W.J., T.M. Spencer, M.L. Mangano, T.R. Adams, R.J. Daines, W.G. Start, J.V. O'Brien, S.A. Chambers, W.R. Adams Jr., N.G. Willetts, T.B. Rice, C.J. Mackey, R.W. Krueger, A.P. Kausch, P.G. Lemaux, 1990. Transformation of maize cells and regeneration of fertile transgenic plants. Plant Cell 2: 603-618.

Grafi G., B.A. Larkins, 1995. Endoreduplication in maize endosperm. Involvement of M phase-promoting factor inhibition and induction of S phase-related kinases. Science 269: 1262-1264.

Green C.E., R.L. Phillips, 1975. Plant regeneration from tissue cultures of maize. Crop Sci. 15: 417-421.

Greyson R.I., Z. Yang, R.A. Bouchard, J.R.H. Frappier, B.G. Atkinson, E. Banasikowska, D.B. Walden, 1996. Maize seedlings show cell-specific responses to heat shock as revealed by expression of RNA and protein. Develop. Genet. 18: 244-253.

Grotewold E., P. Athma, T. Peterson, 1991. Alternatively spliced products of the maize P gene encode proteins with homology to the DNA-binding domain of *myb*-like transcription factors. Proc. Natl. Acad. Sci. USA 88: 4587-4591.

Guan L., J.G. Scandalios, 1993. Charcterization of the catalase antioxidant defense gene *Cat1* of maize, and its developmentally regulated expression in transgenic tobacco. Plant J. 3: 527-536.

Guan L., J.G. Scandalios, 1995. Developmentally related responses of maize catalase genes to salicylic acid. Proc. Natl. Acad. Sci. USA 92: 5930-5934.

Guan L., A.N. Polidoros, J.G. Scandalios, 1996. Isolation, characterization and expression of the maize *Cat2* catalase gene. Plant Mol. Biol. 30: 913-924.

Guo M., D. Davis, J.A. Birchler, 1996. Dosage effects on gene expression in a maize ploidy series. Genetics 142: 1349-1353.

Hallauer A.R., J.B. Miranda Fo., 1988. Quantitative genetics in maize breeding. 2nd ed. Iowa State University Press, Ames, IA.

Han C.D., E.H. Coe Jr., R.A. Martienssen, 1992. Molecular cloning and characterization of *iojap* (*ij*), a pattern striping gene of maize. EMBO J. 11: 4037-4046.

Hannah L.C., 1997. Starch synthesis in the maize seed. In: B.A. Larkins and I.K. Vasil (Eds) Cellular and Molecular Biology of the Plant Seed Development. Kluwer Acad. Publ., Dordrecht (in press).

Hansen J.D., J. Pyee, Y. Xia, T.J. Wen, D.S. Robertson, P.E. Kolattukudy, B.J. Nikolau, P.S. Schnable, 1997. The *glossy1* locus of maize and an epidermis-specific cDNA from *Kleinia odora* define a class of receptor-like proteins required for the normal accumulation of cuticular waxes. Plant Physiol. 113: 1091-1100.

Hanson M.A., B.S. Gaut, A.O. Stec, S.I. Fuerstenberg, M.M. Goodman, E.H. Coe, J.F. Doebley, 1996. Evolution of anthocyanin biosynthesis in maize kernels: the role of regulatory and enzymatic loci. Genetics 143: 1395-1407.

Hardeman K.J., V.L. Chandler, 1993. Two maize genes are each targeted predominantly by distinct classes of Mu elements. Genetics 135: 1141-1150.

Harper L., M. Freeling, 1996. Studies on early leaf development. Curr. Opinion Biotech. 7: 139-144.

Harris L.J., K. Currie, V.L. Chandler, 1994. Large tandem duplication associated with a Mu2 insertion in *Zea mays* B-Peru gene. Plant Mol. Biol. 25: 817-828.

Hartings H., N. Lazzaroni, A. Spada, R. Thompson, F. Salamini, M. Motto, J. Palau, N. Di Fonzo, 1989. The *b-32* protein from endosperm: characterization of genomic sequences. Maize Gen. Coop. Newslet. 63: 29-30.

Hartings H., C. Spilmont, N. Lazzaroni, V. Rossi, F. Salamini, R.D. Thompson, M. Motto, 1991a. Molecular analysis of the *Bg-rbg* transposable element system of *Zea mays* L. Mol. Gen. Genet. 227: 91-96.

Hartings H., V. Rossi, N. Lazzaroni, R.D. Thompson, F. Salamini, M. Motto, 1991b. Nucleotide sequence of the *Bg-rbg* transposable element of *Zea mays* L. *MAYDICA* 36: 355-359.

Hartings H., N. Lazzaroni, V. Rossi, M. Motto, 1996. Distribution of sequences related to the *Bg* transposable element of maize in *Zea* and related genera. Theor. Appl. Genet. 92: 696-701.

Helentjaris T., 1993. Implications for conserved genomic structure among plant species. Proc. Natl. Acad. Sci. USA 90: 8308-8313.

Helentjaris T., D. Weber, S. Wright, 1988. Identification of the genomic locations of duplicate nucleotide sequences in maize by analysis of restriction fragment length polymorphisms. Genetics 118: 356-363.

Helentjaris T., G. King, M. Slocum, C. Siedenstrang, S. Wegman, 1985. Restriction fragment length polymorphisms as probes for plant diversity and their development as tools for applied plant breeding. Plant Mol. Biol. 5: 109-118.

Helentjaris T., M. Slocum, S. Wright, A. Schaefcr, J. Nienhuis, 1986. Construction of genetic linkage maps in maize and tomato using restriction fragment length polymorphisms. Theor. Appl. Genet. 72: 761-769.

Hershberger R.J., C.A. Warren, V. Walbot, 1991. Mutator activity in maize correlates with the presence and expression of the *Mu* transposable element *Mu9*. Proc. Natl. Acad. Sci. USA 88: 10198-10202.

Hershberger R.J., M.I. Benito, K.J. Hardeman, C.A. Warren, V.L. Chandler, V. Walbot, 1995. Characterization of the major transcripts encoded by the regulatory mudr transposable element of maize. Genetics 140: 1087-1098.

Hibberd K.A., C.E. Green, 1982. Inheritance and expression of lysine plus threonine resistance selected in maize tissue culture. Proc. Natl. Aad. Sci. USA 79: 559-563.

Hibberd K.A., T. Walter, C.E. Green, B.G. Gengenbach, 1980. Selection and characterization of a feedback-insensitive tissue culture of maize. Planta 148: 183-187.

Hoecker U., I.K. Vasil, D.R. McCarty, 1995. Integrated control of seed maturation and germination programs by activator and repressor functions of *viviparous-1* of maize. Gene & Develop. 9: 2459-2469.

Holt D.C., V.J. Lay, E.D. Clarke, A. Dinsmore, I. Jepson, S.W.J. Bright, A.J. Greenland, 1995. Characterization of the safener-induced glutathione S-transferase isoform II from maize. Planta 196: 295-302.

Hoisington D., E.H. Coe Jr., M.G. Neuffer, 1988. Genelist and linkage map of maize (*Zea mays* L.). Maize Genet. Coop. Newsl. 62: 125-149.

Hong K.S., T.E. Richter, J.L. Bennetzen, S.H. Hubert, 1993. Complex duplications in maize lines. Mol. Gen. Genet. 239: 115-121.

Hooker A.L., 1985. Corn and sorghum rusts. In The Cereal Rusts, Diseases, Distribution, Epideliology and Control. Vol. 2, edited by A.P. Roelfs, WRBA Press. Academic Press, San Diego, CA, pp. 208-229.

Hooker A.L., D.R. Smith, S.M. Lim, J.B. Beckett, 1970. Reaction of corn seedlings with male-sterile cytoplasm *Helminthosporium maydis*. Plant Dis. Rep. 54: 708-712.

Hsia A.P., P.S. Schnable, 1996. DNA sequence analyses support the role of interrupted gap repair in the origin of the internal deletions of the maize transposon, *MuDR*. Genetics 142: 602-618.

Hu W.M., P.O. Das, J. Messing, 1995. *Zeon-1*, a member of a new maize retrotransposon family. Mol. Gen. Genet. 248: 471-480.

Huang B.Q., W.F. Sheridan, 1994. Female gamethophyte development in maize. Microtubular organization and embryo sac polarity. Plant Cell 6: 845-861.

Huang B.Q., W.F. Sheridan, 1996. Embryo sac development in the maize *indeterminate gametophyte1* mutant: abnormal nuclear behavior and defective microtubule organization. Plant Cell 8: 1391-1407.

Hueros G., S. Varotto, F. Salamini, R.D. Thompson, 1995. Molecular characterization of *BET1*, a gene expressed in the endosperm transfer cells of maize. The Plant Cell 7: 747-757.

Hulbert S.H, J.L. Bennetzen, 1991. Recombination at the *Rp1* locus of maize. Mol. Gen. Genet. 226: 377-382.

Iltis H.H., 1983. From teosinte to maize: the catastrophic sexual transmutation. Science 222: 886-894.

Iltis H.H., J.F. Doebley, R. Guzman, B. Pazy, 1979. *Zea diploperennis* (*Gramineae*): A new teosinte from Mexico. Science 203: 186-188.

Ishida Y., H. Saito, S. Ohta, Y. Hiei, T. Komari, T. Kumashiro, 1996. High efficiency transformation of maize (*Zea mays* L.) mediated by *Agrobacterium tumefaciens*. Nature Biotech. 14: 745-750.

Jackson D., B. Veit, S. Hake, 1994. Expression of maize *knotted1* related homeobox genes in the shoot apical meristem predicts patterns of morphogenesis in the vegetative shoot. Development 120: 405-413.

James M.G., D.S. Robertson, A.M. Myers, 1995. Characterization of the maize gene *sugary1*, a determinant of starch composition in kernels. Plant Cell 7: 417-429.

Jensen P.J., R.S. Bandurski, 1994. Metabolism and synthesis of indole-3-acetic acid (IAA) in *Zea mays*. Plant Physiol. 106: 343-351.

Jepson I., V.J. Lay, D.C. Holt, S.W.J. Bright, A.J. Greenland, 1994. Cloning and characterization of maize herbicide safener-induced cDNAs encoding subunits of glutathione S-transferase isoforms I, II and IV. Plant Mol. Biol. 26: 1855-1866.

Jin Y.K., J.L. Bennetzen, 1994. Integration and nonrandom mutation of a plasma membrane proton ATPase gene fragment within the *Bs1* retroelement of maize. Plant Cell 6: 1177-1186.

Johal G.S., S.P. Briggs, 1992. Reductase activity encoded by the *HM1* disease resistance gene in maize. Science 258: 985-987.

Johns M.A., J. Mottinger, M. Freeling, 1985. A low copy number copia-like transposon in maize. EMBO J. 4: 1093-1102.

Keith C.S., D.O. Hoang, B.M. Barrett, B. Feigelman, M.C. Nelson, H. Thai, C. Baysdolfer, 1993. Partial sequence-analysis of 130 randomly selected maize cDNA clones. Plant Physiol. 101: 329-332.

Keller J., D.G. Jones, E. Harper, E. Lim, F. Carland, E.J. Ralston, H.K. Dooner, 1993a. Effects of gene dosage and sequence modification on the frequency and timing of transposition of the maize element *Activator Ac* in tobacco. Plant Mol. Biol. 21: 157-170.

Keller J., E. Lim, H.K. Dooner, 1993b. Preferential transposition of *Ac* to linked sites in *Arabidopsis*. Theor. Appl. Gen. 86: 585-588.

Kermicle J.L., 1969. Androgenesis conditioned by a mutation in maize. Science 116: 1422-1424.

Kermicle J.L., W.B. Eggleston, M. Alleman, 1995. Organization of paramutagenicity in *R-stippled* in maize. Genetics 141: 361-372.

Kernodle S.P., J.G. Scandalios, 1996. A comparison of the structure and function of the highly homologous maize antioxidant Cu/Zn superoxide dismutase genes, *Sod4*and *Sod4A*. Genetics 144: 317-328.

Kerstetter R., E. Vollbrecht, B. Lowe, B. Veit, J. Yamagushi, S. Hake, 1994. Sequence-analysis and expression patterns divide the maize *knotted1*-like homeobox genes into 2 classes. Plant Cell 6: 1877-1887.

Kindiger B., C. Curtis, J.B. Beckett, 1991. Adjacent II segregation products in B-A translocations of maize. Genome 34: 592-602.

Klein J., H. Saedler, P. Huijser, 1996. A new family of DNA binding proteins includes putative transcriptional regulators of the *Antirrhinum majus* floral meristem identity gene Squamosa. Mol. Gen. Genet. 250: 7-16.

Klinge B., W. Werr, 1995. Transcription of the *Zea mays homeobox* (*ZmHox*) genes is activated early in embryogenesis and restricted to meristems of the maize plant. Dev. Genetics 16: 349-357.

Klinge B., B. Uberlacker B., C. Korfhage, W. Werr, 1996. *Zmhox* - a novel class of maize homeobox genes. Plant Mol. Biol. 30: 439-453.

Klösgen R.J., A. Gierl, Z. Schwarz-Sommer, H. Saedler, 1986. Molecular analysis of the waxy locus of Zea maize. Mol. Gen. Genet. 203: 237-244.

Korth K.L., C.S. Levings II, 1993. Baculovirus expression of the maize mitochondrial protein URF13 confers insecticidal activity in cell cultures and larvae. Proc. Natl. Acad. Sci. USA 90: 3388-3392.

Korth K.L., C.I. Kaspi, J.N. Siedow, C.S. Levings III, 1991. URF13, a maize mitochondrial pore-forming protein, is oligomeric and has a mixed orientation in *Escherichia coli* plasma membranes. Proc. Natl. Acad. Sci. USA 88: 10865-10869.

Knowles R.V., F. Srienc, R.L. Phillips, 1990. Endoreduplication of nuclear DNA in the developing maize endosperm. Dev. Gen. 11: 125-132.

Knowles R.V., M.D. McMullen, G. Yerk, R.L. Phillips, S. Kraemer, F. Srienc, 1992. Endosperm mitotic activity and endoreduplication in maize affected by defective kernel mutations. Genome 35: 68-77.

Koziel M.G. G.L. Beland, C. Bowman, N.B. Carozzi, R. Crenshaw, L. Crossland, J. Dawson, N. Desai, M. Hill, S. Kadwell, K. Launis, K. Lewis, D. Maddox, K. MCPherson, M.R. Meghji, E. Merlin, R. Rhodes, G.W. Warren M. Wright, S.V. Evola, 1993. Field performance of elite transgenic maize plants expressing an insecticidal protein derived from *Bacillus thuringiensis*. Bio/Technology 11: 194-200.

Kranz E., H. Lörz, 1993. *In vitro* fertilization with isolated, single gametes results in zygotic embryogenesis and fertile maize plants. Plant Cell 5: 739-746.

Kuchareck T.A., I. Kommedahl, 1966. Kernel infection and corn stalk and rot caused by *Fusarium moniliforme*. Phytopathology 56: 983-984.

Kunze R., P. Starlinger, 1989. The putative transposase of transposable element *Ac* from *Zea mays* L. interacts with subterminal sequences of *Ac*. EMBO J. 8: 3177-3185.

Kunze R., U. Stochaj, J. Laufs, P. Starlinger, 1987. Transcription of the transposable element *Activator* (*Ac*) of *Zea mays* L. EMBO J. 6: 1555-1563.

Kunze R., S. Kühn, T. Jones, S.R. Scofield, 1993. Somatic and germinal activities of maize *Activator* (*Ac*) transposase mutants in transgenic tobacco. The Plant J. 8: 45-54.

Kyozuka J., M. Olive, W.J. Peacock, E.S. Dennis, K. Shimamoto, 1994. Promoter elements required for developmental expression of the maize *Adh1* gene in transgenic rice. Plant Cell 6: 799-810.

Laetsch W.M., 1974. The C4 syndrome: A structural analysis. Annu. Rev. Plant Physiol. 25: 27-52.

Lamoureux G.L., D.G. Rusness, 1989. The role of glutathione and glutathione-S-transferases in pesticide metabolism, selectivity, and mode of action in plants and insects. In Coenzymes and Cofactors. Vol. 3. Glutathione: Chemical, Biochemical and Medical Aspects. Part B, edited by D. Dolphin, R. Poulson, O. Avramovic. J. Wiley, New York, pp. 153-196.

Langdale J.A, C.A. Kidner, 1994. Bundle-sheath defective, a mutation that disrupts cellular-differentiation in maize leaves. Development 120: 673-681.

Langdale J.A., B. Lane, M. Freeling, T. Nelson, 1989. Cell lineage analysis of maize bundle sheath and mesophyll cells. Dev. Biol. 133: 128-129.

Langdale J.A., L.N. Hall, R. Roth, 1995. Control of cellular differentiation in maize leaves. Philos. Trans. R. Soc. Lond. (Biol.) 350: 53-57.

Laughnan J.R., S. Gabay-Laughnan, 1983. Cytoplasmic male sterility in maize. Ann. Rev. Genet. 17: 27-48.

Lechelt C., T. Peterson, A. Laird, J. Chen, S.L. Dellaporta, E. Dennis, W.J. Peacock, P. Starlinger, 1989. Isolation and molecular analysis of the maize *P* locus. Mol. Gen. Genet. 219: 225-234.

Lee B.H., A.L. Hooker, W.A. Russell, J.G. Dickson, A.L. Flangas, 1963. Genetic relationships of alleles on chromosome 10 for resistance to *Puccinia sorghi* in 11 corn lines. Crop Sci. 3: 24-26.

Levings C.S. III, D.R. Pring, 1976. Restriction endonuclease analysis of mitochondrial DNA from normal and Texas cytoplasmic male-sterile maize. Science 193: 158-160.

Lewin R., 1983. A naturalist of the genome. Science 222: 402-405.

Li Z.H., P.D. Matthews, B. Burr, E.T. Wurtzel, 1996. Cloning and characterization of a maize cDNA encoding phytoene desaturase, an enzyme of the carotenoid biosynthetic pathway. Plant Mol. Biol. 30: 269-279.

Lin B.-Y., 1978. Structural modifications of the female gametophyte associated with the *indeterminate gametophyte (ig1)* mutant in maize. Can. J. Cytol. 20: 249-257.

Lin B.-Y., 1981. Megagametogenetic alterations associated with the *indeterminate gametophyte (ig1)* mutant in maize. Rev. Bras. Biol. 41: 557-563.

Lisch D., P. Chomet, M. Freeling, 1995. Genetic characterization of the Mutator system in maize: behavior and regulation of Mu transposons in a minimal line. Genetics 139: 1777-1796.

Liu C.N., I. Rubenstein, 1993. Transcriptional characterization of an alpha zein gene cluster in maize. Plant Mol. Biol. 22: 323-336.

Liu L., D.S. Maillet, J.R.H. Frappier, D.B. Walden, B.G. Atkinson, 1995. Characterization, chromosomal mapping, and expression of different polyubiquitin genes in tissues from control and heat-shocked maize seedlings. Biochem. Cell Biol. 73: 19-30.

Livini C., P. Ajmone Marsan, A.E. Melchinger, M.M. Messmer, M. Motto, 1992. Genetic diversity of maize inbred lines within and among heterotic groups revealed by RFLPs. Theor. Appl. Genet. 84: 17-25.

Lohmer S., M. Maddaloni, M. Motto, N. Di Fonzo, H. Hartings, F. Salamini, R.D. Thompson, 1991. The maize regulatory locus *opaque-2* encodes a DNA-binding protein which activates transcription of the b-32 gene. EMBO J. 10: 617-624.

Lohmer S., M. Motto, M. Maddaloni, F. Salamini, R.D. Thompson, 1993. Translation of the mRNA of the maize transcriptional activator *Opaque-2* is inhibited by upstream open reading frames present in the leader sequence. Plant Cell 5: 65-73.

Ludwig S.R., L.F. Habera, S.L. Dellaporta, S.R. Wessler, 1989. *Lc*, a member of the maize *R* gene family responsible for tissue-specific anthocyanin production, encodes a protein similar to transcriptional activators and contains the *myc* homology region. Proc. Natl. Acad. Sci. USA 86: 7092-7096.

Ludwig R.S., B. Bowen, L. Beach, S.R. Wessler, 1990. A regulatory gene as a novel visible marker for maize transformation. Science 247: 449-450.

Lugert T., W. Werr, 1994. A novel DNA-binding domain in the Shrunken inibitor-binding protein (IBP1). Plant Mol. Biol. 25: 493-506.

Lund G., P. Ciceri, A. Viotti, 1995a. Maternal-specific demethylation and expression of specific alleles of zein genes in the endosperm of *Zea mays* L. Plant J. 8: 571-581.

Lund G., O.P. Das, J. Messing, 1995b. Tissue-specific DNase I-sensitive sites of the maize P gene and their changes upon epimutation. Plant J. 7: 797-807.

Luo D., R. Carpenter, C. Vincent, L. Copsey, E. Coen, 1995. Origin of floral asymmetry in *Antirrhinum*. Nature 383: 794-799.

Lusardi M.C., G. Neuhaus-Url, I. Potrykus, G. Neuhaus, 1994. An approach towards genetically engineered cell fate mapping in maize using the *Lc* gene as a visible marker: transactivation capacity of *Lc* vectors in differentiated maize cells and microinjection of *Lc* vectors into somatic embryos and shoot apical meristems. Plant J. 5: 571-582.

Maas C., S. Schaal, W. Werr, 1990. A feedback control element near the transcription start site of the maize Shrunken gene determines promoter activity. EMBO J. 9: 3447-3452.

Maddaloni M., L. Barbieri, S. Lohmer, F. Salamini, R. Thompson, 1991. Characterization of an endosperm-specific developmentally regulated protein synthesis inhibitor from maize seeds. J. Genet. & Breed. 45: 377-380.

Maddaloni M., G. Donini, C. Balconi, E. Rizzi, P. Gallusci, F. Forlani, S. Lohmer, R. Thompson, F. Salamini, M. Motto, 1996. The transcriptional activator Opaque-2 controls the expression of a cytosolic form of pyruvate orthophosphate dikinase-1 in maize endosperms. Mol. Gen. Genet. 250: 647-654.

Maddaloni M., F. Forlani, V. Balmas, G. Donini, L. Stasse, L. Corazza, M. Motto, 1997. Tolerance to the fungal pathogen *Rhizoctonia solani* AG4 of transgenic tobacco expressing the maize ribosome-inactivating protein b-32. Trans. Res. (in press).

Mangelsdorf P.C., 1974. Corn, Its Origin, Evolution and Improvement. Harvard Univ. Press, Cambridge, Mass.

Mannervik B., U.H. Danielson, 1988. Glutathione transferases: structure and catalytic activity. Crit. Rev. Biochem. 21: 283-337.

Mao Z.Y., R. Paiva, A.L. Kriz, J.A. Juvik, 1995. Dehydrin gene-expression in normal and viviparous embryos of *Zea mays* during seed development and germination. Plant Physiol. Biochem. 33: 649-653.

Mariani C., M. De Beucheleer, J. Truettner, J. Leemans, R.B. Goldberg, 1990. Induction of male sterility in plants by a chimeric ribonuclease gene. Nature 347: 737-741.

Marocco A., A. Santucci, S. Cerioli, M. Motto, N. Di Fonzo, R. Thompson, F. Salamini, 1991. Three high-lysine mutations control the level of ATP-binding HSP70-like proteins in the maize endosperm. Plant Cell 3: 507-515.

Marrs K.A., M.R. Alfenito, A.M. Lloyd, V. Walbot, 1995. A glutathione S-transferase involved in vacuolar transfer encoded by the maize gene *Bronze-2*. Nature 375: 397-400.

Masson P., N. Fedoroff, 1989. Mobility of the maize *Suppressor-mutator* element in transgenic tobacco cells. Proc. Natl. Acad. Sci. USA: 86: 2219-2223.

Masson P., R. Surosky, J. Kingsbury, N.V. Fedoroff, 1987. Genetic and molecular analysis of the *Spm*-dependent *a-m2* alleles of the maize *a* locus. Genetics 177: 117-137.

Masson P. G. Rutherford, J. Banks, N. Fedoroff, 1989. Essential large transcripts of the maize *Spm* transposable element are generated by alternative splicing. Cell 58: 755-765.

McCarty D.R., 1995. Genetic-control and integration of maturation and germination pathways in seed development. Annu. Rev. Plant Physiol. Plant Mol. Biol. 46: 71-93.

McCarty D.R., J.R. Shaw, L.C. Hannah, 1986. Th cloning, genetic mapping, and expression of the *constitutive sucrose synthase* locus of maize. Proc. Natl. Acad. Sci. USA 83: 9099-9103.

McCarty D.R., C.B. Carson, M. Lazar, S.C. Simonds, 1989a. Transposable element induced mutations of the *viviparous-1* gene of maize. Dev. Genet. 10: 273-281.

McCarty D.R., C.B. Carson, P.S. Stinard, D.S. Robertson, 1989b. Molecular analysis of *viviparous-1*: An abscisic acid-insensitive mutant of maize. Plant Cell 1: 523-532.

McCarty D.R., T. Hattori, C.B. Carson, V. Vasil, M. Lazar, I.K. Vasil, 1991. The *viviparous-1* developmental gene of maize encodes a novel transcriptional activator. Cell 66: 895-905.

McClintock B., 1932. A correlation of ring-shaped chromosomes with variegation in *Zea mays*. Proc. Natl. Acad. Sci. USA 18: 677-681.

McClintock B., 1945. Cytogenetic studies of maize and Neurospora. Carnegie Inst. Wash. Year Book 44: 108-112.

McClintock B., 1954. Mutations in maize and chromosomal aberrations in neurospora. Carnegie Inst. Wash. Year Book 53: 254-260.

McClintock B., 1978. Development of the maize endosperm as revealed by clones. In The clonal basis of development, edited by S. Subtelny, I.M. Sussex. Academic Press, New York, pp. 217.

McClintock B., 1984. The significance of responses of the genome to challenge. Science 226: 792-801.

McDaniel C.N., R.S. Poethig, 1988. Cell-lineage patterns in the shoot apical meristem of the germinating maize embryo. Planta 175: 13-22.

McLaughlin M., V. Walbot, 1987. Cloning of a mutable *bz2* allele of maize by transposon tagging and differential hybridization. Genetics 117: 771-776.

Meeley R.B., J.D. Walton, 1991. Enzymatic detoxification of HC toxin, the host-selective cyclic peptide from *Cochilobolus carborum*. Plant Physiol. 97: 1080-1086.

Meeley R., S. Briggs, 1995. Reverse genetics for maize. Maize Gen. Coop. Newsletter 69: 67-82.

Meeley R.B., G.S. Johal, S.P. Briggs, J.D. Walton, 1992. A biochemical phenotype for a disease resistance gene of maize. Plant Cell 4: 71-77.

Mehta A.D., J.E. Flaherty, G.A. Payne, R.S. Boston, 1995. Investigation of an antifungal activity of a maize ribosome-inactivating protein. Mol. Biol. of Cell 6: 315-325.

Meinke D.W., 1991. Perspectives on genetic analysis of plant embryogenesis. Plant Cell 3: 857-866.

Melchinger A.E., M.M. Messmer, M. Lee, W. Woodman, K.R. Lamkey, 1991. Diversity and relationships among U.S. maize inbreds revealed by restriction fragment length polymorphisms. Crop Sci. 31: 669-678.

Mena M., M.A. Mandel, D.R. Lerner, M.F. Yanofsky, 1995. A characterization of the mads-box gene family in maize. Plant J. 8: 845-854.

Menssen A., S. Höhmann, W. Martin, P.S. Schnable, P.A. Peterson, H. Saedler, A. Gierl, 1990. The *En/Spm* transposable element of *Zea mays* contains splice sites at the termini generating a novel intron from a dSpm element in the *A2* gene. EMBO J. 9: 3051-3057.

Miao S.H., D.R. Duncan, J.M. Widholm, 1986. Selection of lysine plus threonine and 5-methyltryptophan resistance in maize tissue culture. In IV Intl. Congress of Plant Tissue and Cell Culture, edited by D.A. Somers, B.G. Gengenbach, D.D. Biesboer, W.P. Hackett, C.E. Green. Minneapolis, MN, p. 380.

Michel D., H. Hartings, S. Lanzini, M. Michel, M. Motto, G.R. Riboldi, F. Salamini, H.P. Döring, 1995. Insertion mutations at the maize *Opaque-2* locus induced by transposable element families *Ac*, *En/Spm* and *Bg*. Mol. Gen. Genet. 248: 287-292.

Miles D., 1994. The role of high chlorophyll fluorescence photosynthetic mutants in the analysis of chloroplast thylakoid membrane assembly and function. *MAYDICA* 39: 35-45.

Miller M.E., P.S. Chourey, 1992. The maize invertase-deficient *miniature-1* seed mutation is associated with aberrant pedicel and endosperm development. Plant Cell 4: 297-302.

Montanelli C., N. Di Fonzo, R. Marotta, M. Motto, C. Soave, F. Salamini, 1984. Occurrence and behavior of the components of the *o2-m(r)-Bg* system of maize controlling elements. Mol. Gen. Genet. 197: 209-218.

Moose S.P., P.H. Sisco, 1994. *Glossy15* controls the epidermal juvenile-to adult phase transition in maize. Plant Cell 6: 1343-1355.

Moose S.P., P.H. Sisco, 1996. *Glossy15*, an *APETALA2*-like gene from maize that regulates leaf epidermal cell identity. Genes & Develop. 10: 3018-3027.

Moreno M.A., J. Chen, I. Greenblatt, S.L. Dellaporta, 1992. Reconstitutional mutagenesis of the maize *P* gene by short-range *Ac* transpositions. Genetics 131: 939-956.

Moreno M.A., L.C. Harper, R.W. Krueger, S.L. Dellaporta, M. Freeling, 1997. *Liguleless1* encodes a nuclear-localized protein required for induction of ligules and auricles during maize leaf organogenesis. Genes and Develop. 11: 616-628.

Mottinger J.P., M.A. Johns, M. Freeling, 1984. Mutations Of The *Adh1* Gene In Maize Following Infection With Barley stripe mosaic virus. Mol.Gen. Genet. 195: 367-369.

Motto M., N. Di Fonzo, C. Soave, F. Salamini, 1987. Chromosomal and genic instabilities originating from revertant derivatives of the *o2m(r)* mutable allele of maize. *MAYDICA* 32: 249-272.

Motto M., M. Maddaloni, G. Ponziani, M. Brembilla, R. Marotta, N. Di Fonzo, C. Soave, R. Thompson, F. Salamini, 1988. Molecular cloning of the *o2-m5* allele of *Zea mays* using transposon marking. Mol. Gen. Genet. 212: 488-494.

Motto M., N. Di Fonzo, H. Hartings, M. Maddaloni, F. Salamini, C. Soave, R.D. Thompson, 1989a. Regulatory genes affecting maize storage protein synthesis. Oxford Surv. Plant Mol. Cell Biol. 6: 87-114.

Motto M., R.D. Thompson, G. Ponziani, C. Soave, M. Maddaloni, H. Hartngs, F. Salamini, 1989b. Transpositional ability of the *Bg-rbg* system of maize transposable elements. *MAYDICA* 34: 107-122.

Mozer T.J., D.C. Tiemeier, E.G. Jaworski, 1983. Purification and characterization of corn glutathione *S*-transferase. Biochemistry 22: 1068-1072.

Müller-Neumann M., J.I. Yoder, P. Starlinger, 1984. The DNA sequence of the transposable element *Ac* of *Zea mays* L. Mol. Gen. Genet. 198: 19-24.

Murray M., J. Cramer, Y. Ma, D. West, J. Romero-Severson, I. Pitas, S. De Mars, L. Vilbrandt, J. Kirshman, R. McLeester, J. Schilz, J. Lotzer, 1988. Agrigenetics maize RFLP linkage map. Maize Genet. Coop. Newsl. 62: 89-91.

Murry L.E., L.G. Elliot, S.A. Capitant, A. West, K.K. Hanson, L. Scarafia, S. Johnston, C. DeLuca-Flaherty, S. Nichols, D. Cunanan, P.S. Dietrich, I.J. Mettler, S. Dewald, D.A. Warnick, C. Rhodes, R.M. Sinibaldi, K.J. Brunke, 1993. Transgenic corn plants expressing MDMV strain B coat protein are resistant to mixed infections of maize dwarf mosaic virus and maize chlorotic mottle virus. Bio/ Technology 11: 1559-1564.

Nelson O.E., 1980. Genetic control of polysaccharide and storage protein synthesis in the endosperm of barley, maize and sorghum. Adv. Cereal Sci. Tech. 3: 41-71.

Nelson O.E., A.J. Ullstrup, 1964. Resistance to leaf spot in maize. J. Heredity 55: 195-199.

Nelson O.E., D. Pan, 1995. Starch synthesis maize endosperms. Ann. Rev. Plant Physiol. and Plant Mol. Biol. 46: 475-494.

Nelson T., J.A. Langdale, 1992. Developmental genetics of C4 photoynthesis. Annu. Rev. Plant Physiol. Plant Biol. 43: 25-47.

Nelson T., J.A. Langdale, 1993. C4 photosynthetic genes and their expression patterns during leaf development. In: Control of Plant Gene Expression, edited by D.P.S. Verma. Caldwell N.J. Telford Press, pp. 259-274.

Neuffer M.G., W.F. Sheridan, 1980. Defective kernel mutations of maize. I. Genetic and lethality studies. Genetics 95: 929-944.

Nussaume L., K. Harrison, V. Klimyuk, R. Martienssen, V. Sundaresan, J.D.G. Jones, 1995. Analysis of splice donor and acceptor site function in a transposable gene trap derived from the maize element Activator. Mol. Gen. Genet. 249: 91-101.

O'Reilly C., N.S. Shepherd, A. Pereira, Z. Schwartz-Sommer, I. Bertram, D.S. Robertson, P.A. Peterson, H. Saedler, 1985. Molecular cloning of the *a1* locus of *Zea mays* using the transposable elements *En* and *Mu1*. EMBO J. 4: 877-882.

Oishi K.K., M. Freeling, 1988. A new *Mu* element from a Robertson's *Mutator* line. In Plant Transposable Elements, edited by O. Nelson. Plenum Press, New York, pp. 289-291.

Olive M.R., W.J. Peacock, E.S. Dennis, 1991. The anaerobic responsive element contains two GC-rich sequences essential for binding a nuclear protein and hypoxic activation of the maize *Adh1* promoter. Nucl. Acids Res. 19: 7053-7060.

Olive M.R., J.C. Walker, K. Singh, E.S. Dennis, W.J. Peacock, 1990. Functional properties of the anaerobic responsive element of the maize *Adh1* gene. Plant Mol. Biol. 15: 593-604.

Paiva R., A.L. Kriz, 1994. Effect of abscisic-acid on embryo-specific gene-expression during normal and precocious germination in normal and viviparous maize (*Zea mays*) embryos. Planta 192: 332-339.

Pan D., O.E. Nelson, 1984. A debranching enzyme deficiency in endosperm of the *sugary-1* mutants of maize. Plant Physiol. 74: 324-328.

Paterson A.H., Y.R. Lin, Z. Li, K.F. Schertz, J.F. Doebley, S. Pinson, S. Liu, J.W. Stansel, J.E. Irvine, 1995. Convergent domestication of cereal crops by independent mutations at corresponding genetic loci. Science: 269: 1714-1718.

Patterson G.I., C.J. Thorpe, V.L. Chandler, 1993. Paramutation, an allelic interaction, is associated with a stable and heritable reduction of transcription of the maize *b* regulatory gene. Genetics 135: 881-894.

Patterson G.I., K.M. Kubo, T. Shroyer, V.L. Chandler, 1995. Sequences required for paramutation of the maize *b* gene map to a region containing the promoter and upstream sequences. Genetics 140: 1389-1406.

Paul A.L., R.J. Ferl, 1991. *In vivo* footprinting reveals unique *cis* elements and different modes of hypoxic induction in maize. *Adh1*, *Adh2*. Plant Cell 3: 159-168.

Paz-Ares J., D. Ghosal, U. Wienand, P.A. Peterson, H. Saedler, 1987. The regulatory *c1* locus of *Zea mays* encodes a protein with homology to *myb* proto-oncogene products and with structural similarities to transcriptional activators. EMBO J. 6: 3353-3358.

Pedersen K., J. Devereux, D.R. Wilson, E. Sheldon, B.A. Larkins, 1982. Cloning and sequence analysis reveal structural variation among related zein genes in maize. Cell 29: 1015-1026.

Pereira A., H. Saedler, 1989. Transcriptional behavior of the maize *En/Spm* element in transgenic tobacco. EMBO J. 8: 1315-1321.

Pereira A., H. Cuypers, A. Gierl, Zs. Schwarz-Sommer, H. Saedler, 1986. Molecular analysis of the En/Spm transposable element system of *Zea mays*. EMBO J. 5: 835-841.

Perrot G.H., K.C. Cone, 1989. Nucleotide sequence of the maize *R-S* gene. Nucleic Acids Res. 17: 8003.

Peschke V.M., M.M. Sahs, 1994. Characterization and expression of transcripts induced by oxygen deprivation in maize (*Zea mays* L.). Plant Physiol. 104: 387-394.

Peterson P.A., 1953. A mutable *pale-green* locus in maize. Genetics 38: 682-683.

Peterson P.A., 1965. A relationship between the *Spm* and *En* control systems in maize. Am. Nat. 99: 391-398.

Peterson P.A., 1976. Gene repression and the evolution of unisexuality in the maize spikelet. *MAYDICA* 21: 157-164.

Peterson P.A., 1987. Mobile elements in plants. CRC Crit. Rev. Plant Sci. 6: 105-208.

Peterson P.A., 1993. Transposable elements in maize - their role in creating plant genetic-variability. Adv. Agron. 51: 79-124.

Peterson P.A., 1995. Genetic-analysis of the functions of the transposable element *En* in *Zea mays* - limited transposase elicits a differental response on reporter alleles. Genetics 141: 1135-1145.

Poethig R.S., 1984. Cellular parameters of leaf morphogenesis in maize and tobacco. In Contemporary Problems of Plant Anatomy, edited by R.A. White, W.C. Dickinson. Academic Press, New York, pp. 235-258.

Poethig R.S., E.J. Szymkowiak, 1995. Clonal analysis of leaf development in maize. *MAYDICA* 40: 67-76.

Phinney B.O., 1956. Growth response of single-gene dwarf mutants in maize to gibberellic acid. Proc. Natl. Acad. Sci. USA 42: 185-189.

Phinney B.O., C.A. West, 1960. Gibberellins and the growth of flowering plants. In Developing Cell Systems and Their Control, edited by D. Rudnick. Ronald Press, New York, pp. 71-92.

Phinney B.O., C. Spray, 1982. Chemical genetics and the gibberelin pathway in *Zea mays* L. In Plant Growth Substances, edited by P.F. Wareing. Academic Press, New York, pp. 101-110.

Pohlman R.F., N.V. Fedoroff, J. Messing, 1984. The nucleotide sequence of the maize controlling element *Activator*. Cell 37: 635-643.

Pollak P.E., K. Hansen, J.D. Astwood, L.P. Taylor, 1995. Conditional male-fertility in maize. Sexual Plant Rep. 8: 231-241.

Pring D.R., C.S. Levings III, 1978. Heterogeneity of maize cytoplasmic genomes among male-sterile cytoplasms. Genetics 89: 121-136.

Prioli L.M., M.R. Söndahl, 1989. Plant regeneration and recovery of fertile plants from protoplasts of maize (*Zea mys* L.). Bio/Technology 7: 589-594.

Pryor A., 1987. The origin and structure of fungal disease resistance in plants. Trends Genet. 3: 157-161.

Qin M., A.H. Ellingboe, 1990. A transcript identified by *MuA* of maize is associated with *Mutator* activity. Mol. Gen. Genet. 224: 357-363.

Qin M., D.S. Robertson, A.H. Ellingboe, 1991. Cloning of the *Mutator* transposable element *MuA2*, a putative regulator of somatic mutability of the *a1-Mum2* allele in maize. Genetics 129: 845-854.

Radicella J.P., D. Turks, V.L. Chandler, 1991. Cloning and nucleotide sequence of cDNA encoding *B-Peru*, a regulatory protein of the anthocyanin pathway in maize. Plant Mol. Biol. 17: 127-130.

Ragot M., P.H. Sisco, D.A. Hoisington, C.W. Stuber, 1995. Molecular-marker-mediated characterization of favorable exotic alleles at quantitative trait loci in maize. Crop Sci. 35: 1306-1315.

Raina R., M. Schläppi, N. Fedoroff, 1995. Interaction between cis-elements and the trans-factors in regulating the epigenetic activity of the maize *Spm* transposable element. J. Cell. Biochem. 59: 176-191.

Randolph L.F., 1976. Contributions of wild relatives of maize to the evolutionary history of domesticated maize: a synthesis of divergent hypotheses. I. Econ. Bot. 30: 321-345.

Redinbaugh M.G., M. Sabre, J.G. Scandalios, 1990. The distribution of catalase activity, isozyme protein, and transcript in the tissues of developing maize seedlings. Plant Physiol. 92: 375-380.

Reid J.B., 1986. Gibberellin mutants. In Plant Gene Research, edited by P.J. King, A.D. Blonstein. Springer-Verlag, New York, pp. 1-34.

Rhoads D.M., C. Kaspi, C.S. Levings II, J.N. Siedow, 1994. Structural charcterization of URF13, the maize mitochondrial protein receptor that causes CMS and susceptibility to fungal pathogens. Annual Meeting of the American Society of Plant Physiologists, Portland, Oregon, USA, July 30-August 3. p. 105.

Richter T.E., T.J. Pryor, J.L. Bennetzen, S.H. Hulbert, 1995. New rust resistance specificities associated with recombination in the Rp1 complex in maize. Genetics 141: 373-381.

Robertson D.S., 1955. The genetics of vivipary in maize. Genetics 40: 745-760.

Robertson D.S., 1978. Characterization of a mutator system in maize. Mutat. Res. 51: 21-28.

Robertson D.S., 1985. Differential activity of the maize mutator Mu at different loci and different cell lineages. Mol. Gen. Genet. 200: 9-13.

Ronchi A., K. Petroni, C. Tonelli, 1995. The reduced expression of endogenous duplications (REED) in the maize *R* gene family is mediated by DNA methylation. EMBO J. 14: 5318-5328.

Rossi V., M. Motto, L. Pellegrini, 1997. The *Opaque-2* regulatory protein and other endosperm factors specifically bind to methylated sequences mimicking the *in vivo* methylation state of the O_2 promoter in maize endosperm. Plant Cell (in press).

Rossini L., P. Me, C. Frova, K. Hein, M. Sari-Gorla, 1995. Molecular analysis and mapping of two genes encoding maize glutathione S-transferases (GST I and GST II). Mol. Gen. Genet. 248: 535-539.

Russell W.A., 1984. Agronomic performance of maize cultivars representing different eras of breeding. *MAYDICA* 29: 375-390.

Russell W.A., 1991. Genetic improvement of maize yield. Adv. Agron. 46: 246-298.

Sachs M.M., M. Freeling, R. Okimoto, 1980. The anaerobic proteins of maize. Cell 20: 761-767.

Sachs M.M., C.C. Subbaiah, I.N. Saab, 1996. Anaerobic gene-expression and flooding tolerance in maize. J. Exp. Bot. 47: 1-15.

Salamini F., 1980. Genetic instability at the *Opaque-2* locus of maize. Mol. Gen. Genet. 179: 497-507.

Salamini F., 1981. Controlling elements at the *Opaque-2* locus of maize: their involvement in the origin of spontaneous mutations. Cold Spring Harbor Symp. Quant. Biol. 45: 467-476.

SanMiguel P., A. Tikhonov, Y.K. Jin, N. Motchoulskaia, D. Zakharov, A. Melake-Berhan, P. S. Springer, K.J. Edwards, M. Lee, Z. Avramova, J.L. Bennetzen, 1996. Nested retrotransposons in the intergenic regions of the maize genome. Science 274: 765-768.

Sanz-Alferez S., T.E. Richter, S.H. Hulbert, J.L. Bennetzen, 1995. The *Rp3* disease resistance gene of maize: mapping and characterization of introgressed alleles. Theor. Appl. Genet. 91: 25-32.

Saxena K.M.S., A.L. Hooker, 1968. On the structure of a gene for disease resistance in maize. Proc. Natl. Acad. Sci. USA 68: 1300-1305.

Scanlon M., R.G. Scheneeberger, M. Freeling, 1996. The maize mutant narrow sheath fails to establish leaf margin identity in a meristematic domain. Development 117: 985-1000.

Scanlon M.J., P.S. Stinard, M.G. James, A.M. Myers, D.S. Robertson, 1994. Genetic analysis of 63 mutations affecting maize kernel development isolated from *Mutator* stocks. Genetics 136: 281-294.

Schläppi M., R. Raina, N. Fedoroff, 1994. Epigenetic regulation of the maize *Spm* transposable element: novel activation of a methylated promoter by TnpA. Cell 77: 427-437.

Schläppi M., R. Raina, N. Fedoroff, 1995. Epigenetic regulation of the maize Spm transposable element-interaction between element-encoded proteins. J. Cell. Biochem. 48: 177-185.

Schmidt R.J., F.A. Burr, B. Burr, 1987. Transposon tagging and molecular analysis of the maize regulatory locus *opaque-2*. Science 238: 960-963.

Schmidt R.J., M.J. Aukerman, F.A. Burr, B. Burr, 1990. Maize regulatory gene *opaque-2* encodes a protein with a leucine zipper motif that binds to zein DNA. Proc. Natl. Acad. Sci. USA 87: 46-50.

Schnable P.S., P.A. Peterson, H. Saedler, 1989. The *bz-rcy* allele of the *Cy* transposable element system of *Zea mays* contains a *Mu*-like element insertion. Mol. Gen. Genet. 217: 459-463.

Schneeberger R.G., P.W. Becraft, S. Hake, M. Freeling, 1995. Ectopic expression of the *knox* homeobox gene *rough sheath1* alters cell fate in the maize leaf. Genes and Development 9: 2292-2304.

Schön C.C., M. Lee, A.M. Melchinger, W.D. Guthrie, W.L. Woodman, 1993. Mapping and characterization of quantitative trait loci affecting resistance against second-generation European corn borer in maize with the aid of RFLPs. Heredity 70: 648-659.

Schultes N.P., T.P. Brutnell, A. Allen, S.L. Dellaporta, T. Nelon, J. Chen, 1996. *Leaf permease1* gene of maize is required for chloroplast development. Plant Cell 8: 463-475.

368 of Maize Mutants

Schwob E., S.Y. Choi, C. Simmons, F. Migliaccio, L. Ilag, T. Hesse, K. Palme, D. Söll, 1993. Molecular analysis of three maize 22 kDa auxin-binding protein genes - transient promoter expression and regulatory regions. Plant J. 4: 423-432.

Schwarz-Sommer Z., L. Leclerq, E. Gobel, H. Saedler, 1987. *Cin4*, an insert altering the structure of the *A1* gene in *Zea mays*, exhibits properties of nonviral retrotransposons. EMBO J. 6: 3873-3880.

Schwarz-Sommer Z., P. Huijser, W. Nacken, H. Saedler, H. Sommer, 1990. Genetic control of flower development by homeotic genes in *Antirrhinum majus*. Science 250: 931-936.

Shaner D.L., P.C. Anderson, 1985. Mechanism of action of the imidazolinones and cell culture selection of tolerant maize. In Biotechnology in plant science: relevance to agriculture in the Nineteen Eighties, edited by M. Zaitlin, P. Day, A. Hollaender. Academic Press, Orlando, Florida, pp. 287-299.

Shaver J.M., D.C. Bittel, J.M. Sellner, D.A. Frisch, D.A. Somers, B.G. Gengenbach, 1996. Single-amino acid substitutions eliminate lysine inhibition of maize dihydrodipicolinate synthase. Proc. Natl. Acad. Sci. USA 93: 1962-1966.

Shen B.O., N. Carneiro, I. Torre-Jerez, B. Stevenson, T. McCreery, T. Helentjaris, C. Baysdorfer, E. Almira, R.J. Ferl, J.E. Habben, B. Larkins, 1994. Partial sequencing and mapping of clones from two maize cDNA libraries. Plant Mol. Biol. 26: 1085-1101.

Shepherd N.S., M.M. Rhoades, E. Dempsey, 1989. Genetic and molecular characterization of *a-mrh-Mrh* a new mutable system of *Zea mays*. Devel. Genet. 10: 507-519.

Shepherd N.S., Zs. Schwarz-Sommer, J. Blumberg Vel Spalve, M. Gupta, U. Wienand, H. Saedler, 1984. Similarity of the Cin1 repetitive family of *Zea mays* to eukaryotic transposable elements. Nature (Lond.) 387: 185-187.

Sheridan W.F., M.G. Neuffer, 1980. Defective kernel mutants of maize. II. Morphological and embryo culture studies. Genetics 95: 945-960.

Sheridan W.F., N.A. Avalkina, I.I. Shamrov, T.B. Batygina, I.N. Golubovskaya, 1996. The *mac1* gene: controlling the commitment to the meiotic pathway in maize. Genetics 142: 1009-1020.

Shillito R.D., G.K. Carswell, C.M. Johnston, J.J. DiMaio, C.T. Harms, 1989. Regeneration of fertile plants from protoplasts of elite inbred maize. Bio/Technology 7: 581-587.

Shoemaker J., D. Zaitlin, J. Horn, S. De Mars, J. Kirshman, J. Pitas, 1992. A comparison of three agrigenetics maize RFLP linkage maps. Maize Genet. Coop. Newsl. 66: 65-69.

Siemenroth A., T. Borner, U. Metzger, 1980. Biochemical studies on the iojap mutant of maize. Plant Physiol. 65: 1108-1110.

Sinha N., S. Hake, M. Freeling, 1985. Genetic and molecular analysis of leaf development. Curr. Top Dev. Biol. 195: 47-80.

Smart C.M., S.E. Hosken, H. Thomas, J.A. Greaves, B.G. Blair, W. Schuch, 1995. The timing of maize leaf senescence and characterization of senescence-related cDNAs. Physiol. Plant. 93: 673-682.

Smith J.S.C., O.S. Smith, 1991. Restriction fragment length polymorphisms can differentiate among U.S. maize hybrids. Crop Sci. 31: 893-899.

Smith L.G., D. Jackson, S. Hake, 1995. Expression of *knotted1* marks shoot meristem formation during maize embryogenesis. Dev. Genetics 16: 344-348.

Spena A., F. Salamini, 1995. Genetic tagging of cells and cell layers for studies of plant development. Methods in Cell Biol. 49: 331-354.

Spray C.R., B.O. Phinney, P. Gaskin, S.J. Gilmour, J. MacMillan, 1994. Internode length in *Zea mays* L. The *dwarf-1* mutation controls the 3ß-hydroxylation of gibberellin A20 to gibberellin A1. Planta 160: 464-468.

Spray C.R., M. Kaboyashi, Y. Suzuki, B.O. Phinney, P. Gaskin, J. MacMillan, 1996. The *dwarf-1* (*dl*) mutant of *Zea mays* blocks three steps in the gibberellin-biosynthetic pathway. Proc. Natl. Acad. Sci. 93: 10515-10518.

Springer P.S., K.J. Edwards, J.L. Bennetzen, 1994. DNA class organization on maize *Adh1* yeast artificial chromosomes. Proc. Natl. Acad. Sci USA 91: 863-867.

Stadler L.J., 1930. Some genetic effects of X-rays in plants. J. Heredity 21: 3-19.

Stapleton A.E., R.L. Phillips, 1993. A fertile field: Maize Genetics 93.The Plant Cell 5: 723-727).

Staskawicz B.J., F.M. Ausubel, B.J. Baker, J.G. Ellis, J.D.G. Jones, 1995. Molecular genetics of plant disease resistance. Science 268: 661-667.

Steffensen D.M., 1968. A reconstruction of cell development in the shoot apex of maize. Ant. J. Bot. 55: 354-369.

Stinard P.S., D.S. Robertson, P.S. Schnable, 1993. Genetic isolation, cloning and analysis of a *Mutator*-induced, dominant antimorph of the maize *amylose extender-1* locus. Plant Cell 5: 1555-1566.

Stirpe F., L. Barbieri, M.G. Battelli, M. Soria, D.A. Lappi, 1992. Ribosome-inactivating proteins from plants: present status and future prospects. Bio/Technology 10: 405-412.

Stuber C.W., 1995. Mapping and manipulating quantitative traits in maize. Trends in Genetics 11: 477-481.

Stuber C.W., S.E. Lincoln, D.W. Wolff, T. Helentjaris, E.S. Lander, 1992. Identification of genetic factors contributing to heterosis in a hybrid from two elite maize inbred lines using molecular markers. Genetics 132: 823-839.

Sturtevant A.H., 1929. The claret mutant type of *Drosophila simulans*. A study of chromosome elimination and cell lineage. Z. Wiss. Zool. 135: 323-356.

Sudupak M.A., J.L. Bennetzen, S.H. Hulbert, 1993. Unequal exchange and meiotic instability of disease-resistance genes in the *Rp1* region of maize. Genetics 133: 119-125.

Sullivan T.D., L.I. Strelow, C.A. Illingworth, R.L. Phillips, O.E. Nelson Jr., 1991. The maize *brittle-1* locus: molecular characterization based on DNA clones isolated usign the *dSpm*-tagged *brittle-1-mutable* allele. Plant Cell 3: 1337-1348.

Tacke E., C. Korfhage, D. Michel, M. Maddaloni, M. Motto, S. Lanzini, F. Salamini, H-P.Döring, 1995. Transposon tagging of the maize *Glossy-2* locus with the transposable element *En/Spm*. The Plant J. 8: 907-917.

Talbert L.E., G.I. Patterson, V.L. Chandler, 1989. *Mu* transposable elements are structurally diverse and distributed throughout the genus *Zea*. J. Mol. Evol. 29: 28-39.

Taliercio E., S. Shanker, J.H. Choi, P.S. Chourey, 1995. Molecular aspects of cell wall invertase in developing kernels of maize. Plant Physiol. (suppl.) 108: 182.

Taylor L.P., V. Walbot, 1987. Isolation and characterization of a 1.7 kb transposable element from a *Mutator* line of maize. Genetics 117: 297-307.

Theissen G., T. Strater, A. Fischer, H. Saedler, 1995. Structural characterization, chromosomal localization and phylogenetic evaluation of 2 pairs of agamous-like mads-box genes from maize. Gene 156: 155-166.

Theres N., T. Scheele, P. Starlinger, 1987. Cloning of the *Bz2* locus of *Zea mays* using the transposable element *Ds* as a gene tag. Mol. Gen. Genet. 209: 193-197.

Tilney-Basset R.A.E., 1986. Plant Chimeras. Arnold, London.

Timmermans M.C.P., O.P. Das, J. Messing, 1996. Characterization of a meiotic crossover in maize identified by a restriction fragment length polymorphism-based method. Genetics 143: 1771-1783.

Torti G., L. Manzocchi, F. Salamini, 1986. Free and bound indole-acetic acid is low in the endosperm of the maize mutant *defective endosperm-B18*. Theor. Appl. Genet. 72: 602-605.

Touzet P., R.G. Winkler, T. Helentjaris, 1995. Combined genetic and physioloical analysis of a locus contributing to quantitative variation. Theor. Appl. Genet. 91: 200-205.

Turcich M.P., A. Bokhari-Riza, D.A. Hamilton, C. He, W. Messier, C.B. Stewart, J.P. Mascarenhas, 1996. PREM-2, a copia-type retroelement in maize is expressed preferentially in early microspores. Sexual Plant Repr. 9: 65-74.

Ueberlacker B., B. Klinge, W. Werr, 1996. Ectopic expression of the maize homeobox genes *ZmHox1a* or *ZmHox1B* causes pleiotropic alterations in the vegetative and floral development of transgenic tobacco. Plant Cell 8: 349-362.

Uhrig H., A. Marocco, H.-P. Döring, F. Salamini, 1997. The clonal origin of the lateral meristem generating the ear shoot of maize. Plant 201: 9-17.

Vasil V., I.K. Vasil, 1987. Formation of callus and somatic embryos from protoplasts of a commercial hybrid of maize (*Zea mays* L.). Theor. Appl. Genet. 73: 793-798.

Vasil V., W.R. Marcotte, L. Rosenkrans, S.M. Cocciolone, I.K. Vasil, R.S. Quatrano, D.R. McCarty, 1995. Overlap of *viviparous1* (*vp1*) and abscisic acid response elements in the em promoter - g-box elements are sufficient but not necessary for *vp1* transactivation. Plant Cell 7: 1511-1518.

Voelker R., A. Barkan, 1995. Nuclear genes required for post-translational steps in the biogenesis of the chloroplast cytochrome *b-6f* complex in maize. Mol. Gen. Genet. 249: 507-514.

Vollbrecht E., B. Veit, N. Sinha, S. Hake, 1991. The developmental gene Knotted-1 is a member of a maize homeobox gene family. Nature 350: 241-243.

Wagner V.T., Y.C. Song, E. Matthys-Rochon, C. Dumas, 1989. Observations on the isolated embryo sacs of *Zea mays* L. Plant Sci. 59: 127-132.

Walbot V., C. Warren, 1990. DNA methylation in the *Alcohol dehydrogenease-1* gene of maize. Plant Mol. Biol. 15: 121-125.

Walker E.L., T.P. Robbins T.E. Bureau, J.L. Kermicle, S.L. Dellaporta, 1995. Transposon-mediated chromosomal rearrangements and gene duplications in the formation of the maize *R-r* complex. EMBO J. 14: 2350-2363.

Wallace H.A., W.L. Brown, 1956. Corn and its early fathers. Michigan State University Press.

Walsh T.A., A.E. Morgan, T.D. Hey, 1991. Characterization and molecular cloning of a proenzyme form of a ribosome inactivating protein from maize. J. Biol. Chem. 266: 23422-23427.

Walton J.D., 1991. Genetics and biochemistry of toxin synthesis in *Cochliobolus* (*Helminthosporium*). In Molecular Industrial Mycology, edited by S. Leong, R.M. Berka. M. Dekker, New York, pp. 225-249.

Walton J.D., F.R. Holden, 1988. Properties of two enzymes involved in the biosynthesis of the fungal pathogenicity factor HC-toxin. Mol. Plant-Microbe Interact. 1: 128-134.

Wang L., M. Heinlein, R. Kunze, 1996. Methylation pattern of *Activator* transposase binding sites in maize endosperm. The Plant Cell 8: 747-758.

Weatherwax P., 1955. History and origin of corn. I. Early history of corn and theories as to its origin. In Corn and Corn Improvement, edited by G.F. Sprague. Academic Press, New York, pp. 1.

Weber D., T. Helentjaris, 1989. Mapping RFLP loci in maize using B-A translocations. Genetics 121: 583-590.

Weil C.F., S. Marillonnet, B. Burr, S.R.Wessler, 1992. Changes in state of the *wx-m5* allele of maize are due to intragenic transposition of Ds. Genetics 130: 175-186.

Werr W., W.B. Frommer, C. Maas, P. Starlinger, 1985. Structure of the sucrose synthase gene on chromosome 9 of *Zea mays* L. EMBO J. 4: 1373-1380.

Wiegand R.C., D.M. Shah, T.J. Mozer, E.I. Harding, J. Diaz-Collier, C. Saundres, E.G. Jaworski, D.C. Tiemeier, 1986. Mesenger RNA encoding a glutathione-S-transferase responsible for herbicide tolerance in maize is induced in response to safener treatment. Plant Mol. Biol. 7: 235-243.

Wienand U., U. Weydemann, U. Niesbach-Kloesgen, P.A. Peterson, H. Saedler, 1986. Molecular cloning of the *c2* locus of *Zea mays*, the gene coding for chalcone synthase. Mol. Gen. Genet. 203: 202-207.

Williams C.G., M.M. Goodman, C.W. Stuber, 1995. Comparative recombination distances among *Zea mays* L. inbreds, wide crosses and interspecific hybrids. Genetics 141: 1573-1581.

Willman M.R., S.M. Schroll, T.K. Hodges, 1989. Inheritance of somatic embryogenesis and plantlet regeneration from primary (type 1) callus in maize. In Vitro Cell. & Dev. Biol. 25: 95-100.

Winkler R.G., T. Helentjaris, 1995. The maize *dwarf3* gene encodes a cytochrome p450-mediated TI early step in gibberellin biosynthesis. Plant Cell 7: 1307-1317.

Wise R.P., C.L. Dill, P.S. Schnable, 1996. *Mutator*-induced mutations of the *rf1* nuclear fertility restorer of T-cytoplasm maize alter the accumulation of *T-urf13* mitochondrial transcripts. Genetics 143: 1383-1394.

Wise R.P., A.E. Fliss Jr., D.R. Pring, B.G. Gengenbach, 1987. *urf13-T* of T cytoplasm maize mitochondria encodes a 13-kDa polypeptide.Plant Mol. Biol. 9: 121-126.

Wright A.D., M.B. Sampson, M. G. Neuffer, L. Michalczuk, J. P. Slovin, J.D. Cohen, 1991. Indole-3-acetic acid biosynthesis in the mutant maize *orange pericarp*, a tryptophan auxotroph. Science 254: 998-1000.

Wright A.D., C.A. Moehlenkamp, G.H. Perrot, M.G. Neuffer, K.C. Cone, 1992. The maize auxotrophic mutant orange pericarp is defective in duplicate genes for tryptophan synthase-beta. Plant Cell 4: 711-719.

Xu X., A.P. Hsia, L. Zhang, B.J. Nikolau, P.S. Schnable, 1995. Meiotic recombination break points resolve at high rates at the 5' end of a maize coding sequence. Plant Cell 7: 2151-2161.

Yanofsky M.F., H. Ma, J.L. Bowman, G.N. Drews, K.A. Feldmann, E.M. Meyerowitz, 1990. The protein encoded by the *Arabidopsis* homeotic gene agamous resembles transcription factors. Nature 346: 35-39.

Yu L.X., C.M. Lazarus, 1991. Structure and sequence of an auxin-binding protein gene from maize (*Zea mays* L.). Plant Mol. Biol. 16: 925-930.

Zehr B.E., M.E. Williams, D.R. Duncan, J.M. Widholm, 1987. Somaclonal variation in the progeny of plants regenerated from callus cultures of seven inbred lines of maize. Can. J. Bot. 65: 491-499.

Zhu D., J.D. Scandalios, 1994. Differential accumulation of manganese-superoxide dismutase transcripts in maize in response to abscisic acid and high osmoticum. Plant Physiol. 106: 173-178.

Zhu D., J.D. Scandalios, 1995. The maize mitochondrial MnSODs encoded by multiple genes are localized in the mitochondrial matrix of transformed yeast cells. Free Rad. Biol. Medicine 18: 179-183.

Index